JACARANDA
MATHS QUEST

GENERAL MATHEMATICS 11 FOR QUEENSLAND

UNITS 1 & 2 | SECOND EDITION

MARK BARNES

PAULINE HOLLAND

REVIEWED BY

Judi Dau | Frances Healy | Tanya Parnwell | Emilia Sinton

jacaranda
A Wiley Brand

Second edition published 2024 by
John Wiley & Sons Australia, Ltd
Level 4, 600 Bourke Street, Melbourne, Vic 3000

First edition published 2018

Typeset in 10.5/13 pt TimesLTStd

© John Wiley & Sons Australia, Ltd 2024

The moral rights of the authors have been asserted.

ISBN: 978-1-394-26967-9

Front cover images: TWINS DESIGN STUDIO/Adobe Stock Photos, veekicl/Adobe Stock Photos, mrhighsky/Adobe Stock Photos, sapunkele/Adobe Stock Photos, NARANAT STUDIO/Adobe Stock Photos, Kullaya/Adobe Stock Photos, vectorplus/Adobe Stock Photos, WinWin/Adobe Stock Photos, valeriya_dor/Adobe Stock Photos, Ludmila/Adobe Stock Photos, Анастасия Трофимова/Adobe Stock Photos, katarinalas/Adobe Stock Photos, izzul fikry (ijjul)/Adobe Stock Photos, Tatsiana/Adobe Stock Photos, nadiinko/Adobe Stock Photos

Illustrated by various artists, diacriTech and Wiley Composition Services

Typeset in India by diacriTech

A catalogue record for this book is available from the National Library of Australia

Printed in Singapore
M130181_010824

The publisher of this series acknowledges and pays its respects to Aboriginal Peoples and Torres Strait Islander Peoples as the traditional custodians of the land on which this resource was produced.

This suite of resources may include references to (including names, images, footage or voices of) people of Aboriginal and/or Torres Strait Islander heritage who are deceased. These images and references have been included to help Australian students from all cultural backgrounds develop a better understanding of Aboriginal and Torres Strait Islander Peoples' history, culture and lived experience.

It is strongly recommended that teachers examine resources on topics related to Aboriginal and/or Torres Strait Islander Cultures and Peoples to assess their suitability for their own specific class and school context. It is also recommended that teachers know and follow the guidelines laid down by the relevant educational authorities and local Elders or community advisors regarding content about all First Nations Peoples.

All activities in this resource have been written with the safety of both teacher and student in mind. Some, however, involve physical activity or the use of equipment or tools. **All due care should be taken when performing such activities**. To the maximum extent permitted by law, the authors and publisher disclaim all responsibility and liability for any injury or loss that may be sustained when completing activities described in this resource.

The publisher acknowledges ongoing discussions related to gender-based population data. At the time of publishing, there was insufficient data available to allow for the meaningful analysis of trends and patterns to broaden our discussion of demographics beyond male and female gender identification.

Contents

online only

Problem-solving and modelling task guide

UNIT 1 MONEY, MEASUREMENT, ALGEBRA AND LINEAR EQUATIONS   1

TOPIC 1 CONSUMER ARITHMETIC

TOPIC 2 SHAPE AND MEASUREMENT

online only

PRACTICE ASSESSMENT 1
Problem-solving and modelling task

PRACTICE ASSESSMENT 2
Unit 1 Examination

online only

PRACTICE ASSESSMENT 3
Unit 2 Examination

PRACTICE ASSESSMENT 4
Units 1 & 2 Examination

Learning with learnON

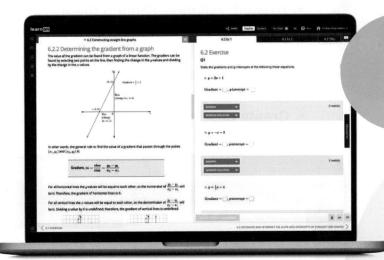

Everything you need
for your students to succeed

JACARANDA MATHS QUEST
GENERAL MATHEMATICS 11
UNITS 1 AND 2 FOR QUEENSLAND |
SECOND EDITION

Developed by expert teachers for students

Tried, tested and trusted. The completely revised and updated second edition of *Jacaranda Maths Quest General Mathematics 11 Units 1 & 2 for Queensland* continues to focus on helping teachers achieve learning success for every student — ensuring no student is left behind and no student is held back.

Because both what and how students learn matter

Learning is personal

Whether students need a challenge or a helping hand, you'll find what you need to create engaging lessons.

Whether in class or at home, students can get unstuck and progress! Scaffolded lessons with detailed worked examples and difficult concepts are supported by teacher-led video eLessons. Automatically marked, differentiated question sets are all supported by detailed worked solutions. And brand-new exam-style questions support in-depth skill acquisition in every lesson.

Learning is effortful

Learning happens when students push themselves. With learnON, Australia's most powerful online learning platform, students can challenge themselves, build confidence and ultimately achieve success.

Learning is rewarding

Through real-time results data, students can track and monitor their own progress and easily identify areas of strength and weakness.

And for teachers, Learning Analytics provide valuable insights to support student growth and drive informed intervention strategies.

Learn online with Australia's most

Everything you need for each of your lessons in one simple view

- Trusted, syllabus-aligned theory
- Engaging, rich multimedia
- All the teacher support resources you need
- Deep insights into progress
- Immediate feedback for students
- Create custom assignments in just a few clicks.

Practical teaching advice and ideas for each lesson provided in teachON

Each lesson linked to content points from the QCAA General Mathematics 2025 General senior syllabus

Reading content and rich media including embedded videos and interactivities

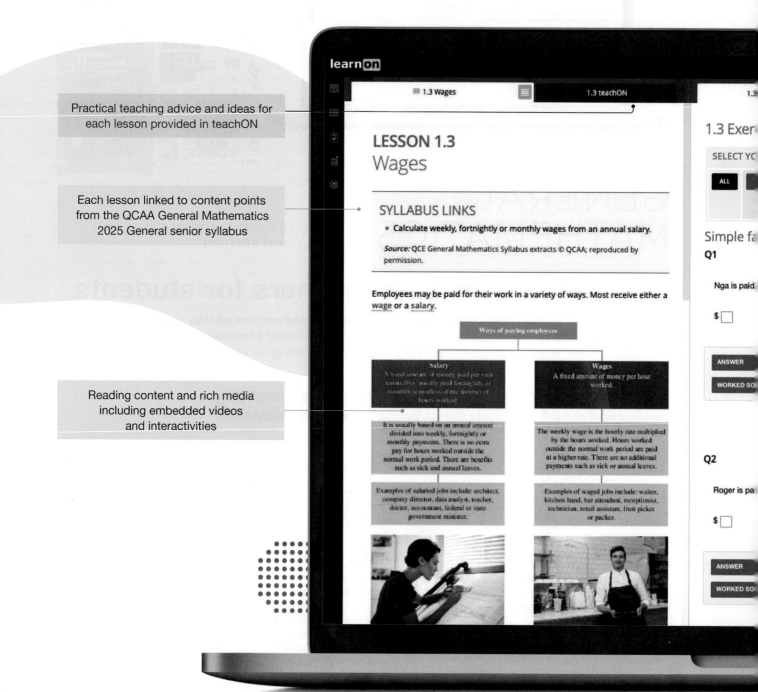

learn**on**

1.3 Wages   1.3 teachON   1.3

LESSON 1.3
Wages

SYLLABUS LINKS
- Calculate weekly, fortnightly or monthly wages from an annual salary.

Source: QCE General Mathematics Syllabus extracts © QCAA; reproduced by permission.

Employees may be paid for their work in a variety of ways. Most receive either a wage or a salary.

Ways of paying employees

Salary
A fixed amount of money paid per year annually, usually paid fortnightly or monthly regardless of the number of hours worked.

Wages
A fixed amount of money per hour worked.

It is usually based on an annual amount divided into weekly, fortnightly or monthly payments. There is no extra pay for hours worked outside the normal work period. There are benefits such as sick and annual leaves.

The weekly wage is the hourly rate multiplied by the hours worked. Hours worked outside the normal work period are paid at a higher rate. There are no additional payments such as sick or annual leaves.

Examples of salaried jobs include: architect, company director, data analyst, teacher, doctor, accountant, federal or state government minister.

Examples of waged jobs include: waiter, kitchen hand, bar attendant, receptionist, technician, retail assistant, fruit picker or packer.

1.3 Exer

SELECT YC

ALL

Simple fa
Q1

Nga is paid

$

ANSWER

WORKED SO

Q2

Roger is pa

$

ANSWER

WORKED SO

powerful learning tool, learnON

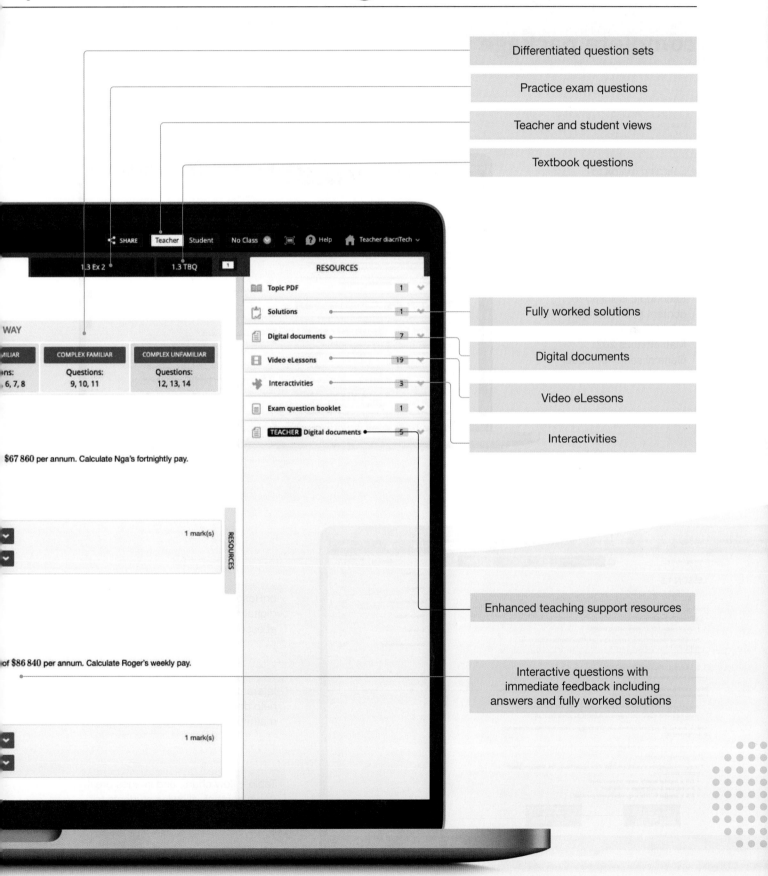

Differentiated question sets

Practice exam questions

Teacher and student views

Textbook questions

Fully worked solutions

Digital documents

Video eLessons

Interactivities

Enhanced teaching support resources

Interactive questions with immediate feedback including answers and fully worked solutions

SHARE Teacher Student No Class Help Teacher diacriTech

1.3 Ex 2   1.3 TBQ   1

RESOURCES

Topic PDF   1
Solutions   1
Digital documents   7
Video eLessons   19
Interactivities   3
Exam question booklet   1
TEACHER Digital documents   5

WAY

FAMILIAR	COMPLEX FAMILIAR	COMPLEX UNFAMILIAR
ns: 6, 7, 8	Questions: 9, 10, 11	Questions: 12, 13, 14

$67 860 per annum. Calculate Nga's fortnightly pay.

1 mark(s)

of $86 840 per annum. Calculate Roger's weekly pay.

1 mark(s)

Online, these new editions are the **complete package**

Trusted Jacaranda theory, plus tools to support teaching and make learning more engaging, personalised and visible.

Learning matrix to monitor student's confidence level throughout topics.

Each topic is linked to content points from the QCAA General Mathematics 2025 General senior syllabus.

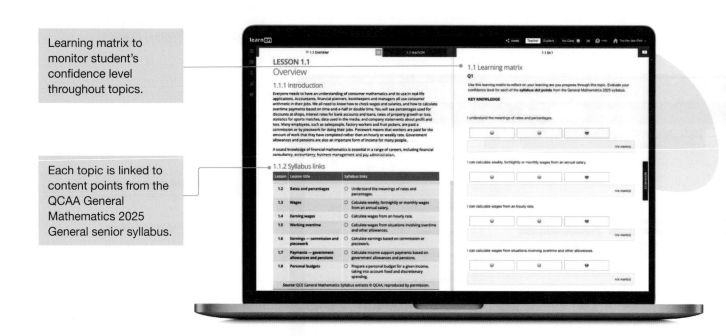

onResources link to targeted digital resources including video eLessons and weblinks.

Interactive glossary terms help develop and support mathematical literacy.

Tables, flow charts and images break down content, allowing students to understand complex concepts.

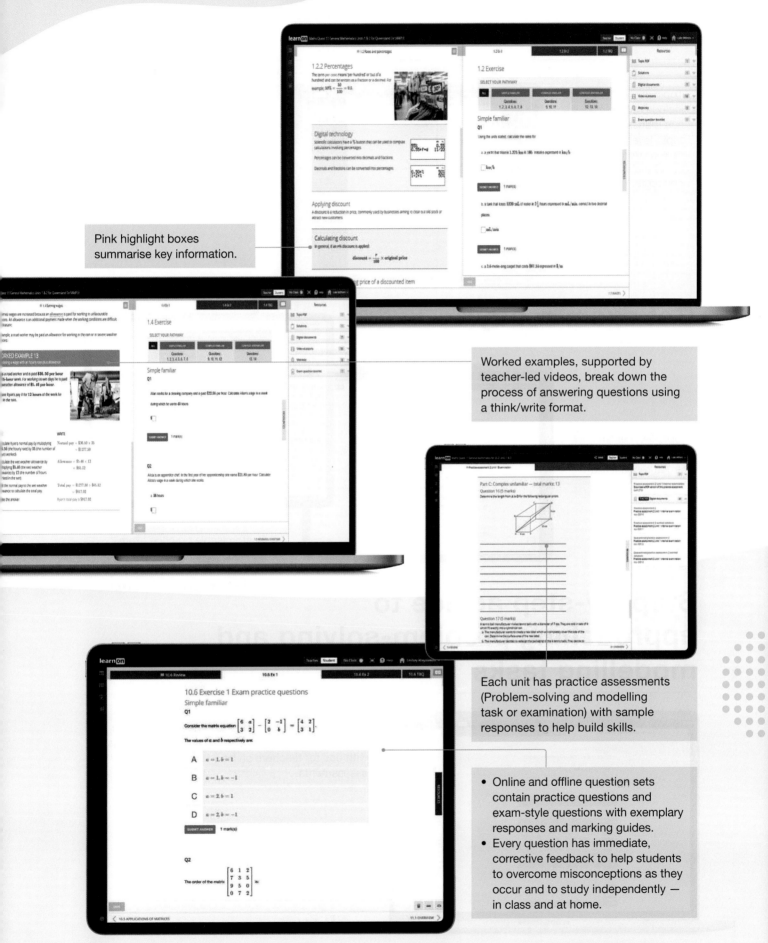

Pink highlight boxes summarise key information.

Worked examples, supported by teacher-led videos, break down the process of answering questions using a think/write format.

Each unit has practice assessments (Problem-solving and modelling task or examination) with sample responses to help build skills.

- Online and offline question sets contain practice questions and exam-style questions with exemplary responses and marking guides.
- Every question has immediate, corrective feedback to help students to overcome misconceptions as they occur and to study independently — in class and at home.

Topic and Unit reviews

Topic and Unit reviews include online summaries and topic-level and unit review exercises that cover multiple concepts.

Get exam-ready!

Topic-level exam questions are structured just like the exams.

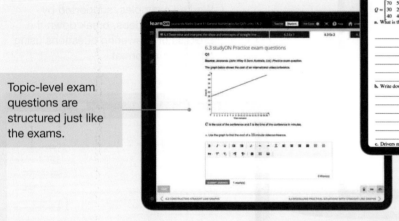

Customisable practice exam question booklets are available in every topic to build student competence and confidence.

Step-by-step advice to approaching problem-solving and modelling tasks

Step-by-step student guide on how to approach and complete problem-solving and modelling tasks with tips for teachers on how to create good assessments.

Teaching with learnON

learn on
one resource

PLAN · TEACH · LEARN · ASSESS · ANALYSE

ONGOING SUPPORT · CLASS SET-UP · INTEGRATIONS

Enhanced teacher support resources, including:

- work programs and syllabus grids
- teaching advice and additional activities
- quarantined topic tests (with solutions)
- Quarantined PSMTs and examinations
  Custom exam-builder with question differentiation (SF/CF/CU) question filters

Customise and assign

A testmaker enables you to create custom tests from the complete bank of thousands of questions (including past QCAA exam questions in year 12).

Reports and results

Data analytics and instant reports provide data-driven insights into performance across the entire course.

Show students (and their parents or carers) their own assessment data in fine detail. You can filter their results to identify areas of strength and weakness.

Acknowledgements

The authors and publisher would like to thank the following copyright holders, organisations and individuals for their assistance and for permission to reproduce copyright material in this book.

Images

• © Alessandro Biascioli/Adobe Stock Photos: **458** • © AntonioDiaz/Adobe Stock Photos: **30** • © Bruce/Adobe Stock Photos: **88** • © Daniela/Adobe Stock Photos: **263** • © Danko/Adobe Stock Photos: **234** • © Dilok/Adobe Stock Photos: **4, 79** • © Drobot Dean/Adobe Stock Photos: **39, 217** • © DusanJelicic/Adobe Stock Photos: **270** • © dusanpetkovic1/Adobe Stock Photos: **42** • © evgeniia_1010/Adobe Stock Photos: **6** • © Flamingo Images/Adobe Stock Photos: **34** • © frantisek hojdysz/Adobe Stock Photos: **359** • © Friends Stock/Adobe Stock Photos: **18** • © Ivan Traimak/Adobe Stock Photos: **29** • © JackF/Adobe Stock Photos: **89** • © Jade Maas/peopleimages.com/Adobe Stock Photos: **42** • © J Bettencourt/peopleimages.com/Adobe Stock Photos: **263** • © Jelena/Adobe Stock Photos: **267** • © Jim Ekstrand/Adobe Stock Photos: **518** • © Kaspars Grinvalds/Adobe Stock Photos: **438** • © Kevin/Adobe Stock Photos: **481** • © korkeng/Adobe Stock Photos: **3, 51, 93, 155, 201, 225, 241, 283, 331, 391, 437, 495** • © Kirsten D/peopleimages.com/Adobe Stock Photos: **274** • © LIGHTFIELD STUDIOS/Adobe Stock Photos: **266** • © lordn/Adobe Stock Photos: **35, 40** • © Lumos sp/Adobe Stock Photos: **11** • © Maksym/Adobe Stock Photos: **264** • © marbenzu/Adobe Stock: **356** • © micromonkey/Adobe Stock Photos: **459** • © mila103/Adobe Stock Photos: **231** • © milkovasa/Adobe Stock Photos: **226** • © Monkey Business/Adobe Stock Photos: **425** • © MyMicrostock/Stocksy/Adobe Stock Photos: **473** • © myphotobank.com.au/Adobe Stock Photos: **274** • © nickolya/Adobe Stock Photos: **235** • © Nomad_Soul/Adobe Stock Photos: **43** • © Pawel Pajor/Adobe Stock Photos: **519** • © Pixel-Shot/Adobe Stock Photos: **11, 324** • © Quality Stock Arts/Adobe Stock Photos **73** • © Rawpixel.com/Adobe Stock Photos: **487** • © rufous/Adobe Stock Photos: **33** • © Scanrail/Adobe Stock Photos: **56** • © Sergey Ryzhov/Adobe Stock Photos: **238** • © sorapop/Adobe Stock Photos: **37** • © Stockbym/Adobe Stock Photos: **81** • © stocksolutions/Adobe Stock Photos: **442** • © tampatra/Adobe Stock Photos: **81** • © Thanate/Adobe Stock Photos: **235** • © TheSupporter/Adobe Stock Photos: **273** • © Urbanscape/Adobe Stock Photos: **59** • © VetalStock/Adobe Stock Photos: **231** • © volff/Adobe Stock Photos: **10** • © Wollwerth Imagery/Adobe Stock Photos: **431** • © Yaroslav Astakhov/Adobe Stock Photos: **23** • © Action Plus Sports Images/Alamy Stock Photo: **41** • © A. Astes/Alamy Stock Photo: **135** • © Gary Dyson/Alamy Stock Photo: **123** • © Ingo Schulz/Alamy Stock Photo: **407** • © Alistair Berg/Getty Images: **24** • © Bettmann/Getty Images: **242** • © Bloomberg/Getty Images: **53** • © kali9/Getty Images: **303** • © kledge/Getty Images: **265** • © mareciok/Getty Images: **408** • © Monty Rakusen/Getty Images: **21** • © Olga_Danylenko/iStock/Getty Images: **418** • © Photodisc/Getty Images: **484** • © PeopleImages/E+/Getty Images, Ground Picture/Shutterstock: **12** • © Replace "Offstocker/istockphoto" with "offstocker/istock/Getty Images": **80** • © Baris Simsek/iStockphoto: **52** • © Jacob Wackerhausen/iStockphoto: **34** • © Krystian Kaczmarski/iStockphoto: **419** • © 06photo/Shutterstock: **285** • © abimages/Shutterstock: **55** • © Albert Michael Cutri/Shutterstock: **512** • © Aleksandar Mijatovic/Shutterstock: **129** • © Alena Ozerova/Shutterstock: **318** • © Alex Bogatyrev/Shutterstock: **479** • © Alex Tsuper/Shutterstock: **350** • © Alexey Boldin/Shutterstock: **78** • © AlexLMX/Shutterstock: **231** • © Anastasios71/Shutterstock: **378** • © Andras Deak/Shutterstock: **471** • © Andy Dean Photography/Shutterstock: **25, 176** • © Anggara dedy/Shutterstock: **193** • © Anna Kucherova/Shutterstock: **300** • © ArtFamily/Shutterstock: **393** • © bbevren/Shutterstock: **35** • © Bianda Ahmad Hisham/Shutterstock: **82** • © Big Pants Production/Shutterstock: **198** • © Brian A Jackson/Shutterstock: **89** • © CandyBox Images/Shutterstock: **54, 506** • © Christian Bertrand/Shutterstock: **221** • © Curioso.Photography/Shutterstock: **65** • © Danny Xu/Shutterstock: **136** • © David Acosta Allely/Shutterstock: **284** • © David Papazian/Shutterstock: **350** • © DeiMosz/Shutterstock: **58** • © Djem/Shutterstock: **504** • © Dmitrijs Mihejevs/Shutterstock: **269** • © Dmitry Kalinovsky/Shutterstock: **23** • © Dream79/Shutterstock: **41** • © ekler/Shutterstock: **212** • © Eliyahu Yosef Parypa/Shutterstock: **9** • © Empirephotostock/Shutterstock: **300** • © erashov/Shutterstock: **208** • © Eugene Onischenko/Shutterstock: **299** • © Evgeniya Uvarova/Shutterstock: **305** • © Federico Rostagno/Shutterstock: **43** • © freesoulproduction/Shutterstock: **5** • © g0d4ather/Shutterstock: **76** • © garetsworkshop/Shutterstock: **6**

• © Gingerss/Shutterstock: **86** • © Goncharov_Artem/Shutterstock: **25** • © Gorodenkoff/Shutterstock: **156** • © Henny van Roomen/Shutterstock: **60** • © homydesign/Shutterstock: **367, 374** • © ibreakstock/Shutterstock: **68** • © imagedb.com/Shutterstock: **519** • © irabel8/Shutterstock: **316** • © itor/Shutterstock: **218** • © Ivan Cholakov/Shutterstock: **392** • © James Flint/Shutterstock: **298** • © JAY KRUB/Shutterstock: **450** • © Jazper/Shutterstock: **146** • © joephotostudio/Shutterstock: **480** • © Juan_Gomez/Shutterstock: **213** • © Karramba Production/Shutterstock: **22** • © KKulikov/Shutterstock: **134** • © Kletr/Shutterstock: **232** • © koya979/Shutterstock: **78, 233** • © Krakenimages.com/Shutterstock: **401** • © Le Quang Photo/Shutterstock: **235** • © Leftleg/Shutterstock: **60** • © Libor Fousek/Shutterstock: **420** • © Lissandra Melo/Shutterstock: **315** • © littlenySTOCK/Shutterstock: **431** • © Lotus_studio/Shutterstock: **407** • © LP Design/Shutterstock: **449** • © MarcelClemens/Shutterstock: **202** • © Maridav/Shutterstock: **266** • © marijaf/Shutterstock: **419** • © Mariya Volik/Shutterstock: **304** • © Martin M303/Shutterstock: **332** • © Marynka Mandarinka/Shutterstock: **145** • © Matej Kastelic/Shutterstock: **13** • © max blain/Shutterstock: **472** • © Maxx-Studio/Shutterstock: **467** • © Maya Kruchankova/Shutterstock: **268** • © MedstockPhotos/Shutterstock: **135** • © Michael Kraus/Shutterstock: **60** • © michaeljung/Shutterstock: **27** • © Millionstock/Shutterstock: **299** • © Minerva Studio/Shutterstock: **64** • © Monkey Business Images/Shutterstock: **34, 38** • © muzsy/Shutterstock: **220** • © myphotobank.com.au/Shutterstock: **452** • © niderlander/Shutterstock: **304** • © Nils Versemann/Shutterstock: **57** • © Nuchylee/Shutterstock: **157** • © Oaklizm/Shutterstock: **8** • © Olga Danylenko/Shutterstock: **81** • © OoddySmile Studio/Shutterstock: **193** • © Ozgur Coskun/Shutterstock: **60** • © Passion Images/Shutterstock: **85** • © Paul Brennan/Shutterstock: **367** • © Peshkova/Shutterstock: **183** • © Peter Turansky/Shutterstock: **193** • © Phovoir/Shutterstock: **65** • © PL Petr Lerch/Shutterstock: **135** • © Pot of Grass Productions/Shutterstock: **55** • © Poznyakov/Shutterstock: **320** • © Radu Bercan/Shutterstock: **53, 218** • © Rafael Lavenere/Shutterstock: **315** • © Rashevskyi Viacheslav/Shutterstock: **209** • © Rawpixel.com/Shutterstock: **398** • © ricochet64/Shutterstock: **145** • © Robert Biedermann/Shutterstock: **394** • © ronstik/Shutterstock: **73** • © rosypatterns/Shutterstock: **304** • © rukxstockphoto/Shutterstock: **60** • © Ruslan Semichev/Shutterstock: **267** • © Ruth Black/Shutterstock: **234** • © Sarah Cates/Shutterstock: **366** • © Sergey Maksimov/Shutterstock: **323** • © Shvaygert Ekaterina/Shutterstock: **266** • © sirtravelalot/Shutterstock: **310** • © snake3d/Shutterstock: **272** • © SpeedKingz/Shutterstock: **30, 314** • © Stefan Holm/Shutterstock: **273** • © Stephen Coburn/Shutterstock: **17** • © Steve Mann/Shutterstock: **262** • © StockLite/Shutterstock: **303** • © stockyimages/Shutterstock: **316, 511** • © style_TTT/Shutterstock: **203** • © Sunny_baby/Shutterstock: **302** • © Syda Productions/Shutterstock: **318** • © Take A Pix Media/Shutterstock: **398** • © Teia/Shutterstock: **184** • © Thinglass/Shutterstock: **26** • © Thomas M Perkins/Shutterstock: **502** • © Timothy Epp/Shutterstock: **375** • © Tom Wang/Shutterstock: **301** • © Tony Alt/Shutterstock: **136** • © Totsapon Phattaratharnwan/Shutterstock: **64** • © Trong Nguyen/Shutterstock: **447** • © Tyler Olson/Shutterstock: **29** • © Vector Department/Shutterstock: **60** • © Vectorpocket/Shutterstock: **135** • © Wesley Walker/Shutterstock: **360** • © Wollertz/Shutterstock: **302** • © Woman working in workshop: **32** • © xshot/Shutterstock: **457** • © Yellow duck/Shutterstock, pattang/Shutterstock, Denis Tabler/Shutterstock, Nattika/Shutterstock, rangizzz/Shutterstock: **440** • © Yuriy Golub/Shutterstock: **496** • © zstock/Shutterstock: **77** • © John Wiley & Sons Australia/Photo by Renee Bryon: **292**

Text

• © Source: Data from Services Australia. Youth Allowance for Students and Australian Apprentices. Retrieved from https://www.servicesaustralia.gov.au/how-much-youth-allowance-for-students-and-apprentices-you-can-get?context=43916: **30** • © Source: Data from Services Australia. Austudy. Retrieved from https://www.servicesaustralia.gov.au/how-much-austudy-you-can-get?context=22441: **30** • © Source: Data from Services Australia. Age pension. Retrieved from https://www.servicesaustralia.gov.au/how-much-age-pension-you-can-get?context=22526: **30** • © Queensland Curriculum and Assessment Authority: **4**

UNIT 1 Money, measurement, algebra and linear equations

Source: General Mathematics Senior Syllabus 2024 © State of Queensland (QCAA) 2024; licensed under CC BY 4.0.

1 Consumer arithmetic — earning and budgeting

LESSON SEQUENCE

Fully worked solutions for this chapter are available online.

EXAM PREPARATION

Access exam-style questions in every lesson, available online.

on Resources

Solutions	Solutions — Chapter 1 (sol-1242)
Exam questions	Exam question booklet — Chapter 1 (eqb-0258)
Digital documents	Learning matrix — Chapter 1 (doc-41467)
	Chapter summary — Chapter 1 (doc-41468)

LESSON
1.1 Overview

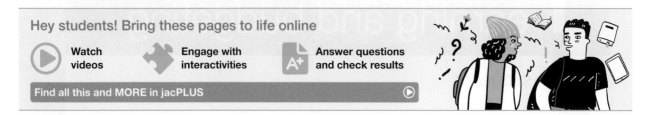
1.1.1 Introduction

Everyone needs to have an understanding of consumer mathematics and its use in real-life applications. Accountants, financial planners, bookkeepers and managers all use consumer arithmetic in their jobs. We all need to know how to check wages and salaries, and how to calculate overtime payments based on time-and-a-half or double time.

You will see percentages used for discounts at shops, interest rates for bank accounts and loans, rates of property growth or loss, statistics for sports matches, data used in the media, and company statements about profit and loss. Many employees, such as salespeople, factory workers and fruit pickers, are paid a commission or by piecework for doing their jobs. Piecework means that workers are paid for the amount of work that they have completed rather than an hourly or weekly rate. Government allowances and pensions are also an important form of income for many people.

1.1.2 Syllabus links

Lesson	Lesson title	Syllabus links
1.2	**Rates and percentages**	◯ Understand the meanings of rates and percentages.
1.3	**Wages**	◯ Calculate weekly, fortnightly or monthly wages from an annual salary.
1.4	**Earning wages**	◯ Calculate wages from an hourly rate.
1.5	**Working overtime**	◯ Calculate wages from situations involving overtime and other allowances.
1.6	**Earnings — commission and piecework**	◯ Calculate earnings based on commission or piecework.
1.7	**Payments — government allowances and pensions**	◯ Calculate income support payments based on government allowances and pensions.
1.8	**Personal budgets**	◯ Prepare a personal budget for a given income, taking into account fixed and discretionary spending.
		◯ Use a spreadsheet to display examples of the above computations when multiple or repeated computations are required, e.g. preparing a wage sheet displaying the weekly earnings of workers in an organisation, preparing a budget, investigating the potential cost of owning and operating a car over a year.

Source: General Mathematics Senior Syllabus 2024 © State of Queensland (QCAA) 2024; licensed under CC BY 4.0.

LESSON
1.2 Rates and percentages

SYLLABUS LINKS

- Understand the meaning of rates and percentages.

Source: General Mathematics Senior Syllabus 2024 © State of Queensland (QCAA) 2024; licensed under CC BY 4.0.

1.2.1 Definition of rates

Rates are used to compare two related quantities. A rate is a measure of change between two different units. Examples of rates are a car's speed in kilometres per hour (km/h), the number of loaves of bread a baker makes in a day (loaves/day) or the cost of a concert ticket for each person (cost/person). A rate is calculated per unit or per item.

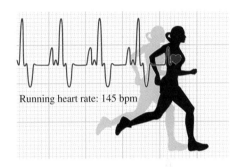

Running heart rate: 145 bpm

If a rider travels at a constant speed of 20 km/h, then they will travel 20 km in 1 hour, 40 km in 2 hours, 60 km in 3 hours and so on.

The units used for the rate depend on the units used to measure each quantity.

WORKED EXAMPLE 1 Identifying rates

After a school assembly, 689 students leave the assembly hall in 13 minutes. Determine the rate in students per minute.

THINK	WRITE
1. Identify the two quantities: number of students and time. The time is measured in minutes.	The quantities are 689 students and 13 minutes.
2. Write the rate as a fraction in terms of number of students over number of minutes.	$\dfrac{689 \text{ students}}{13 \text{ minutes}}$
3. Simplify.	$\dfrac{689 \text{ students}}{13 \text{ minutes}} = 53$ students per minute
4. Write the answer, including the units.	Students leave the assembly hall at a rate of 53 students per minute.

---

eles-3036

WORKED EXAMPLE 2 Calculating a rate in km/h

Calculate the rate (in km/h) at which you are moving if you are on a bus that travels 11.5 km in 12 minutes.

THINK	WRITE
1. Identify the two quantities: distance and time. As the question asks for the answer in km/h, convert the time quantity units from minutes to hours.	The quantities are 11.5 km and 12 minutes. $\dfrac{12}{60} = \dfrac{1}{5}$ or 0.2 hours
2. Write the rate as a fraction in terms of number of kilometres over number of hours.	$\dfrac{11.5 \text{ km}}{0.2 \text{ h}}$

▶

3. Simplify.

$$\frac{11.5 \text{ km}}{0.2 \text{ h}} = 57.5 \text{ km/h}$$

4. Write the final answer, including the units.

You are travelling at 57.5 km/h.

WORKED EXAMPLE 3 Calculating a rate of pay per hour

James works as a barista at the local café and is paid $99 for 6 hours. Calculate his rate of pay per hour.

THINK	WRITE
1. Identify the two quantities: money and time.	The quantities are $99 and 6 hours.
2. Write the rate as a fraction in terms of money over the number of hours.	$\dfrac{99}{6}$
3. Simplify.	$\dfrac{99}{6} = \$16.5$ per hour
4. Write the answer.	James is paid $16.50 per hour.

1.2.2 Percentages

The term **per cent** means 'per hundred' or 'out of a hundred' and can be written as a fraction or a decimal.

For example, $50\% = \dfrac{50}{100} = 0.5$.

Digital technology

Scientific calculators have a % button that can be used to compute calculations involving percentages.

Percentages can be converted into decimals and fractions.

Decimals and fractions can be converted into percentages.

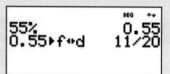

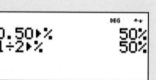

Applying discount

A discount is a reduction in price, commonly used by businesses aiming to clear out old stock or attract new customers.

Calculating discount

In general, if an $r\%$ discount is applied:

$$\text{discount} = \frac{r}{100} \times \text{original price}$$

Calculating the selling price of a discounted item

- **Method 1**
  Use the percentage remaining after the percentage discounted has been subtracted from 100%; that is, if an item for sale has a 10% discount, then the price must be 90% of the marked price.
- **Method 2**
  The new sale price of the item can be solved by calculating the amount of the discount, then subtracting the discount from the marked price.

  For example, see the two different methods used to calculate the sale price on a pair of shoes marked $95 if a 10% discount is given.

Method 1

$$\begin{aligned}\text{Sale price} &= 90\% \text{ of } \$95 \\ &= \$85.50\end{aligned}$$

Method 2

$$\begin{aligned}\text{Discount} &= \frac{r}{100} \times \text{original price} \\ &= \frac{10}{100} \times 95.00 \\ &= \$9.50 \\ \text{Sale price} &= \text{marked price} - \text{discount} \\ &= \$95.00 - \$9.50 \\ &= \$85.50\end{aligned}$$

In other words, reducing the price by 10% is the same as multiplying by 90% or $(100 - 10)\%$.

Increasing or decreasing a quantity by $x\%$

To decrease a quantity by $x\%$, multiply by $(100 - x)\%$.

To increase a quantity by $x\%$, multiply by $(100 + x)\%$.

Note: **Convert the percentage to a decimal or fraction before multiplying.**

WORKED EXAMPLE 4 Calculating a percentage increase

Calculate an increase to $76 by 15%.

THINK	WRITE
1. The original percentage is always 100%, so an increase of 15% means a total of 115%.	$(100 + 15)\% = 115\%$
2. Write 115% as a fraction and then express it as a decimal.	$115\% = \frac{115}{100}$ $= 1.15$
3. Multiply the amount $76 by 1.15.	$76 \times 1.15 = 87.40$
4. Write the answer.	$87.40

eles-6361

WORKED EXAMPLE 5 Calculating the sales price

Sarah bought a car for $7500 and sold it 4 years later for 30% less than she paid for it. Calculate the price she sold the car for.

THINK	WRITE
1. The original percentage is always 100%, so a discount of 30% means a total of 70%.	$(100 - 30)\% = 70\%$
2. Write 70% as a fraction and then express it as a decimal.	$70\% = \dfrac{70}{100}$ $= 0.70 \text{ or } 0.7$
3. Multiply the amount $7500 by 0.7.	$7500 \times 0.7 = 5250$
4. Write the answer.	Sarah sold it for $5250.

1.2.3 Percentage increase and decrease

Percentage increase and decrease can be used to calculate sale prices, discounts, profits and many other quantities. It is calculated as a percentage of the original amount.

Percentage change

The formula for calculating the percentage increase/decrease is:

$$\text{Percentage increase/decrease} = \frac{\text{amount of increase/decrease}}{\text{original amount}} \times 100$$

WORKED EXAMPLE 6 Calculating the percentage discount

Ramon bought a laptop in a sale for $774.40. If the original price was $968, calculate the percentage discount.

THINK	WRITE
1. Calculate the amount of the discount by subtracting $774.40 from $968.	Discount $= \$968 - \$774.40$ $= \$193.60$
2. Calculate the percentage discount by using the formula: percentage discount $= \dfrac{\text{amount of discount}}{\text{original amount}} \times 100$.	Percentage discount $= \dfrac{193.60}{968} \times 100$ $= 20$
3. Write the answer as a percentage.	The percentage discount is 20%.

When a large number of values are being considered in a problem involving percentages, spreadsheets or other technologies can be useful to help carry out most of the associated calculations.

For example, a spreadsheet can be set up so that entering the original price of an item will automatically calculate several different percentage increases for comparison.

	A	B	C	D	E
1	Original		Increase by:		
2	price	+5%	+8%	+12%	+15%
3	$100.00	$105.00	$108.00	$112.00	$115.00
4	$150.00	$157.50	$162.00	$168.00	$172.50
5	$200.00	$210.00	$216.00	$224.00	$230.00
6	$300.00	$315.00	$324.00	$336.00	$345.00
7	$450.00	$472.50	$486.00	$504.00	$517.50

Original price	increased by			
	0.05	0.08	0.12	0.15
100	=A3*1.05	=A3*1.08	=A3*1.12	=A3*1.15
150	=A4*1.05	=A4*1.08	=A4*1.12	=A4*1.15
200	=A5*1.05	=A5*1.08	=A5*1.12	=A5*1.15
300	=A6*1.05	=A6*1.08	=A6*1.12	=A6*1.15
350	=A7*1.05	=A7*1.08	=A7*1.12	=A7*1.15

Exercise 1.2 Rates and percentages

learn on

1.2 Exercise	1.2 Exam questions on

Simple familiar	Complex familiar	Complex unfamiliar
1, 2, 3, 4, 5, 6, 7, 8, 9	10, 11, 12	13, 14

These questions are even better in jacPLUS!
• Receive immediate feedback
• Access sample responses
• Track results and progress

Find all this and MORE in jacPLUS ▶

Simple familiar

1. **WE1** Using the units stated, calculate the rates for:
   a. 4-kg bag of apples that cost $23.60 expressed in $/kg
   b. a tank that loses 1320 mL of water in 2 hours expressed in mL/h
   c. a 3.6-metre-long carpet that costs $67.14 expressed in $/m
   d. a basketball player who has scored a total of 833 points in 68 games expressed in points/game.

2. Calculate the following rates when the units are changed as indicated. Where necessary, give answers correct to 2 decimal places.
   a. 1.5 m/s to km/h
   b. 60 km/h to m/s
   c. 65 cents per gram to $/kg
   d. $5.65 per kilogram to cents per gram

3. **WE2** Calculate the rate (in km/h) that you are moving if you are in a passenger aircraft that travels 1770 km in 100 minutes.

4. **WE3** Michael works as an apprentice chef and is paid $424.80 for 36 hours. Determine his rate of pay per hour.

5. **WE4** Calculate the following increases:
   a. $35 by 8%
   c. $142.85 by 22.15%
   b. $96 by 12.5%
   d. $42 184 by 0.285%.

6. Calculate the following decreases:
   a. $54 by 16%
   c. $102.15 by 32.15%
   b. $7.65 by 3.2%
   d. $12 043 by 0.0455%.

7. **WE5** A clothing shop is closing down and offers a discount of 25% on all items. Calculate how much Jade would pay for a pair of jeans that were originally $80.

8. **WE6** Determine the percentage discount if a piece of silverware has a price tag of $168 at a market, but the seller is bartered down and sells it for $147.

9. The following graph shows the change in the price of gold (in US dollars per ounce) from 27 July to 27 August.

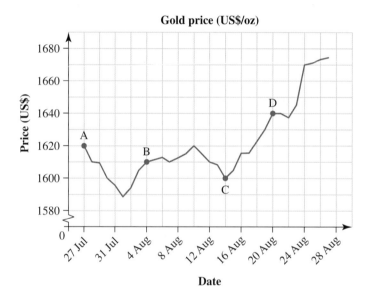

Gold price (US$/oz)

a. Calculate the percentage change from:
   i. the point marked A to the point marked B
   ii. the point marked C to the point marked D.
b. Calculate the percentage change from the point marked A to the point marked D.

Complex familiar

10. The price of a bottle of wine was originally $19.95. After it received an award for wine of the year, the price was increased by 12.25%. Twelve months later the price was reduced by 15.5%.
    Calculate the percentage change of the final price from the original price.

11. A student's test results in Mathematics are shown in the table.

	Test 1	Test 2	Test 3	Test 4
Mark	$\dfrac{14}{21}$	$\dfrac{42}{60}$	$\dfrac{14.5}{20}$	$\dfrac{26}{35}$
Percentage				

Complete the table and calculate the overall percentage increase or decrease in the student's marks from the first test to the last test.

12. The ladder for the top four teams in the A-League is shown in the following table.

Team	Win	Loss	Draw	Goals for	Goals against
1. Western Sydney Wanderers	18	6	3	41	21
2. Central Coast Mariners	16	5	6	48	22
3. Melbourne Victory	13	9	5	48	45
4. Adelaide United	12	10	5	38	37

Use a spreadsheet to:

a. express the win, loss and draw columns as a percentage of the total games played, correct to 2 decimal places

b. express the goals for as a percentage of the total goals, correct to 2 decimal places.

Complex unfamiliar

13. a. An advertisement for bedroom furniture states that you save $55 off the recommended retail price when you buy it for $385. Calculate by what percentage the price has been reduced.

   b. If another store was advertising the same furniture for 5% less than the sale price of the first store, calculate the percentage the price has been reduced from the recommended retail price.

14. A house originally purchased for $320 000 is sold to a new buyer at a later date for $377 600.
The new buyer pays a deposit of 15% and borrows the rest from a bank. They are required to pay the bank 5% of the total amount borrowed each year.
If they purchased the house as an investment, calculate how much they should charge in rent per month in order to fully cover their bank payments.

Fully worked solutions for this chapter are available online.

LESSON
1.3 Wages

SYLLABUS LINKS

• Calculate weekly, fortnightly or monthly wages from an annual salary.

Source: General Mathematics Senior Syllabus 2024 © State of Queensland (QCAA) 2024; licensed under CC BY 4.0.

1.3.1 Wages and salaries

Employees may be paid for their work in a variety of ways. Most receive either a **wage** or a **salary**.

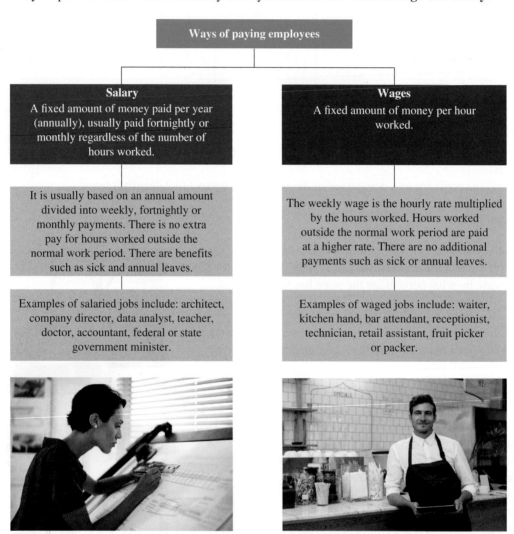

Ways of paying employees

Salary
A fixed amount of money paid per year (annually), usually paid fortnightly or monthly regardless of the number of hours worked.

Wages
A fixed amount of money per hour worked.

It is usually based on an annual amount divided into weekly, fortnightly or monthly payments. There is no extra pay for hours worked outside the normal work period. There are benefits such as sick and annual leaves.

The weekly wage is the hourly rate multiplied by the hours worked. Hours worked outside the normal work period are paid at a higher rate. There are no additional payments such as sick or annual leaves.

Examples of salaried jobs include: architect, company director, data analyst, teacher, doctor, accountant, federal or state government minister.

Examples of waged jobs include: waiter, kitchen hand, bar attendant, receptionist, technician, retail assistant, fruit picker or packer.

Key points

• **Normal working hours in Australia are *38 hours* per week.**
• **There are *52 weeks* in a year.**
• **There are *26 fortnights* in a year (this value will be slightly different for a leap year).**
• **There are *12 months* in a year.**

WORKED EXAMPLE 7 Calculating the fortnightly pay

Dimitri works as an accountant and receives an annual salary of $83 460. Calculate the pay that Dimitri is paid each fortnight.

THINK	WRITE
1. There are 26 fortnights in a year, so we divide $83 460 by 26.	Fortnightly pay = $83 460 ÷ 26
2. Evaluate.	= $3210
3. Write the answer.	Dimitri's fortnightly pay is $3210.

WORKED EXAMPLE 8 Calculating the annual salary

Grace is a solicitor who is paid $7500 per month. Calculate Grace's annual salary.

THINK	WRITE
1. There are 12 months in a year, so multiply $7500 (monthly pay) by 12.	Annual salary = $7500 × 12
2. Evaluate.	= $90 000
3. Write the answer.	Grace's annual salary is $90 000.

eles-6364

WORKED EXAMPLE 9 Calculating an hourly rate of pay

Charlotte works as a laboratory technician and is paid an annual salary of $71 560. If Charlotte works an average of 42 hours per week, calculate her equivalent hourly rate of pay.

THINK	WRITE
1. Calculate the weekly pay by dividing the salary by 52.	Weekly pay = $71 560 ÷ 52 = $1376.15
2. Calculate the equivalent hourly rate by dividing the weekly pay by 42.	Hourly rate = $1376.15 ÷ 42 = $32.77
3. Write the answer.	Charlotte's hourly rate of pay is $32.77 per hour.

1.3.2 Wages and inflation

Inflation measures how much more expensive a set of goods and services has become over a certain period, usually a year. It is an increase in the price of goods and services and a decrease in the value of our money.

Inflation is a rate that is expressed as a percentage and in Australia is called the consumer price index (CPI). As inflation increases, the spending power of a set amount of money will decrease.

For example, if the cost of a loaf of bread was $4.00 and rose with inflation, then in five years it might cost $4.50. As inflation gradually decreases the spending power of the dollar, peoples' salaries often increase in line with inflation. This increase counterbalances the decreasing spending power of money.

eles-6365

WORKED EXAMPLE 10 Calculating a weekly salary with a CPI increase

George received a weekly salary of $1050 in 2022. The rate of inflation was 5.4% in 2023. Calculate his weekly salary at the end of 2023 if it increased with the CPI.

THINK	WRITE
1. An increase in inflation of 5.4% means a total of 105.4%.	$(100 + 5.4)\% = 105.4\%$
2. Write 105.4% as a fraction and then express it as a decimal.	$105.4\% = \dfrac{105.4}{100}$ $= 1.054$
3. Multiply the weekly salary $1050 by 1.054.	$1050 \times 1.054 = 1106.70$
4. Write the answer.	His salary will be $1106.70.

on Resources

Digital documents   SkillSHEET Converting units of time (doc-10849)
                    SkillSHEET Multiplying and dividing a quantity (money) by a whole number (doc-10850)
                    SkillSHEET Payroll calculations (doc-1439)

Exercise 1.3 Wages

 learn on

1.3 Exercise	1.3 Exam questions on

Simple familiar	Complex familiar	Complex unfamiliar
1, 2, 3, 4, 5, 6, 7, 8, 9	10, 11, 12, 13	14

These questions are even better in jacPLUS!
• Receive immediate feedback
• Access sample responses
• Track results and progress

Find all this and MORE in jacPLUS ▶

Simple familiar

1. **WE7** Nga is paid a salary of $67 860 per annum. Calculate Nga's fortnightly pay.

2. Roger is paid a salary of $86 840 per annum. Calculate Roger's weekly pay.

3. Frieda is paid a salary of $84 000 per annum. Calculate Frieda's monthly pay.

4. Wendy works as an office secretary and is paid a salary of $76 440 per annum. Calculate Wendy's pay if she is paid:

   a. weekly   b. fortnightly

5. **WE8** Maxine is paid a salary. She receives $1060 per week. Calculate Maxine's annual salary.

6. Thao receives $2250 per fortnight. Calculate Thao's annual salary.

7. **MC** Determine which of the following people receives the greatest salary.
   A. Goran, who receives $1060 per week   B. Bryan, who receives $2150 per fortnight
   C. Wayne, who receives $4660 per month   D. Chris, who receives $2200 per fortnight

8. **WE9** Fiona receives a salary of $44 116.80 per annum. If Fiona works an average of 40 hours per week, calculate the equivalent hourly rate of pay.

9. **WE10** Tina receives a weekly salary of $890. Tina receives a salary increase equal to the rate of inflation. If the rate of inflation is 5.5%, determine her new salary.

Complex familiar

10. Use a spreadsheet to complete the table below for food production employees.

Annual salary	Weekly pay	Fortnightly pay	Monthly pay
$156000			
$93600			
$62400			
$71760			
$109200			

11. Use a spreadsheet to calculate the wages for each of the following eight employees.

Employee	Hourly pay rate	Hours worked per week	Wages per week	Wages per annum
A. Richardson	$35.50	20		
S. Provis	$21.75	35		
N. Liu	$27.45	8		
W. Naidoo	$29.55	28		
G. Riddell	$35.75	16		
C. Ve	$23.60	38		
O. Chang	$31.00	26		
D. Evans	$22.00	13		

12. Jade receives a salary of $66 000 per annum and works an average of 36 hours each week. Calculate the hourly rate to which Jade's salary is equivalent. Give your answer correct to the nearest cent.

13. Lisa is on an annual salary of $71 552. Letitia is on a wage and is paid $16.00 per hour.
    If Lisa works an average of 42 hours per week, calculate whether Lisa or Letitia receives the better rate of pay.

Complex unfamiliar

14. Garry earns $85 000 per year while his friend Henry earns $37.00 per hour. Calculate the number of hours that Henry will need to work each week to earn more money than Garry does.

Fully worked solutions for this chapter are available online.

LESSON
1.4 Earning wages

SYLLABUS LINKS

- Calculate wages from an hourly rate and other allowances.

Source: General Mathematics Senior Syllabus 2024 © State of Queensland (QCAA) 2024; licensed under CC BY 4.0.

1.4.1 Wages paid at an hourly rate

A **wage** is paid at an hourly rate. The wage for each week is calculated by multiplying the normal rate of pay by the number of hours worked during that week.

WORKED EXAMPLE 11 Calculating a weekly wage

Sadiq works as a mechanic and is paid $37.50 per hour.
Calculate Sadiq's wage in a week during which he works 38 hours.

THINK	WRITE
1. Multiply $37.50 (the hourly rate) by 38 (the number of hours worked).	Wage $= \$37.50 \times 38$ $= \$1425.37$
2. Write the answer.	Sadiq's wage for the week is $1425.37.

To compare two people's wages, we need to consider the number of hours each has worked. Wages are compared by looking at the hourly rate. To calculate the hourly rate of an employee, we need to divide the wage by the number of hours worked.

WORKED EXAMPLE 12 Calculating the hourly rate given the weekly wage

Georgina works 42 hours as a computer technician.
Her wage for the week totalled $2123.10.
Calculate Georgina's hourly rate of pay.

THINK	WRITE
1. Divide $2123.10 (the wage) by 42 (the number of hours worked).	Hourly rate = $2123.10 ÷ 42 = $50.55
2. Write the answer.	Georgina's hourly rate of pay is $50.55.

1.4.2 Allowance

Sometimes wages are increased because an **allowance** is paid for working in unfavourable conditions. An allowance is an additional payment made when the working conditions are difficult or unpleasant.

For example, a road worker may be paid an allowance for working in the rain or in severe weather conditions.

WORKED EXAMPLE 13 Calculating a wage with an hourly rate plus allowance

Ryan is a road worker and is paid $36.50 per hour
for a 35-hour week. For working on wet days he is
paid a wet weather allowance of $5.46 per hour.
Calculate Ryan's pay if for 12 hours of the week he
works in the rain.

THINK	WRITE
1. Calculate Ryan's normal pay by multiplying $36.50 (the hourly rate) by 35 (the number of hours worked).	Normal pay = $36.50 × 35 = $1277.50
2. Calculate the wet weather allowance by multiplying $5.46 (the wet weather allowance) by 12 (the number of hours worked in the wet).	Allowance = $5.46 × 12 = $65.52
3. Add the normal pay to the wet weather allowance to calculate the total pay.	Total pay = $1277.50 + $65.52 = $817.02
4. Write the answer.	Ryan's total pay is $817.02.

Exercise 1.4 Earning wages

learnon

1.4 Exercise	1.4 Exam questions on

Simple familiar	Complex familiar	Complex unfamiliar
1, 2, 3, 4, 5, 6, 7, 8, 9, 10	11, 12	13, 14

These questions are even better in jacPLUS!
- Receive immediate feedback
- Access sample responses
- Track results and progress

Find all this and MORE in jacPLUS ▶

Simple familiar

1. **WE11** Allan works for a cleaning company and is paid $22.95 per hour. Calculate Allan's wage in a week during which he works 40 hours.

2. Alicia is an apprentice chef. In the first year of her apprenticeship she earns $21.80 per hour. Calculate Alicia's wage in a week during which she works:

   a. 36 hours
   b. 48 hours
   c. 42.5 hours.

3. Domonic is a fully qualified refrigeration mechanic. He earns $43.50 per hour. Calculate Domonic's wage in a week during which he works:

   a. 32 hours
   b. 37 hours
   c. 44.5 hours.

4. **MC** Select the worker who earns the highest wage for the week.

   A. Dylan, who works 35 hours at $27.00 per hour
   B. Lachlan, who works 37 hours at $25.86 per hour
   C. Connor, who works 38 hours at $25.34 per hour
   D. Cameron, who works 40 hours at $24.38 per hour

5. **WE12** Calculate the hourly rate of a person who works 40 hours for a wage of $896.00.

6. Julie earns $22.84 per hour. Calculate the number of hours worked by Julie during a week in which she is paid $845.08.

7. **MC** Select the worker who is paid at the highest hourly rate.

   A. Melissa, who works 35 hours for $732.90
   B. Belinda, who works 36 hours for $752.40
   C. April, who works 38 hours for $799.52
   D. Nicole, who works 40 hours for $839.32

8. **WE13** Ingrid works as an industrial cleaner and is paid $34.60 per hour for a 35-hour working week. When Ingrid is working with toxic substances, she is paid an allowance of $1.08 per hour.
   Calculate Ingrid's pay if she works with toxic substances all week.

9. Copy and complete the table below for these casual workers and their wages.

Name	Wage	Hours worked	Hourly rate
Brent	$832.32	36	
Sandra	$1077.60	40	
Ann	$739.26	37	
Chris	$1627.92		$19.38
Anna	$462.60		$15.42
Toni	$1553.44		$20.44

10. **MC** Select the person who worked the greatest number of hours.
    A. Su-Li, who earned $979.32 at $30.60 per hour
    B. Denise, who earned $1152 at $28.80 per hour
    C. Vera, who earned $666.40 at $19.04 per hour
    D. Camille, who earned $1414.50 at $34.50 per hour

Complex familiar

11. Katherine works as a casual waitress. Casual workers earn
    20% more per hour than full-time workers to compensate for
    their lack of holidays and sick leave. A full-time waitress earns
    $32.50 per hour.
    Calculate Katherine's wage in a week during which she works
    6 hours on Saturday and 7 hours on Sunday.

12. Richard works as an electrician and is paid $120.94 per hour for
    a 38-hour week. When he has to work at heights, he is paid a
    $3.50 per hour 'height allowance'.
    Calculate Richard's pay during a week in which he spends 15 hours
    working at heights.

Complex unfamiliar

13. Rema works as a tailor and earns $29.50 per hour. She works 37 hours per week. Zhong is Rema's assistant
    and earns $27.60 per hour. Determine the least time Zhong must work if he is to earn more money than
    Rema does.

14. Tamarin works 38 hours per week at $27.80 per hour. Zoe earns the same amount each week as Tamarin
    does, but Zoe works a 40-hour week. Calculate Zoe's hourly rate of pay.

Fully worked solutions for this chapter are available online.

LESSON
1.5 Working overtime

SYLLABUS LINKS

- Calculate wages from situations involving overtime.

Source: General Mathematics Senior Syllabus 2024 © State of Queensland (QCAA) 2024; licensed under CC BY 4.0.

1.5.1 Overtime and penalties

Overtime is paid when an employee works more than the regular hours of work.

When an employee works overtime, a higher rate is paid. This higher rate of pay is called a **penalty rate**. The rate is normally calculated at either:

1. **time and a half**, which means that the person is paid $1\frac{1}{2}$ times the usual rate of pay, or

2. **double time**, which means that the person is paid twice the normal rate of pay.

A person may also be paid these overtime rates for working at unfavourable times such as at night or during weekends.

Calculating overtime

The overtime hourly rate is usually a multiple of the regular hourly rate. Some examples of overtime rates include:
- **1.5 × regular hourly wage (*time and a half*)**
- **2 × regular hourly wage (*double time*)**
- **2.5 × regular hourly wage (*double time and a half*)**

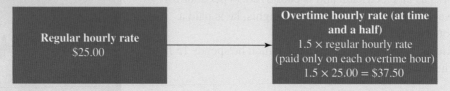

Regular hourly rate $25.00 → Overtime hourly rate (at time and a half) 1.5 × regular hourly rate (paid only on each overtime hour) 1.5 × 25.00 = $37.50

WORKED EXAMPLE 14 Calculating an hourly rate at time and a half

Gustavo is paid $25.70 per hour in his job as a childcare worker. Calculate Gustavo's hourly rate when he is being paid for overtime at time and a half.

THINK	WRITE
1. Multiply $25.70 (the normal hourly rate) by 1.5 (the overtime factor for time and a half). (*Note:* A half is expressed in decimal form as 0.5.)	Time-and-a-half rate $= \$25.70 \times 1.5$ $= \$38.55$
2. Write the answer.	Gustavo's hourly rate when he is being paid for overtime at time and a half is $38.55.

WORKED EXAMPLE 15 Calculating the amount earned at time and a half

Adrian works as a shop assistant and his normal rate of pay is $29.90 per hour. Calculate the amount that Adrian earns for 6 hours work on Saturday, when he is paid time and a half.

THINK	WRITE
1. Multiply $29.90 (the normal pay rate) by 1.5 (the overtime factor) and by 6 (hours worked at time and a half).	Pay = $29.90 \times 1.5 \times 6$ $\qquad = \$269.10$
2. Write the answer.	Adrian will earn $269.10.

When we need to calculate the total pay for a week that involves overtime, we need to calculate the normal pay and then add the amount earned for any overtime.

eles-6368

WORKED EXAMPLE 16 Calculating the weekly pay with overtime

Natasha works as a waitress and is paid $27.80 per hour for a 38-hour week. Calculate Natasha's pay in a week during which she works 5 hours at time and a half in addition to her regular hours.

THINK	WRITE
1. Calculate Natasha's normal pay.	Normal pay = 27.80×38 $\qquad = \$1056$
2. Calculate Natasha's pay for 5 hours at time and a half.	Pay at time and a half = $27.80 \times 1.5 \times 5$ $\qquad = \$208.50$
3. Add the normal pay and the time-and-a-half pay together.	Total pay = $1056 + $208.50 $\qquad = \$1264.90$
4. Write the answer.	Natasha's total pay is $1264.90.

Some examples of jobs will have more than one overtime rate to consider, and some will require you to calculate how many hours have been worked at each rate.

eles-6369

WORKED EXAMPLE 17 Calculating the weekly pay with different overtime rates

Graeme is employed as a car assembly worker and is paid $31.10 per hour for a 36-hour week. If Graeme works overtime, the first 6 hours are paid at time and a half and the remainder at double time.
Calculate Graeme's pay in a week during which he works 45 hours.

THINK	WRITE
1. Calculate the number of hours overtime Graeme worked.	Overtime = 45 − 36 $\qquad = 9$ hours
2. Of these nine hours, calculate how much was at time and a half and how much was at double time.	Time and a half = 6 hours Double time = 3 hours

3. Calculate Graeme's normal pay.

Normal pay $= \$31.10 \times 36$
$= \$1119.60$

4. Calculate what Graeme is paid for 6 hours at time and a half.

Pay at time and a half $= \$31.10 \times 1.5 \times 6$
$= \$279.90$

5. Calculate what Graeme is paid for 3 hours at double time.

Pay at double time $= \$31.10 \times 2 \times 3$
$= \$186.60$

6. Calculate Graeme's total pay by adding the time-and-a-half and double-time payments to his normal pay.

Total pay $= \$1119.60 + \$279.90 + \$186.60$
$= \$1586.10$

7. Write the answer.

Graeme's total pay is $1586.10.

 Resources

 Digital document SkillSHEET Multiplying and dividing a quantity (money) by a fraction (doc-10851)

Exercise 1.5 Working overtime

learn on

1.5 Exercise	**1.5 Exam questions**

Simple familiar	Complex familiar	Complex unfamiliar
1, 2, 3, 4, 5, 6, 7, 8	9, 10, 11, 12	13, 14

These questions are even better in jacPLUS!
- Receive immediate feedback
- Access sample responses
- Track results and progress

Find all this and MORE in jacPLUS ⊙

Simple familiar

1. **WE14** Reece works in a restaurant and is paid a normal hourly rate of $25.60. Calculate the amount Reece earns each hour when he is being paid time and a half.

2. Gareth works in a warehouse and is normally paid $26.70 per hour. For working on a Sunday, he is paid time and a half and on a public holiday at double time. Calculate Gareth's hourly rate of pay on:

   a. a Sunday   **b.** a public holiday.

3. **WE15** Ben works in a hotel and is paid $26.40 per hour. Calculate the total amount Ben will earn for an 8-hour shift on Saturday when he is paid at time and a half.

4. **MC** Ernie works as a chef and is paid $29.90 per hour. Select Ernie's hourly rate when he is paid time and a half for overtime.

   A. $14.95
   B. $22.43
   C. $44.85
   D. $59.80

5. **MC** Stephanie works in a florist's shop and is paid $29.40 per hour. Calculate how much more Stephanie will earn for 8 hours work at time and a half than she would at normal pay rates.

A. $14.70
B. $44.10
C. $117.60
D. $352.80

6. Copy and complete the table below.

Name	Ordinary rate	Normal hours	Time-and-a-half hours	Double-time hours	Total pay
W. Clark	$18.60	38	4	—	
A. Hurst	$19.85	37	—	6	
S. Gannon	$24.50	38	5	2	
G. Dymock	$36.23	37.5	4	1.5	
D. Colley	$34.90	36	6	8	

7. **WE16** Rick works 37 hours at normal time each week and receives $22.40 per hour. Calculate Rick's pay in a week when, in addition to his normal hours, he works 4 hours overtime at time and a half.

8. Grant works as a courier and is paid $32.04 per hour for a 35-hour working week. Calculate Grant's pay for a week during which he works 4 hours at time and a half and 2 hours at double time in addition to his regular hours.

Complex familiar

9. Eric works on the wharves unloading containers. Eric is paid $24.20 per hour. Calculate the number of hours at time and a half that Eric will have to work to earn the same amount of money that he will earn in 9 hours at normal pay rates.

10. Jenny is a casual worker at a motel. The normal rate of pay is $40 per hour. Jenny works 8 hours on Saturday for which she is paid time and a half. On Sunday, she works 6 hours for which she is paid double time.
    Determine how many hours Jenny would need to work at her normal rate of pay to earn the same amount of money as she earns on Saturday and Sunday combined.

11. **WE17** Steven works on a car assembly line and is paid $40 per hour for a 36-hour working week. The first 4 hours overtime he works each week is paid at time and a half with the rest paid at double time. Calculate Steven's earnings for a week in which he works 43 hours.

12. Zac works in a supermarket. He is paid at an ordinary rate of $24 per hour. If Zac works more than 8 hours on any one day, the first two hours are paid at time and a half and the rest at double time.
    Calculate Zac's pay if the hours worked each day are:
    Monday — 8 hours
    Tuesday — 9 hours
    Wednesday — 12 hours
    Thursday — 7 hours
    Friday — 10.5 hours.

Complex unfamiliar

13. Patricia works a 35-hour week and is paid $29.60 per hour. Any overtime that Patricia does is paid at time and a half. Patricia wants to work enough overtime so that she earns more than $1200 each week.
    Calculate the minimum number of hours that Patricia will need to work to earn this amount of money.

14. Megan works a 38-hour week and for any extra time she is paid at time and a half. When she worked a 45-hour week, she received $1358.
    Calculate her earning for a week in which she worked 40 hours.

Fully worked solutions for this chapter are available online.

LESSON
1.6 Earnings — commission and piecework

SYLLABUS LINKS

- Calculate earnings based on commission or piecework.

Source: General Mathematics Senior Syllabus 2024 © State of Queensland (QCAA) 2024; licensed under CC BY 4.0.

1.6.1 Commission

Commission is paid to a salesperson to motivate them to sell more products. For example, when real estate agents sell houses, they are paid a commission. It is calculated as a percentage of the total value of the goods sold.

In addition to commission, most salespeople receive a fixed amount per week, called a **retainer**. A retainer does not depend on sales.

WORKED EXAMPLE 18 Calculating the commission paid on weekly sales figure

Jack is a computer salesman who is paid a commission of 12% of all sales. Calculate the commission that Jack earns in a week if he makes sales to the value of $15 000.

THINK	WRITE
1. Convert 12% to a decimal.	$12\% = \dfrac{12}{100}$ $= 0.12$
2. Calculate 12% of $15 000.	$\text{Commision} = 0.12 \times \$15\,000 = \$1800$
3. Write the answer.	Jack earns $1800 commission.

WORKED EXAMPLE 19 Calculating the weekly wage when adding a sales commission to a retainer

Peter works in a menswear store. He earns 7.5% commission on all sales on top of his retainer of $600 per week. Calculate Peter's wage in a week when his sales are $7400.

THINK	WRITE
1. Convert 7.5% to a decimal.	$7.5\% = \dfrac{7.5}{100} = 0.075$
2. Calculate 7.5% of 7400.	$\text{Commission} = 0.075 \times 7400$ $= 555$
3. Calculate the total wage by adding the retainer to the commission.	$\text{Peter's wage} = \text{Retainer} + \text{commission}$ $= 600 + 555$ $= 1155$
4. Write the answer.	Peter's wage is $1155.

eles-6371

WORKED EXAMPLE 20 Calculating the amount earned from commission only

A real estate agent is paid commission on her sales at the following rate:
- **1.75% on the first $830 000**
- **2.5% on the balance of the sale price.**

Calculate the commission earned on the sale of a property for $935 000.

THINK	WRITE
1. Calculate 1.75% of $830 000.	Commission 1 = 1.75% of $830 000 $$= \frac{\$1.75}{100} \times 830\,000$$ $$= \$145\,25$$
2. Calculate the balance of the sale.	Balance = $935 000 − $830 000 $$= \$105\,000$$
3. Calculate 2.5% of $105 000.	Commission 2 = 2.5% of $105 000 $$= \frac{2.5}{100} \times 105\,000$$ $$= \$2625$$
4. Add up each portion to calculate the commission.	Total commission = $14 525 + $2625 $$= \$17\,150$$
5. Write the answer.	The real estate agent is paid a total commission of $17 150.

1.6.2 Piecework

Piecework means that workers are paid for the amount of work that they have completed rather than an hourly or weekly pay rate. Permanent employees are not paid by piecework; it is usually factory workers, fruit pickers and packers who are paid piece rates.

WORKED EXAMPLE 21 Calculating the amount paid for letterbox drops

Sian delivers flyers to letterboxes and is paid $25 for 1000 brochures.

Calculate how much she earned for delivering 3500 brochures.

THINK	WRITE
1. Calculate the number of thousands of brochures delivered.	$$\frac{3500}{1000} = 3.5$$
2. Multiply by $25 to calculate what Sian is paid.	Pay = 3.5 × 25 $$= 87.5$$
3. Write the answer.	Sian earns $87.50.

eles-6372

WORKED EXAMPLE 22 Calculating the amount paid for machine work

Michelle works as a machine worker in a factory. She is paid $2.75 each for the first 150 items that she sews and then $3.35 per item thereafter.
Calculate how much she get paid if she sews 295 items.

THINK	WRITE
1. Determine how much she is paid for 150 items.	Payment 1 $= 150 \times 2.75 = 412.50$
2. Calculate how many items more than 150 she sews.	Number of items more than $150 = 295 - 150 = 145$
3. Determine how much she is paid for 145 items.	Payment 2 $= 145 \times 3.35$ $= 485.75$
4. Calculate how much she is paid for 295 items.	Total payment $= 412.50 + 485.75$ $= 898.25$
5. Write the answer.	Michelle is paid $898.25.

WORKED EXAMPLE 23 Comparing commission contracts

Bob and Sanjeev are part-time employees of a mobile phone company. Bob is paid $600 plus 2.4% of sales per week and Sanjeev is on a commission-only contract, paid 22% of sales per week.
a. Use a spreadsheet to compare their salaries for sales up to $2500. Determine who earns more money for sales up to $2500.
b. Determine for what value of sales it is better to work on a commission-only contract.

THINK

1. Set up headings and value amounts of sales in $500 increments to $2500.

	A	B	C
1	Sales	Retainer + 2.4% Commission	22% Commission
2	0		
3	500		
4	1000		
5	1500		
6	2000		
7	2500		

2. Enter formula in cell B2 to calculate a salary of $600 plus 2.4% commission on sales and fill down.

	A	B	C
1	Sales	Retainer + 2.4% Commission	22% Commission
2	0	=600+0.024*A2	
3	500		
4	1000		
5	1500		
6	2000		
7	2500		

	A	B	C
1	Sales	Retainer + 2.4% Commission	22% Commission
2	0	600	
3	500	612	
4	1000	624	
5	1500	636	
6	2000	648	
7	2500	660	

3. Enter formula in cell C2 to calculate a salary of 22% commission on sales and fill down.

	A	B	C
1	Sales	Retainer + 2.4% Commission	22% Commission
2	0	600	=0.22*A2
3	500	612	
4	1000	624	
5	1500	636	
6	2000	648	
7	2500	660	

	A	B	C
1	Sales	Retainer + 2.4% Commission	22% Commission
2	0	600	0
3	500	612	110
4	1000	624	220
5	1500	636	330
6	2000	648	440
7	2500	660	550

4. Write the answer to part **a.**

A retainer of $600 plus 2.4% commission on sales is better for sales up to $2500, so Bob earns more money.

5. Complete the sales column for amounts up to $5000.

	A	B	C
1	Sales	Retainer + 2.4% Commission	22% Commission
2	0	600	0
3	500	612	110
4	1000	624	220
5	1500	636	330
6	2000	648	440
7	2500	660	550
8	3000	672	660
9	3500	684	770
10	4000	696	880
11	4500	708	990
12	5000	720	1100

6. Write the answer to part **b.**

Sales of $3500 or more earn a better wage when 22% commission is paid.

Exercise 1.6 Earnings — commission and piecework

1.6 Exercise	**1.6 Exam questions**

Simple familiar	Complex familiar	Complex unfamiliar
1, 2, 3, 4, 5, 6, 7, 8, 9, 10, 11	12, 13, 14	N/A

These questions are even better in jacPLUS!
- Receive immediate feedback
- Access sample responses
- Track results and progress

Find all this and MORE in jacPLUS ▶

Simple familiar

1. **WE18** Kylie is an insurance salesperson. She is paid 8% of the value of any insurance that she sells. Calculate the amount that Kylie is paid for selling insurance to the value of $25 000.

2. **MC** Ursula is a computer software salesperson. Ursula's sales total $105 000 and she is paid a commission of 0.8%. Calculate how much Ursula receives in commission.

A. $105 **B.** $840 **C.** $1050 **D.** $8400

3. **MC** Asif is a sales representative for a hardware firm. Asif earns $870 commission on sales of $17 400. Calculate the rate of commission Asif receives.

A. 0.05%

B. 0.5%

C. 5%

D. 10%

4. **WE19** Stanisa is a car salesman who is paid a retainer of $250 per week plus a commission of 2% of any sales he makes. Calculate Stanisa's pay in a week where his sales total $35 000.

5. **MC** In a group of sales representatives, each representative has $10 000 in sales. Identify who earns the most money.

A. Averil, who is paid a commission of 8%

B. Bernard, who is paid $250 plus 6% commission

C. Cathy, who is paid $350 plus 4% commission

D. Darrell, who is paid $540 plus 2.5% commission

6. **WE20** A real estate agent charges commission at the following rate:
   - 1.75% on the first $830 000
   - 2.5% on the balance of the sale price.

   Calculate the commission charged on the sale of a property for $870 000.

7. **WE21** Matthew delivers pamphlets to local letterboxes. He is paid $21.80 per thousand pamphlets delivered. Calculate what Matthew will be paid for delivering 15 000 pamphlets.

8. Julia works after school at a car yard detailing cars. If Julia is paid $10.85 per car, calculate what she will earn in an afternoon when she details 7 cars.

9. Keith is a taxi driver. He is paid $3.00 plus $1.60 per kilometre. Calculate the amount Keith will earn for a journey of:

   a. 5 km

   b. 15.5 km

10. **WE22** Hamish makes leather belts for a local farmers' market. He is paid $4.50 for each belt for the first 50 belts and $5.10 thereafter. Determine what his income is for a day in which he produces 68 belts.

11. A production line worker is paid $3.45 for each of the first 75 toasters assembled then $4.90 per toaster thereafter. Calculate how much she earns if she produces 120 toasters.

12. Jade sells cosmetics, and the company pays her a fixed weekly wage plus 2.5% commission on all sales she makes. Last week she sold $3000 worth of cosmetics and her total pay was $900. Determine how much she would expect to earn in a week when she sold $4050 worth of cosmetics.

13. Charlie works in a caryard as a detailer. Charlie is paid $11.60 per car. In an afternoon, Charlie can detail 15 cars.
If Charlie could finish in 6 hours, calculate the hourly rate of pay he would earn.

14. **WE23** Tasha and Neesha work at a bridal store. Tasha is paid $700 plus 3.8% of sales per week. Neesha is on a commission-only contract and is paid 27% of sales per week.

   a. Use a spreadsheet to compare their salaries for sales up to $8000. Identify who earns more money for sales up to $8000.
   b. Determine the value of sales for which it is better to work on a commission-only contract.

Fully worked solutions for this chapter are available online.

LESSON
1.7 Payments — government allowances and pensions

SYLLABUS LINKS

- Calculate income support payments based on government allowances and pensions.

Source: General Mathematics Senior Syllabus 2024 © State of Queensland (QCAA) 2024; licensed under CC BY 4.0.

1.7.1 Youth Allowance

Centrelink is an Australian government agency that provides income support and other payments to Australians. Various government allowances and benefits are available for students, families, carers, unemployed people, retirees and disabled people. These allowances include youth allowance, age pension, unemployment benefits, disability payments and other welfare benefits. The amount of the benefit is affected by age, income and assets.

Youth Allowance is financial help for young people who fall into one of the following categories:
- 16 to 21 and looking for full-time work
- 18 to 24 and studying full-time
- 16 to 24 and doing a full-time Australian Apprenticeship
- 16 to 17 and independent or needing to live away from home to study
- 16 to 17, studying full time and have completed Year 12 or equivalent.

To be eligible, people must also meet Australian residence rules, complete an approved course or apprenticeship, and satisfy income and assets tests or the parental means test.

If eligible for Youth Allowance, a claim must be submitted to Centrelink and, if approved, maximum fortnightly payments can be received at the following rates.

Your circumstances	Your maximum fortnightly payment
Single, no children, younger than 18 years, and live at your parent's home	$395.30
Single, no children, younger than 18 years, and need to live away from your parent's home to study, train or look for work	$639.00
Single, no children, 18 years or older and live at parent's home	$455.20
Single, no children, 18 years or older and need to live away from parent's home	$639.00
Single, with children	$806.50
Member of a couple, with no children	$639.00
Member of a couple, with children	$691.80

Note: The payment table is subject to change and was correct at the time of writing.

Your parents' annual income affects how much Youth Allowance you can claim. If your parents' income is less than a certain amount, around $60 000, then you may be able to claim the full allowance.

If your parents' income is more than the threshold amount, the allowance is reduced by 20 cents in each dollar over (this is also affected by the number of children in your family group).

1.7.2 Austudy

Austudy is financial assistance offered by the government, if you are:
- at least 25 years old
- a fulltime student in an approved course or a fulltime Australian apprentice or trainee
- under the income and assets test limits.

If you're	the highest payment per fortnight is
single, no children	$639.00
single, with children	$806.00
in a couple, no children	$639.00
in a couple, with children	$691.00

eles-3092

WORKED EXAMPLE 24 Calculating Youth Allowance

Calculate the Youth Allowance payment or Austudy payment for each of the following per fortnight.
a. Ben is 30 years old, a single dad and a fulltime student.
b. Chloe lives at home and is 17 years old. She is completing a carpentry course at TAFE.

THINK

a. Ben would receive Austudy because he is over 18 years old and studying fulltime.

b. Chloe would receive Youth Allowance because she is under 18 years old.

WRITE

a. Because Ben is a single parent, he will be paid at the 'single with children' rate: $806.00.

b. Chloe will be paid at the rate for the category 'single, no children, younger than 18 years, and live at your parent's homes. This is $395.30.

1.7.3 Age Pension

Senior Australian residents aged over 65 years and 6 months whose assets and income come under the government test limits can be eligible for the Age Pension to assist with their living requirements during retirement.

If eligible for the Age Pension, a claim must be submitted to Centrelink and, if approved, maximum fortnightly payments can be received at the following rates.

Per fortnight	Single	Couple each	Couple combined	Couple apart due to ill health
Maximum basic rate	$1002.50	$755.70	$1511.40	$1002.50
Maximum pension supplement	$80.10	$60.40	$120.80	$80.10
Energy supplement	$14.10	$10.60	$21.20	$14.10
Total	$1096.70	$826.70	$1653.40	$1096.70

Note: The payment table is subject to change and was correct at the time of writing.

Income per fortnight is shown below.

Single person

If your income per fortnight is	your pension will reduce by
up to $204	$0
over $204	50 cents for each dollar over $204

Couple living together or apart due to ill health

If your combined income per fortnight is	your combined pension will reduce by
up to $360	$0
over $360	50 cents each dollar over $360

WORKED EXAMPLE 25 Calculating Age Pension

Rhonda is 78 years old and lives alone. She receives an income of $280 per fortnight from her investments. Use the tables above to calculate how much Rhonda would receive from the Age Pension per fortnight if she is entitled to the maximum basic rate.

THINK	WRITE
1. Read the table to identify the maximum basic rate for a single person.	Single age pension = $1002.50
2. Determine the income above $204 that Rhonda receives.	Amount above $204 = $280 − $204 = $76
3. Determine the amount by which her pension is reduced by calculating 50 cents for each dollar over $204.	Reduction = $76 × 0.50 = $38
4. Calculate how much pension Rhonda receives.	$1002.50 − $38 = $964.50
5. Write the answer.	Rhonda would receive $964.50.

Exercise 1.7 Payments — government allowances and pensions

1.7 Exercise	**1.7 Exam questions** on

Simple familiar	Complex familiar	Complex unfamiliar
1, 2, 3, 4, 5, 6, 7	8, 9	10

These questions are even better in jacPLUS!
- Receive immediate feedback
- Access sample responses
- Track results and progress

Find all this and MORE in jacPLUS ▶

Simple familiar

1. **WE24** Sean is 16 years old and a fulltime student. He lives at home with his parents, who are both unemployed. Calculate how much Youth Allowance he is entitled to per fortnight.

2. Grace is 17 years old, completing a nursing degree and living with her father, who does not work. Calculate how much Youth Allowance Grace receives per year.

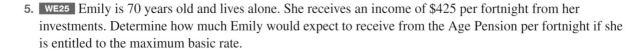

3. Rebecca is a fulltime student living at home. Identify the factors that need to be considered to determine if she is entitled to receive Youth Allowance.

4. **MC** Matt is 16 years old, studies fulltime at high school and lives with his grandparents. Calculate how much he receives in Youth Allowance per week.

   A. $244.10   **B.** $122.05   **C.** $319.50   **D.** $240.25

5. **WE25** Emily is 70 years old and lives alone. She receives an income of $425 per fortnight from her investments. Determine how much Emily would expect to receive from the Age Pension per fortnight if she is entitled to the maximum basic rate.

6. George is 62 years old, lives alone and earns $150 per fortnight. Determine how much Age Pension he receives per fortnight.

7. Mitch is 74 years old and lives with his wife. Determine how much Age Pension he receives per year.

Complex familiar

8. Aycsha qualifies for Youth Allowance and is entitled to $639.00 per fortnight. It is reduced by 50 cents in the dollar for any income earned over $150 per fortnight. Ayesha gets a part-time job as a waitress and earns $200 per fortnight.
   Calculate how much Youth Allowance payment Ayesha receives per year.

9. Jarrod studies fulltime and lives at home with his parents. He is entitled to receive the maximum Youth Allowance per fortnight for a single person over 18 years old.
Jarrod has a part-time job and earns $500 per fortnight. His payment is reduced by 60 cents for every dollar he earns over $250.
Calculate how much Youth Allowance payment Jarrod receives per fortnight.

Complex unfamiliar

10. Angus is married with children and receives the maximum Youth Allowance per fortnight. His payments are reduced by 50 cents in the dollar for any income earned over $150 per fortnight.
He started a part-time job earning $25.00 per hour.
Determine how many hours per fortnight he can work without affecting his Youth Allowance payment.

Fully worked solutions for this chapter are available online.

LESSON
1.8 Personal budgets

SYLLABUS LINKS

- Prepare a personal budget for a given income, taking into account fixed and discretionary spending.
- Use a spreadsheet to display examples of the above computations when multiple or repeated computations are required, e.g. preparing a wage sheet displaying the weekly earnings of workers in an organisation, preparing a budget, investigating the potential cost of owning and operating a car over a year.

Source: General Mathematics Senior Syllabus 2024 © State of Queensland (QCAA) 2024; licensed under CC BY 4.0.

1.8.1 Expenses

A **budget** is a list of all planned income and costs. A budget can be prepared for long periods of time, but for individuals it is usually a monthly plan.

In order to create a monthly budget, monthly income and expenses need to be known. Monthly income is $\frac{1}{12}$ of the total annual income. The expenses in a personal budget can be divided into two major categories: fixed expenses and variable or discretionary expenses.

Fixed expenses may include rent or mortgage, medical insurance, car registration and other regular payments that must be paid and can't be varied.

Expenses that are non-essential and/or can be controlled or varied are called discretionary expenses and include items such as food, entertainment and clothing. To reduce your expenses in order to save more money, you would look at reducing your discretionary expenses.

A weekly, monthly or yearly budget can be prepared. Expenses may be calculated weekly (such as food), monthly (health insurance), quarterly (electricity bills) or yearly (car registration). Depending on the budget duration, all expenses should be converted to weekly, monthly or yearly amounts.

The following table of conversion will help you in preparing a budget.

Purpose	Convert from	Convert to	Operation
Weekly budget	Monthly cost	Weekly cost	$\times 12$, then $\div 52$
	Yearly cost	Weekly cost	$\div 52$
Monthly budget	Weekly cost	Monthly cost	$\times 52$, then $\div 12$
	Yearly cost	Monthly cost	$\div 12$
Yearly budget	Weekly cost	Yearly cost	$\times 52$
	Monthly cost	Yearly cost	$\times 12$

eles-6375

WORKED EXAMPLE 26 Preparing a budget

Karla's fortnightly budget is as follows.

Income		Expenses	
Salary after tax	**$2050**	**Rent of 1-bedroom flat**	**$760**
Dividends from shares	**$23**	**Electricity**	**$140**
		Phone and internet	**$80**
		Health insurance	**$100**
		House contents insurance	**$20**
		Car registration	**$20**
		Car insurance	**$30**
		Fuel	**$70**
		Food	**$200**
		Clothing	**$80**
		Entertainment	**$150**
		Gym membership	**$100**
		Miscellaneous	**$40**
Total:	**$2073**	**Total:**	**$1770**

Use the table to answer the following.
a. Calculate the total of the fixed expenses. Note that Karla views health insurance, house contents insurance and car insurance as important expenses, and their regular payments are fixed.
b. Calculate the total of the discretionary expenses.
c. Calculate the amount available for saving per fortnight.
d. If Karla wishes to take a vacation and travel to New Zealand (estimated cost $3500), calculate for how many weeks she will have to save.

THINK	WRITE
a. Identify and calculate the total of the fixed expenses.	**a.** Fixed expenses:

Rent	$760
Health insurance	$100
House contents insurance	$20
Car registration	$20
Car insurance	$30
Electricity	$140
Phone and internet	$80
Fuel	$70

Total expenses $= 760 + 100 + 20 + 20 + 30 + 140 + 80 + 70 = \$1220$

b. Calculate the total of the discretionary expenses by subtracting fixed expenses from the total.

b. Total of discretionary expenses
$$= \text{total expenses} - \text{total of fixed expenses}$$
$$= \$1770 - \$1220$$
$$= \$550$$

c. Calculate fortnightly savings by subtracting expenses from the income.

c. Fortnightly savings
$$= \text{fortnightly income} - \text{fortnightly expenses}$$
$$= \$2073 - \$1770$$
$$= \$303$$

d. Calculate the number of weeks required to save for the holiday.

d. Karla saves $303 per fortnight, which is $151.50 per week.
To save $3500 at the rate of $151.50 per week: $3500 \div 151.50 = 23.1$ or 24 weeks (rounded up).
Karla needs to save for 24 weeks.

Exercise 1.8 Personal budgets

1.8 Exercise	1.8 Exam questions on

Simple familiar	Complex familiar	Complex unfamiliar
1, 2, 3, 4, 5, 6, 7	8, 9, 10	11

These questions are even better in jacPLUS!
- Receive immediate feedback
- Access sample responses
- Track results and progress

Find all this and MORE in jacPLUS ⊙

Simple familiar

1. Explain the difference between fixed and discretionary expenses.

2. **MC** Select the item that can be considered as a fixed cost.
   A. Speeding ticket
   B. Soccer club membership
   C. Holiday to Thailand
   D. Visit to the dentist

3. Identify as many discretionary expenses as you can that an average 17-year-old would have in a month.

4. **WE26** The table below shows the fortnightly budget for a couple with one school-aged child.

Income		Expenses	
		Mortgage repayments	$1543
		Rates	$46
		Building insurance	$30
		Contents insurance	$20
		Electricity	$200
		Telephones and internet	$125
		Car registration	$35
		Car insurance	$50
		Health insurance	$150
		School fees	$300
		Food	$500
		Clothing	$120
		Entertainment	$150
		Miscellaneous	$100
		Maintenance and repairs	$50
		Petrol	$200
Income after tax:	$4102	Total:	$3619

Use the table to calculate:

a. the total of fixed expenses
b. the total of discretionary expenses
c. the total weekly savings
d. the number of weeks needed for the family to save enough for a trip to Bali (estimated cost $4000).

5. Complete the following table of income and expenses to determine the monthly savings for Phoenix.

Annual salary	Monthly bank interest	Monthly fixed expenses	Monthly discretionary expenses	Monthly savings
$73 700	$784	$3623	$2563	

6. Determine the total monthly cost from the following list of fixed and variable costs. The variable costs represent the amounts spent in a month.

Fixed costs			Variable costs	
Item	Frequency	Amount	Item	Amount
Rent	Weekly	$205	Food	$494
Health insurance	Monthly	$89	Clothing	$205
Vehicle registration	Yearly	$540	Entertainment	$123
Vehicle insurance	Yearly	$499	Car repairs	$72

7. Sally's monthly electricity bill is $250. Calculate how much she should allow for electricity in her weekly budget.

Complex familiar

8. A university student who lives with her parents has the following expenses:
   she pays her parents $100 per week for board and food; a monthly ticket for public transport costs her $80; she spends on average $45 a month on books and stationery; her single health insurance premium is $70.28 a month; entertainment costs her about $120 a month; the university enrolment fee is $900 a year; clothes cost $40 per month; and her mobile phone costs approximately $50 per month.

   a. Use a spreadsheet to prepare a monthly budget if the student's income consists of Austudy (which is $639.00 per fortnight) plus birthday and Christmas presents ($250 per year).
   b. Calculate the amount of money that she can save per month.

9. Use a spreadsheet to determine the total yearly expenses from the following list of fixed and variable expenses. The variable expenses are monthly expenses.

Fixed costs			Variable costs	
Item	Frequency	Amount	Item	Amount
Rent	Weekly	$230	Food	$423
Health insurance	Monthly	$78	Clothing	$107
Vehicle registration	Yearly	$620	Entertainment	$85
Vehicle insurance	Yearly	$389	Car repairs	$325

10. Use the figures in the table below to calculate the total of the weekly expenses.

Item	Cost and period
Rent	$600 per month
Food	$90 per week
Electricity	$420 per 3 months
Gas	$40 per 2 months
Phone	$360 per 3 months
Car registration	$430 per year
Car insurance	$500 per year
Petrol	$50 per week
Health insurance	$175 per 3 months
Contents insurance	$125 per year
Clothes	$100 per month
Entertainment	$80 per month

11. Parents on a kindergarten committee are discussing the budget for the next year.

Their income will come from three sources: a government subsidy of $4200; annual enrolment fees of $1250 per child; and profits from various events.

They estimate that the cake stall will bring in $400, and profits from the two sausage sizzles will yield about $800 each. They also estimate that they will have 22 children enrolled.

The money will be spent as follows: rent of the premises at $500 a month; publishing the newsletter at $240 per quarter; new equipment will be $7200 per year electricity and phone bills at $220 a month; public liability and contents insurance will be $1860 per year. Advertising will cost $30 per month and stationery $250 per year.

A new computer will cost $3500, and $2000 is allowed for unexpected expenses.

Calculate the amount of money that can be saved for future renovations of the kindergarten.

Express your answer to part **b** as a percentage of the annual income.

Fully worked solutions for this chapter are available online.

LESSON
1.9 Review

1.9.1 Summary

doc-41468

Hey students! Now that it's time to revise this chapter, go online to:

 Access the chapter summary

 Review your results

 Practise exam questions

Find all this and MORE in jacPLUS

1.9 Exercise

 learn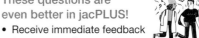on

1.9 Exercise	**1.9 Exam questions** on

Simple familiar	Complex familiar	Complex unfamiliar
1, 2, 3, 4, 5, 6, 7, 8, 9, 10, 11, 12	13, 14, 15, 16	17, 18, 19, 20

These questions are even better in jacPLUS!
- Receive immediate feedback
- Access sample responses
- Track results and progress

Find all this and MORE in jacPLUS

Simple familiar

1. **MC** Identify which of the following is the same as 1.06.

   A. 10.6% **B.** 106% **C.** $1\frac{6}{10}$ **D.** 160%

2. **MC** Australia needs to make 280 runs in 50 overs to beat England. Select the number of runs per over this is.

   A. 6.5
   B. 5
   C. 6
   D. 5.6

3. **MC** Sasha used to weigh 96 kg. After joining a gym he lost 12 kg. The percentage of the original weight lost was:

   A. 8% **B.** 10% **C.** 12% **D.** 12.5%

4. **MC** A jewellery salesperson earns 12% commission on all sales. If she sold two rings, one valued at $400 and the other at $280, select the calculation from the following that *could not* be used to evaluate her commission.

   A. $\frac{12}{100} \times 680$

   B. 680×0.12

   C. $\frac{12}{100} \times 280 + \frac{12}{100} \times 400$

   D. $(400 + 280) \times 12$

5. **MC** Natasha is picking apples on a farm. She is paid $6.50 per basket plus a bonus of $1.80 for every basket after the first ten. Select how much she will be paid on the day when she picks 16 baskets of apples.

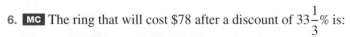

A. $105.80
B. $75.80
C. $114.80
D. $128.80

6. **MC** The ring that will cost $78 after a discount of $33\frac{1}{3}\%$ is:

A. a friendship ring, normally $104
B. a mother of pearl ring, normally $130
C. a sapphire ring, normally $260
D. a Russian band ring, normally $117

7. **MC** Michael earns a retainer of $220 per week plus 7.5% commission on all sales. The total of Michael's sales during the week when he earned $884 was:

A. $6513.33   B. $8913.33   C. $3420   D. $8853.33

8. **MC** Identify which of the following is not a fixed expense.

A. Car insurance   B. Mortgage repayments
C. Council rates   D. Sport expenses

9. **MC** The calculation that *could not* be used to calculate the sale price of a $64 item after a discount of 12.5% is:

A. $\dfrac{12.5 \times 64}{100}$ B. $64 - \dfrac{12.5 \times 64}{100}$ C. $\dfrac{87.5 \times 64}{100}$ D. $\dfrac{7}{8}$ of 64

10. **MC** Stan is paid $23.20 per hour for the first 40 hours, time and a half for a further 3 hours and double time thereafter. If last week Stan worked 10 hours overtime, his total earning for the week were:

A. $1370   B. $1377.60   C. $1357.20   D. $1307.20

11. A mobile phone is sold for $1275. If this represents a 15% reduction from the RRP, calculate the original price.

12. A person has to pay the following bills out of their weekly budget of $1100.

Food	$280
Electricity	$105
Telephone	$50
Petrol	$85
Rent	$320

Giving answers correct to 2 decimal places where necessary:

a. express each bill as a percentage of the total bills
b. express each bill as a percentage of the weekly budget.

Complex familiar

13. Kay is working in a large department store. She earns a retainer of $500 per week for 5 working days plus 10% commission on all sales.
    During a 'facials' promotion week she also earns $50 for each facial that she conducts.

In one day she did 7 facials and sold her clients different products to the total value of $1089.
Calculate Kay's earnings for that day.

14. To make a T-shirt, parts are traced onto the material using a pattern and then cut. A cutter is paid $20 for cutting every 100 standard T-shirts. If a T-shirt contains any extra parts, the cutter is paid an extra $4.50 for each 100 parts.
    Calculate the amount of money that a cutter will earn cutting an order of 360 T-shirts with two extra parts.

15. A power company claims that if you install solar panels for $1800, you will make this money back in savings on your electricity bill in 2 years. If you usually pay $250 per month, calculate what percentage your bill will be reduced by if their claims are correct.

16. The table below shows the share prices of different companies on the first day of two consecutive months. For each company, identify whether there was an increase or decrease in price, calculate the amount of increase/decrease in dollars, and express it as a percentage of the original price (to 2 decimal places).

Company name	Price on 1/9/24	Price on 1/10/24	Increase (I) or decrease (D)?	Change of price ($)	Percentage change
a. BHT	$12.82	$8.19			
b. Super Cole	$5.14	$7.25			
c. SuNatCo	$21.35	$19.00			
d. AMB	$4.70	$5.76			
e. ANX Bank	$9.52	$9.80			
f. Pronto Co	$0.45	$0.61			
g. EIK Gold	$25.40	$29.12			
h. Motors International	$7.80	$7.06			
i. Optocom au	$5.70	$4.98			
j. National Metro	$18.28	$19.15			

Complex unfamiliar

17. Jim works in retail selling clothes, and receives $800 retainer and 12% commission on all sales. Jill works in another store, also selling clothes, and receives $850 retainer and 10% commission.
    Calculate the value of total sales Jim and Jill have to make so that they both receive $1000.

18. Use a spreadsheet to complete the following time sheet and calculate the wages of an employee who earns $22.50 per hour.
    Normal pay is paid for an 8-hour day Monday to Friday. Any overtime on weekdays is paid time and a half for the first 3 hours and double time thereafter. Any hours worked on the weekend are paid at double time.

Name	Time sheet						
Day	On	Off	On	Off			
Monday	7.30	11.30	12.00	4.30			
Tuesday	7.30	11.30	12.00	4.00			
Wednesday	7.30	12.00	12.30	5.30			
Thursday	7.30	11.30	12.30	4.30			
Friday	7.30	11.30	1.00	5.30			
Saturday	8.00	12.00					
Sunday	8.00	11.00					
				Total:			

19. Copy and complete the following table.

Item	Cost price ($)	Percentage discount	Discount ($)	Selling price ($)
a	200	12%		
b	150			142.50
c	98		9.80	
d			16.25	113.75
e		20%		332.80
f		$33\frac{1}{3}\%$	76	

20. The following table shows a list of Rose's expenses.

Item	Cost	Period
Rent	$434	Weekly
Electricity	$500	Quarterly
Phone	$300	Quarterly
Car registration	$840	Yearly
Car insurance	$900	Yearly
Contents insurance	$155	Quarterly
Health insurance	$80	Monthly
Food	$200	Weekly
Sport	$50	Weekly
Entertainment	$80	Weekly
Clothes	$220	Monthly
Holidays	$2400	Yearly

Rose has a part-time job as a receptionist and earns $1070 per week. Use a spreadsheet to produce a budget for Rose and calculate the amount that she can save per year.

Ensure that values in your spreadsheet are rounded to 2 decimal places before you make your final calculation.

Fully worked solutions for this chapter are available online.

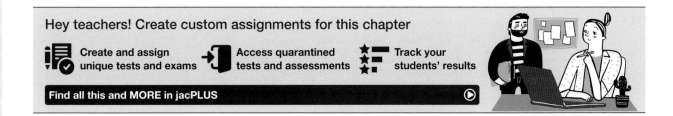

Hey teachers! Create custom assignments for this chapter

Create and assign unique tests and exams

Access quarantined tests and assessments

Track your students' results

Find all this and MORE in jacPLUS

Answers

Chapter 1 Consumer arithmetic — earning and budgeting

1.2 Rates and percentages

1.2 Exercise

1. a. $5.90/kg
   b. 660 mL/h
   c. $18.65/m
   d. 12.25 points/game

2. a. 5.4 km/h
   b. 16.67 m/s
   c. $650/kg
   d. 0.57 cents/g

3. 1062 km/h

4. $11.80/h

5. a. $37.80   b. 108   c. $174.49   d. $42 304.22

6. a. $45.36   b. $7.41   c. $69.31   d. $12 037.52

7. $60

8. 12.5%

9. a. i. Reduction of 0.62%   ii. Increase of 2.5%
   b. Increase of 1.23%

10. Decrease of 5.16%

11.

	Test 1	Test 2	Test 3	Test 4
Mark	$\dfrac{14}{21}$	$\dfrac{42}{60}$	$\dfrac{14.5}{20}$	$\dfrac{26}{35}$
Percentage	66.7%	70%	72.5%	74.29%

11.43%

12. a. See the figure at the foot of the page.*
    b. See the figure at the foot of the page.*

13. a. 12.5%   b. 16.875%

14. $1337.33

1.3 Wages

1.3 Exercise

1. $2610

2. $1670

3. $7000

4. a. $1470   b. $2940

5. $55 120

6. $58 500

7. D

8. $21.21

9. $938.95

10. See the table at the foot of the page.*

---

*12. a.

	A	B	C	D	E	F	G	H	I	J
1									PERCENTAGES	
2	TEAM	Win	Loss	Draw	Goals For	Goals Against	Total Games	% WIN	% LOSS	% DRAW
3	1. Western Sydney Wanderers	18	6	3	41	21	27	66.67	22.22	11.11
4	2. Central Coast Mariners	16	5	6	48	22	27	59.26	18.52	22.22
5	3. Melbourne Victory	13	9	5	48	45	27	48.15	33.33	18.52
6	4. Adelaide United	12	10	5	38	37	27	44.44	37.04	18.52
7										

*12. b.

	A	B	C	D	E	F	G	H
1								PERCENTAGES
2	TEAM	Win	Loss	Draw	Goals For	Goals Against	Total Games	% Goals FOR
3	1. Western Sydney Wanderers	18	6	3	41	21	27	195.24
4	2. Central Coast Mariners	16	5	6	48	22	27	218.18
5	3. Melbourne Victory	13	9	5	48	45	27	106.67
6	4. Adelaide United	12	10	5	38	37	27	102.70
7								

*10.

Annual salary	Weekly pay	Fortnightly pay	Monthly pay
$156 000	$3000	$6000	$13 000
$93 600	$1800	$3600	$7800
$62 400	$1200	$2400	$5200
$71 760	$1380	$2760	$5980
$109 200	$2100	$4200	$9100

11. See the figure at the foot of the page.*

12. $35.26

13. Lisa ($32.76 per hour)

14. 45 hours

1.4 Earning wages

1.4 Exercise

1. $918

2. a. $784.80   b. $1046.40   c. $926.50

3. a. $1392.00   b. $1609.50   c. $1935.75

4. D

5. $22.40

6. 37 hours

7. C

8. $1248.80

9.

Name	Wage	Hours worked	Hourly rate
Brent	$832.32	36	$23.12
Sandra	$1077.60	40	$26.94
Ann	$739.26	37	$19.98
Chris	$1627.92	84	$19.38
Anna	$462.60	30	$15.42
Toni	$1553.44	76	$20.44

10. D

11. $507

12. $4648.22

13. 40 hours

14. $26.41/h

1.5 Working overtime

1.5 Exercise

1. $38.40/h

2. a. $40.05/h   b. $53.40/h

3. $316.80

4. C

5. C

6. See the table at the foot of the page.*

7. $963.20

8. $1441.80

9. 6 hours

10. 24 hours

11. $1920

12. $1236

13. 39 hours

14. $1148

*11.

	Employee	Hourly pay rate	Hours worked per week	Wages per week	Wages per annum
1					
2	A.Richardson	$35.50	20	$710.00	$36,920.00
3	S.Provis	$21.75	35	$761.25	$39,585.00
4	N.Liu	$27.45	8	$219.60	$11,419.20
5	W.Naidoo	$29.55	28	$827.40	$43,024.80
6	G.Riddell	$35.75	16	$572.00	$29,744.00
7	C.Ve	$23.60	38	$896.80	$46,633.60
8	O.Chang	$31.00	26	$806.00	$41,912.00
9	D.Evans	$22.00	13	$286.00	$14,872.00
10					

*6.

Name	Ord. rate	Normal hours	Time-and-a-half hours	Double-time hours	Total pay
W. Clark	$18.60	38	4		$818.40
A. Hurst	$19.85	37		6	$972.65
S. Gannon	$24.50	38	5	2	$1212.75
G. Dymock	$36.23	37.5	4	1.5	$1684.70
D. Colley	$34.90	36	6	8	$2128.90

1.6 Earnings — commission and piecework

1.6 Exercise

1. $2000
2. B
3. C
4. $950
5. B
6. $15 525
7. $327
8. $75.95
9. a. $11   b. 27.80
10. $316.80
11. $479.25
12. $926.25
13. $29.00/h
14. a. See the figure at the foot of the page.*
    b. A commission-only contract is better for sales above $3500 (or equal to $3500).

1.7 Payments — government allowances and pensions

1.7 Exercise

1. $395.30
2. $10 277.80

3. Factors to consider are whether she is in the 16–24 age bracket and what her parents' income is.
4. C
5. $892
6. George doesn't receive the Age Pension as he is under 65 years and 6 months.
7. $19 648.20 per year
8. $15 964 per year
9. $305.20 per fortnight
10. 6 hours

1.8 Personal budgets

1.8 Exercise

1. Fixed expenses are essential expenses such as rent, electricity and gas. Discretionary expenses are non-essential expenses and/or can be controlled or varied, such as entertainment and clothes.
2. D
3. Sample answer: Expenses may include sporting events, entertainment, food, drinks, driving lessons, gym membership and clothing.
4. a. $2699   b. $920   c. $483   d. 17 weeks
5. See the table at the foot of the page.*
6. $1957.91
7. $57.69

*14. a.

	A	B	C
1	Sales	Tasha: $700 + 3.8% Sales	Neesha: 27% Sales
2	0	700	0
3	500	719	135
4	1000	738	270
5	1500	757	405
6	2000	776	540
7	2500	795	675
8	3000	814	810
9	3500	833	945
10	4000	852	1080
11	4500	871	1215
12	5000	890	1350
13	5500	909	1485
14	6000	928	1620
15	6500	947	1755
16	7000	966	1890
17	7500	985	2025
18	8000	1004	2160

*5.

Annual salary	Monthly bank interest	Monthly fixed expenses	Monthly discretionary expenses	Monthly savings
$73 700	$784	$3623	$2563	$739.65

Note: Savings are rounded down to the nearest 5c.

8. **a.** Set up a spreadsheet with the following headings and formulas to calculate the monthly income and expenses. See the figure at the foot of the page.*

   b. $491.72

9. See the figure at the foot of the page.*

10. $418.37

11. The amount of money that can be saved for future renovations is $8930

   The amount of money saved as a percentage of the annual income is 26.5%

1.9 Review

1.9 Exercise

1. B
2. D
3. D
4. D
5. C
6. D
7. B
8. D
9. A
10. C
11. $1500
12. **a.** Food: 33.33%; electricity: 12.5%; telephone: 5.95%; petrol: 10.12%; rent: 38.10%

   b. Food: 25.45%; electricity: 9.55%; telephone: 4.55%; petrol: 7.73%; rent: 29.09%
13. $558.90
14. $104.40
15. Bills will be reduced by 30%.

*8. a.

	A	B	C	D	E
1	**INCOME**	Amount	Period	Monthly cost	
2	Austudy	639	fortnightly	1384.50	
3	Presents	250	yearly	20.83	
4				**Total income**	1405.33
5	**EXPENSES**				
6	Board & food	100	weekly	433.33	
7	Public transport	80	monthly	80.00	
8	Books & stationery	45	monthly	45.00	
9	Health insurance	70.28	monthly	70.28	
10	Entertainment	120	monthly	120.00	
11	Uni enrolment fee	900	yearly	75.00	
12	Clothes	40	monthly	40.00	
13	Mobile phone	50	monthly	50.00	
14				**Total expenses**	913.61

*9.

	A	B	C	D
1	**FIXED COSTS**	Period	Amount	Yearly Amount
2	Rent	weekly	230	11960
3	Health insurance	monthly	78	936
4	Vehicle registration	yearly	620	620
5	Vehicle insurance	yearly	389	389
6				
7	**DISCRETIONARY COSTS**			
8	Food	monthly	423	5076
9	Clothing	monthly	107	1284
10	Entertainment	monthly	85	1020
11	Car repairs	monthly	325	3900
12			TOTAL	25185

16. See the table at the foot of the page.*

17. Jim must have sales of $1666.67. Jill must have sales of $1500.

18. See the figure at the foot of the page.*
    His wages were $1288.13.

19. See the table at the foot of the page.*

20. Set up a spreadsheet with the following headings and enter formulas to calculate the weekly expenditure.

	A	B	C	D
1	Expenses	Cost	Period	Weekly
2	Rent	434	weekly	434.00
3	Electricity	500	quarterly	38.46
4	Phone	300	quarterly	23.08
5	Car registration	840	yearly	16.15
6	Car insurance	900	yearly	17.31
7	Contents insurance	155	quarterly	11.92
8	Health insurance	80	monthly	18.46
9	Food	200	weekly	200.00
10	Sport	50	weekly	50.00
11	Entertainment	80	weekly	80.00
12	Clothes	220	monthly	50.77
13	Holidays	2400	yearly	46.15
14			Total	986.30

Rose can save $4352.40 per year.

*16.

Company name	Price on 1/9/24	Price on 1/10/24	Increase (I) or decrease (D)?	Change of price ($)	% Change
a. BHT	$12.82	$8.19	D	$4.63	36.12%
b. Super Cole	$5.14	$7.25	I	$2.11	41.05%
c. SuNatCo	$21.35	$19.00	D	$2.35	11.01%
d. AMB	$4.70	$5.76	I	$1.06	22.55%
e. ANX Bank	$9.52	$9.80	I	$0.28	2.94%
f. Pronto Co	$0.45	$0.61	I	$0.16	35.56%
g. EIK Gold	$25.40	$29.12	I	$3.72	14.65%
h. Motors International	$7.80	$7.06	D	$0.74	9.49%
i. Optocom au	$5.70	$4.98	D	$0.72	12.63%
j. National Metro	$18.28	$19.15	I	$0.87	4.76%

*18.

	A	B	C	D	E	F	G
1	DAY	Total hours	Hourly rate	Normal pay	Time and a half	Double time	Total weekly wage
2	Monday	8.5	22.5	180	11.25	NA	191.25
3	Tuesday	8	22.5	180	0	NA	180
4	Wednesday	9.5	22.5	180	50.625	NA	230.63
5	Thursday	8	22.5	180	0	NA	180
6	Friday	8.5	22.5	180	11.25	NA	191.25
7	Saturday	4	22.5	NA	NA	180	180
8	Sunday	3	22.5	NA	NA	135	135
9						Total	1288.13

*19.

Item	Cost price ($)	Percentage discount	Discount ($)	Selling price ($)
a	200	12%	24	176
b	150	5%	7.50	142.50
c	98	10%	9.80	88.20
d	130	12.5%	16.25	113.75
e	416	20%	83.20	332.80
f	228	$33\frac{1}{3}\%$	76	152

2 Consumer arithmetic — managing finances

LESSON SEQUENCE

Fully worked solutions for this chapter are available online.

EXAM PREPARATION

Access exam-style questions in every lesson, available online.

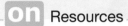 **Resources**

Solutions	Solutions — Chapter 2 (sol-1243)
Exam questions	Exam question booklet — Chapter 2 (eqb-0259)
Digital documents	Learning matrix — Chapter 2 (doc-41470)
	Chapter summary — Chapter 2 (doc-41471)

LESSON
2.1 Overview

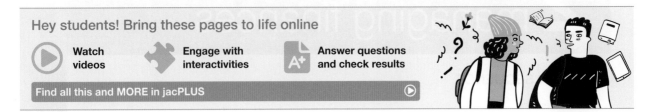
2.1.1 Introduction

Since we live in a commercial world, all consumers should try to understand the mathematics that is around us, from shopping online to buying shares on the share market. When buying online or travelling overseas, we need to understand exchange rates so that we are able to convert between Australian dollars and foreign currencies. Goods and services tax (GST) is paid on most items we buy in Australia and is presently charged at 10%.

When you invest money, you will receive interest, and when you take out a loan, you pay interest. This interest is calculated as a percentage of the money invested or borrowed and may be calculated as simple interest or compound interest.

2.1.2 Syllabus links

Lesson	Lesson title	Syllabus links
2.2	Unit cost	○ Compare prices and values using the unit cost method.
2.3	Mark-ups and discounts	○ Apply percentage increase or decrease in various contexts, e.g. percentage mark-ups and discounts.
2.4	Goods and services tax (GST)	○ Apply percentage increase or decrease in various contexts, e.g. GST.
2.5	Profit and loss	○ Apply percentage increase or decrease in various contexts, e.g. percentage profit and loss.
2.6	Simple interest	○ Apply percentage increase or decrease in various contexts, e.g. simple interest. • $I = Pin$ where I is simple interest, P is principal, i is interest rate per year and n is number of years.
2.7	Compound interest (optional)	○ (This is optional content and is covered more thoroughly in Units 3 & 4.)
2.8	Exchange rates	○ Use currency exchange rates to convert between the Australian dollar and foreign currencies.
2.9	Dividends	○ Calculate the dividend paid on a portfolio of shares, given the dividend yield or dividend paid per share, and compare share values by calculating a price-to-earnings (P/E) ratio. • dividend yield $= \dfrac{\text{dividend}}{\text{share price}} \times 100$ • P/E ratio $= \dfrac{\text{market price per share}}{\text{annual earnings per share}}$

Source: General Mathematics Senior Syllabus 2024 © State of Queensland (QCAA) 2024; licensed under CC BY 4.0.

LESSON
2.2 Unit cost

SYLLABUS LINKS

• Compare prices and values using the unit cost method.

Source: General Mathematics Senior Syllabus 2024 © State of Queensland (QCAA) 2024; licensed under CC BY 4.0.

2.2.1 Unit cost method

Unit cost means the cost per unit of an item. It can be the price per litre, per 100 mL or per 100 g. It is generally used for items sold by weight or volume. It is useful to compare the prices of different sized items and to determine which item is the best buy. Supermarkets in Australia are required to show the unit cost on most products.

It is similar to the process used when simplifying rates. If x items cost $\$y$, divide the cost by x to calculate the price of one item.

$$x \text{ items} = \$y$$

$$1 \text{ item} = \$\frac{y}{x}$$

eles-3162

WORKED EXAMPLE 1 Calculating costs

Calculate the cost per 100 g of pet food if a 1.25-kg box costs $7.50.

THINK	WRITE
1. Identify the cost and the weight. As the final answer is to be referenced in grams, convert the weight from kilograms to grams.	Cost: $7.50 Weight: $1.250 \text{ kg} = 1250 \text{ g}$
2. Calculate the unit cost for 1 gram by dividing the cost by the weight.	$\text{Unit cost} = \dfrac{7.50}{1250}$ $= 0.006$
3. Calculate the cost for 100 g by multiplying the unit cost by 100.	$\text{Cost for } 100 \text{ g} = 0.006 \times 100$ $= 0.60$
4. Write the answer.	Therefore, the cost per 100 g is $0.60.

The unit cost method is used to compare the price per unit of items of different sizes. Items must be in the same units when making comparisons.

Three shampoos are sold in the following quantities.
Brand A: 200 mL for $5.38
Brand B: 300 mL for $5.98
Brand C: 400 mL for $8.04
Determine which shampoo is the best buy.

THINK

1. Calculate the unit cost for each brand by dividing the cost by the number of mL.

2. Write the answer.

WRITE

Unit cost of Brand A $= \dfrac{5.38}{200}$
$= \$0.0269$

Unit cost of Brand B $= \dfrac{5.98}{300}$
$= \$0.0199$

Unit cost of Brand C $= \dfrac{8.04}{400}$
$= \$0.0201$

Brand B: 300 mL of shampoo for $5.98 is the best buy.

Exercise 2.2 Unit cost

learn on

2.2 Exercise	2.2 Exam questions on

Simple familiar	Complex familiar	Complex unfamiliar
1, 2, 3, 4, 5, 6, 7, 8, 9, 10	11, 12, 13	14

These questions are even better in jacPLUS!
- Receive immediate feedback
- Access sample responses
- Track results and progress

Find all this and MORE in jacPLUS ▶

Simple familiar

1. Products (and services) other than supermarket items can be sold with a unit pricing scheme. Identify what unit pricing might be used in the following cases.
   a. Petrol
   b. A lawyer's fee
   c. Hotel accommodation
   d. Lounge room carpet
   e. Floor tiling
   f. Wages at a fast-food restaurant

2. Some products are easy to convert to unit prices. Without a calculator, determine the price per 100 grams.
   a. 1 kg apples at $5.90
   b. 500 g laundry powder at $7.00

3. Without a calculator, calculate the following prices by determining the price per 100 grams.
   a. A 400-g tin of canned peaches at $3.20
   b. 5 kg potatoes at $15.00

4. **WE1** Calculate the cost per 100 grams for:
   a. a 650-g box of cereal costing $6.25
   b. a 350-g packet of biscuits costing $3.25.

5. Calculate the cost per 100 grams for:

   a. a 425-g jar of hazelnut spread costing $6.20
   b. a 550-g container of yoghurt costing $5.69.

6. Calculate the cost:

   a. per litre if a box of 24 cans that each contain 375 mL costs $36.70
   b. per 100 mL if a 4-litre bottle of cooking oil costs $16.75.

7. Calculate the cost:

   a. per kilogram if a 250-g pack of cheese slices costs $7.50
   b. per kilogram if a 400-g frozen chicken dinner costs $7.38.

8. **WE2** Brand H baked beans sell for $1.86 for 400 grams, while Brand E are $1.67 for 350 grams. Determine which brand is the best buy.

9. **MC** Liquids are unit priced according to the price per 100 mL (0.1 litre) instead of per 100 grams. Calculate the unit price of a 1.25 L bottle of cola that sells for $1.87.

   A. $0.187
   B. $0.1496
   C. $0.1558
   D. $0.2338

10. A butcher has the following pre-packed meat specials.

| BBQ lamb chops in packs of 12 for $30.50 |
| Porterhouse steaks in packs of 5 for $36.25 |
| Chicken drumsticks in packs of 11 for $18.70 |

Calculate the price per single unit of each pre-packed meat, correct to the nearest cent.

11. If 6 avocadoes cost $13.50, determine how much 9 avocadoes cost.

12. Complete the following table by calculating the missing entries.

	Item	Size	Selling price	Unit price (per 100 g or 100 mL)
a.	Cheese		$4.78	$1.06
b.	Onions	2.5 kg		$0.23
c.	BBQ sauce	750 g		$0.42
d.	Milk		$3.12	$0.10

13. A particular car part is shipped in containers that hold 2054 items. Each container costs the receiver $8000.

   a. If the car parts are sold for a profit of 15%, determine how much is charged for each.
   b. The shipping company also has smaller containers that cost the receiver $7000, but only hold 1770 items. If the smaller containers are the only ones available, calculate how much the car part seller must charge to make the same percentage profit.

14. Another nearby butcher matched the price per pack of the butcher in question **10** but decided to do a special bundle deal that includes 2 packs of each type of meat from question **10**.
Each bundle is slightly different, and the weights for each are shown in the table below.

Meat	Bundle 1	Bundle 2
BBQ lamb chops	2535 grams	2602 grams
Porterhouse steak	1045 grams	1068 grams
Chicken drumsticks	1441 grams	1453 grams

Determine which Bundle is better value for money and justify the reasonableness of your solution.

Fully worked solutions for this chapter are available online.

LESSON
2.3 Mark-ups and discounts

SYLLABUS LINKS

- Apply percentage increase or decrease in various contexts, e.g. percentage mark-ups and discounts.

Source: General Mathematics Senior Syllabus 2024 © State of Queensland (QCAA) 2024; licensed under CC BY 4.0.

2.3.1 Mark-ups

A **mark-up** is an amount or a percentage by which goods or services are increased. For example, petrol is frequently marked up.

A mark-up is an example of a percentage increase. Remember from Chapter 1: to increase an amount by $x\%$, multiply by $(100 + x)\%$.

WORKED EXAMPLE 3 Calculating mark-ups

The price of petrol was marked up by **8.8%** from the wholesaler to the retailer. If the wholesaler bought it at 170.5 cents per litre, calculate the sale price per litre after the mark-up. Give your answer correct to 2 decimal places.

THINK	WRITE
1. Determine the total percentage of the sale price including the mark-up.	Sale price (%) $= 100\% + 8.8\%$ $= 108.8\%$
2. Convert 108.8% to a decimal.	Sale price (decimal) $= \dfrac{108.8}{100}$ $= 1.088$
3. Determine the sale price by multiplying the original price by 1.088.	Sale price per litre $= 170.5 \times 1.088$ $= 185.504$
4. Write the answer, correct to 2 decimal places.	The sale price is 185.50 cents per litre.

When we know the original amount and the sale price and need to determine the mark-up as a percentage, we use the following formula.

Calculating percentage increase (mark-up)

$$\text{Percentage increase (mark-up)} = \frac{\text{amount of increase}}{\text{original amount}} \times 100$$

WORKED EXAMPLE 4 Determining percentage mark-up (increase)

Cooper bought Bluetooth speakers for $345 and sold them for $500. Calculate the percentage mark-up, correct to the nearest whole number.

THINK	WRITE
1. Determine the amount of the mark-up.	Mark-up $= \$500 - \$345$ $= \$155$
2. Use the percentage increase (mark-up) formula.	Percentage mark-up $= \dfrac{155}{345} \times 100$ $= 44.9\%$
3. Write the answer correct to the nearest whole number.	The percentage mark-up was 45%.

2.3.2 Discounts

A **discount** is an amount of money by which the price of an item is reduced. If expressed as a percentage of the original price, it is called a **percentage discount**.

Calculating percentage decrease (discount)

$$\text{Discount} = \text{original price} - \text{sale price}$$

$$\text{Percentage decrease (discount)} = \frac{\text{decrease (discount)}}{\text{original price}} \times 100\%$$

WORKED EXAMPLE 5 Calculating percentage discount (decrease)

A Playstation game is discounted from \$120 to \$84.

Calculate the percentage discount.

THINK	WRITE
1. Determine the discount in dollars.	Discount = original price − sale price = \$120 − \$84 = \$36
2. Write the formula for the percentage discount.	Percentage discount = $\dfrac{\text{discount}}{\text{original price}} \times 100\%$
3. Substitute the values of the discount and the original price into the formula and evaluate.	Percentage discount = $\dfrac{36}{120} \times 100\%$ = 30%
4. Write the answer.	The game was discounted by 30%.

Sometimes we are required to calculate the original price and are given the sale price and the discount, as shown in Worked example 6.

WORKED EXAMPLE 6 Calculating the original price given the sale price and discount

Aamir bought a baseball cap, in a 20% off everything sale, for \$16. Determine the original price of the cap.

THINK	WRITE
1. The original percentage is 100%, so a discount of 20% means a sale price of 80%.	$(100 - 20)\% = 80\%$ 80% of original price = \$16
2. Write 80% as a fraction and express it as a decimal.	$\dfrac{80}{100}$ of original price = \$16 $\dfrac{80}{100} = 0.80$ or 0.8
3. Divide the sale price of \$16 by 0.8.	0.8 of original price = \$16 Original price = $\dfrac{\$16}{0.8}$ = \$20
4. Write the answer.	The original price was \$20.

Exercise 2.3 Mark-ups and discounts

2.3 Exercise	2.3 Exam questions **on**

Simple familiar	Complex familiar	Complex unfamiliar
1, 2, 3, 4, 5, 6, 7, 8, 9, 10	11, 12, 13	14

These questions are even better in jacPLUS!
- Receive immediate feedback
- Access sample responses
- Track results and progress

Find all this and MORE in jacPLUS ▶

Simple familiar

1. **WE3** The wholesale price of petrol was marked up by 5.5% from 169 cents per litre. Calculate the retail price per litre.

2. A transport company adjusts its charges as the price of petrol changes. Calculate what percentage, correct to 2 decimal places, their fuel costs change by if the price per litre of petrol increases from $1.82 to $1.96.

3. The price of jeans was marked up by 20%. If the original price was $99, calculate the marked-up price.

4. **WE4** Smoothies were increased from $4.80 to $5.45 at the local cafe. Calculate the percentage mark-up, correct to the nearest whole number.

5. Calculate the percentage mark-up on an item that was originally $108.90 and sold for $185.50. Give answer correct to 2 decimal places.

6. **WE5** Determine the percentage discount for each of the following items.
   a. A dress, discounted from $80 to $60
   b. A watch, discounted from $365 to $185

7. Healthway is promoting health and beauty products with a discount. For the item price tag shown, calculate:
   a. the amount of the discount   b. the percentage discount.

   original price ~~99.50~~   *price now* $**49.50** ✓

8. **WE6** Elijah bought some secondhand school books for $280. He got them 60% cheaper than if he purchased them new. Calculate what he would have paid for the books if purchased new.

9. The items shown are from an online electrical store. Next to each item is the retail price and the online price. For each item, calculate:
   i. the discount amount in dollars when the goods are purchased direct
   ii. the percentage discount.

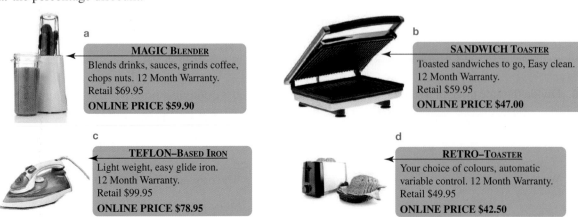

a
MAGIC BLENDER
Blends drinks, sauces, grinds coffee, chops nuts. 12 Month Warranty.
Retail $69.95
ONLINE PRICE $59.90

b
SANDWICH TOASTER
Toasted sandwiches to go, Easy clean.
12 Month Warranty.
Retail $59.95
ONLINE PRICE $47.00

c
TEFLON–BASED IRON
Light weight, easy glide iron.
12 Month Warranty.
Retail $99.95
ONLINE PRICE $78.95

d
RETRO–TOASTER
Your choice of colours, automatic variable control. 12 Month Warranty.
Retail $49.95
ONLINE PRICE $42.50

10. Compare the following. Identify the greatest percentage discount.

   a. A clock, discounted from $47 to $34
   b. A lamp, discounted from $59 to $42

11. A department store announced a 15% discount on every purchase for one day only. Elena decided to use the opportunity to buy new clothes for her daughter.
   She bought a dress normally priced at $29, a 3-piece shorts set (normally $30), pants (normally $16), an embroidered top (normally $18) and sandals (normally $26). Calculate the amount of money Elena was able to save on these purchases by shopping on that day.

12. A carpet company offers a trade discount of 12.5% to a builder for supplying the floor coverings on a new housing estate.
   The builder pays $32 250 for the carpet and charges his customers a total of $35 000. Calculate the percentage discount the customer has received compared to buying direct from the carpet company.

13. The following table shows the mark-ups and discounts applied by a clothing store.

Item	Cost price	Normal retail price (255% mark-up)	Standard discount (12.5% mark-down of normal retail price)	January sale (32.25% mark-down of normal retail price)	Stocktake sale (55% mark-down of normal retail price)
Socks	$1.85				
Shirts	$12.35				
Trousers	$22.25				
Skirts	$24.45				
Jackets	$32.05				
Ties	$5.65				
Jumpers	$19.95				

   Use a spreadsheet to answer these questions.

   a. Identify the calculation that is required in order to determine the stocktake sale price.
   b. Enter the information in your spreadsheet and use it to evaluate the normal retail prices and discount prices for each column as indicated.
   c. Calculate the percentage change between the standard discount price and the stocktake sale price of a jacket.

14. Before the beginning of a winter sale, a shop assistant was asked to reduce the prices of all items in the store by 12.5%. She calculated the new prices and attached new tags to the goods. At the end of the sale she was asked to put the old prices back.
   Unfortunately, the shop assistant had thrown the old tags away as she did not think she would need them again. She decided to add 12.5% to the sale prices. If the shop assistant proceeds in this manner, determine whether she will get back to the original prices.
   Explain your answer with calculations and mathematical reasoning.

Fully worked solutions for this chapter are available online.

LESSON
2.4 Goods and services tax (GST)

SYLLABUS LINKS

- Apply percentage increase or decrease in various contexts, e.g. GST.

Source: General Mathematics Senior Syllabus 2024 © State of Queensland (QCAA) 2024; licensed under CC BY 4.0.

2.4.1 Calculating GST

The **goods and services tax (GST)** is a tax added onto most purchases and services. In Australia it is 10% of the purchase price of an item or service. There are some items and services that are exempt from the GST. These include fresh food, some educational costs and some medical costs. The tax is collected, as part of the total price, at the point of sale.

To calculate the amount of GST payable on an item, we simply calculate 10% of the purchase price. This is another example of a percentage increase.

eles-3095

WORKED EXAMPLE 7 Calculating GST

A cricket bat has a pre-GST price of $127.50. Calculate the GST payable on the purchase of the bat.

THINK	WRITE
Calculate 10% of $127.50.	GST payable $= 10\%$ of $\$127.50$ $= 0.1 \times \$127.50$ $= \$12.75$

Calculating the price with GST

When calculating the price of an item with GST included we multiply by 1.1 (110%), and when calculating the price excluding GST we divide by 1.1 (110%), as follows.

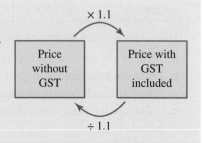

WORKED EXAMPLE 8 Solving problems involving GST

The Besenko family are celebrating the seventh birthday of their daughter, who wants to visit a local theme park. The pre-tax cost of entry for three people is $315.
Calculate how much the Besenkos have to pay to enter the theme park, including the GST.

THINK	WRITE
1. Calculate 110% of $315.	Total cost $= 110\%$ of $\$315$ $= 1.1 \times \$315$ $= \$346.50$
2. Write the answer.	The cost of entry will be $346.50.

WORKED EXAMPLE 9 Calculating the pre-tax price

Calculate the pre-tax price of a car that costs \$31 350, including GST.

THINK	WRITE
1. Total cost is 110% of the price.	Price with GST = 110% of pre-GST price
	$31\,350 = 1.1 \times$ pre-GST price
2. Divide price with GST by 1.1.	Pre-GST price $= \dfrac{31\,350}{1.1}$
3. Write the answer.	The car's pre-tax price was \$28 500.

 Resources

📄 **Digital documents** SkillSHEET Increasing a quantity by a percentage (doc-6902)
SkillSHEET Decreasing a quantity by a percentage (doc-5349)

Exercise 2.4 Goods and services tax (GST)

learn

2.4 Exercise	2.4 Exam questions 

Simple familiar	Complex familiar	Complex unfamiliar
1, 2, 3, 4, 5, 6, 7, 8, 9	10, 11, 12, 13	14

These questions are even better in jacPLUS!
• Receive immediate feedback
• Access sample responses
• Track results and progress

Find all this and MORE in jacPLUS ▶

Simple familiar

1. **WE7** Calculate the GST payable on a book that has a pre-tax price of \$35.60.

2. Calculate the GST payable on each of the following items (prices given are pre-tax):
   a. a bottle of dishwashing liquid at \$2.30
   b. a basketball at \$68.90
   c. a pair of cargo pants at \$98.50
   d. a bus fare at \$1.30.

3. Calculate the GST payable on each of the following items (correct to the nearest cent):
   a. a barbecued chicken with a pre-tax price of \$7.99
   b. a tin of shoe polish with a pre-tax price of \$4.81
   c. a tin of dog food with a pre-tax price of 93c
   d. a pack of toilet rolls with a pre-tax price of \$6.25.

4. **WE8** A pair of sports shoes that cost \$152.50 has the 10% GST added to its cost. Calculate the total cost of the sports shoes.

5. Calculate the total cost of each of the following items after the 10% GST has been added (prices given are pre-tax):
   a. a football jersey priced at \$114.90
   b. a computer game priced at \$29.90
   c. a bunch of flowers priced at \$14.70
   d. a birthday card priced at \$4.95.

6. **WE9** A restaurant bill totals $108.35, including the 10% GST. Calculate the actual price of the meal before the GST was added.

7. A bus fare was $2.09, including the 10% GST. Calculate:
   a. the bus fare without the GST
   b. how much GST was paid.

8. Austin travels to the USA. In the state of Utah the tax is levied at 11%. Calculate what Austin will pay for four nights accommodation in a hotel that charges $102.50 per night.

9. Sachin decides to purchase a new car. The pre-tax cost for the basic model of the car is $30 500. It is an extra $1200 for an automatic car, an extra $1600 for air-conditioning, $1000 for power steering, $600 for metallic paint and $450 for alloy wheels.
   Calculate the cost of each of the following cars, after the 10% GST has been added:
   a. the basic model car
   b. an automatic car with air-conditioning
   c. a car with metallic paint and alloy wheels
   d. a car with all of the above added extras.

Complex familiar

10. Calculate the amount of GST included in an item purchased for a total of:
    a. $34.98
    b. $586.85
    c. $56 367.85
    d. $2.31.

11. Jules is shopping for groceries and buys the following items.
    Bread — $3.30*
    Fruit juice — $5.50*
    Meat pies — $5.80
    Ice-cream — $6.90
    Breakfast cereal — $5.00*
    Biscuits — $2.90
    All prices are listed before GST has been added.
    The items marked with an asterisk (*) are exempt from GST.
    Jules has a voucher that gives him a 10% discount from this shop.
    Calculate how much Jules pays for his groceries.

12. The Australian government is considering raising the GST tax from 10% to 12.5% in order to raise funds and cut the budget deficit.
    The following shopping bill lists all items exclusive of GST.
    1 litre of soft drink — $2.80
    Large bag of pretzels — $5.30
    Frozen lasagne — $6.15
    Bottle of shampoo — $7.60
    Box of chocolate — $8.35
    2 tins of dog food — $1.75 each
    Calculate the amount by which this shopping bill would increase if the rise in GST did go through.
    Note: GST must be paid on all of the items in this bill.

13. Two companies are competing for the same job. Company A quotes a total of $5575 inclusive of GST. Company B quotes $5800 plus GST but offers a 10% reduction on the total price for payment in cash. Determine which is the cheaper offer and by how much.

14. A plumber quotes his clients the cost of any parts required plus $110.50 per hour for his labour, and then adds on the required GST.

    The plumber quotes for a job that requires $250 in parts (excluding GST) and should take 4 hours to complete.

    After attending the job, the plumber realises he was able to complete the job in 3 hours, and therefore only charges the client for 3 hours labour.

    Calculate the discount on the original quote.

Fully worked solutions for this chapter are available online.

LESSON
2.5 Profit and loss

SYLLABUS LINKS

- Apply percentage increase or decrease in various contexts, e.g. percentage profit and loss.

Source: General Mathematics Senior Syllabus 2024 © State of Queensland (QCAA) 2024; licensed under CC BY 4.0.

2.5.1 Calculating profit and loss

When an item is sold for more than it costs, the difference is said to be a **profit**. It is customary to express profit as a percentage of the cost price.

Percentage profit and loss are examples of percentage increase and percentage decrease, respectively.

> **Calculating percentage profit (increase) and loss (decrease)**
>
> $$\text{Profit} = \text{selling price} - \text{cost price}$$
>
> $$\text{Percentage profit} = \frac{\text{profit}}{\text{cost price}} \times 100\%$$
>
> $$\text{Loss} = \text{cost price} - \text{selling price}$$
>
> $$\text{Percentage loss} = \frac{\text{loss}}{\text{cost price}} \times 100\%$$

WORKED EXAMPLE 10 Calculating percentage profit

Calculate the percentage profit on an item that was bought for \$30 and later sold for \$38.

THINK	WRITE
1. Identify the cost price (CP) and the selling price (SP).	CP = \$30; SP = \$38
2. Determine if there is a profit or loss and write the formula. (In this example, SP > CP; therefore, a profit has been made.)	Profit = SP − CP
3. Substitute the values of CP and SP into the formula and evaluate.	Profit = \$38 − \$30 = \$8
4. Write the formula for the percentage profit.	$\text{Percentage profit} = \dfrac{\text{profit}}{\text{CP}} \times 100\%$
5. Substitute the values of profit and CP into the formula and evaluate.	$\text{Percentage profit} = \dfrac{8}{30} \times 100\%$ $\approx 26.67\%$
6. Write the answer.	The percentage profit was 26.67%.

The selling price of an item can be determined if the cost price and profit or loss (x, given as a percentage) are known. The profit (or loss) can considered as a percentage change of the cost price.

> ### Calculating the selling price
>
> **Profit (increase by x): selling price = cost price × $(100 + x)\%$**
>
> **Loss (decrease by x): selling price = cost price × $(100 − x)\%$**
>
> **where x is the profit or loss expressed as a percentage.**

WORKED EXAMPLE 11 Calculating profit

Amira buys a watch online for \$485 and sells it to her friend at a profit of 35%. Calculate how much she sold the watch for.

THINK	WRITE
1. The selling price % is the sum of the cost price 100% and the profit (x) 35%.	Selling price % = $(100 + x)\%$ of \$485 = $(100 + 35)\%$ = 135%
2. Using the formula selling price = cost price × $(100 + x)\%$ multiply the cost price \$485 by 1.35.	Selling price = cost price × 135% = 485 × 1.35 = \$654.75
3. Write the answer.	Amira sold the watch for \$654.75.

To calculate the cost price of an item, when the selling price and profit or loss (x, given as a percentage) are known, we use the formula for calculating the selling price, substitute in the known values and use algebra to solve.

WORKED EXAMPLE 12 Calculating the cost price

A retailer sells a smartphone for \$732, making a profit of 22%. Calculate the wholesale price (cost price) of the smartphone.

THINK	WRITE
1. The selling price % is the sum of the cost price 100% and the profit (x) 22%.	Selling price % $= (100 + x)\%$ $= (100 + 22)\%$ $= 122\%$
2. Write 122% as a decimal.	$122\% = \dfrac{122}{100}$ $122\% = 1.22$
3. Using the formula for selling price.	Selling price = cost price $\times (100 + x)\%$ $732 =$ cost price $\times 1.22$
4. To determine the cost price, divide the selling price \$732 by 1.22.	Cost price $= \dfrac{732}{1.22}$ $= 600$
5. Write the answer.	The wholesale price was \$600.

 Resources

 Interactivity Calculating percentage change (int-6459)

Exercise 2.5 Profit and loss

learn on

2.5 Exercise	2.5 Exam questions on

Simple familiar	Complex familiar	Complex unfamiliar
1, 2, 3, 4, 5, 6, 7, 8	9, 10, 11, 12, 13	14

Simple familiar

1. **WE10** Calculate the percentage profit (to 2 decimal places) for each of the following items.

	Item	Cost price (\$)	Selling price (\$)
a.	Tracksuit	80	139.95
b.	T-shirt	16	22.50
c.	Tennis shoes	49.95	89.95
d.	Tank top	6	9

2. Calculate the percentage profit (to 2 decimal places) for each of the following items.

	Item	Cost price ($)	Selling price ($)
a.	Swimsuit	38	59
b.	Short socks	2	5.95
c.	Training pants	20	29
d.	Tennis skirt	22	36

3. The following goods were sold at a garage sale. Calculate the percentage loss for each of the items, correct to 2 decimal places.

	Item	Cost price ($)	Selling price ($)
a.	Cutlery	40	8
b.	Two bedside lamps	100	22
c.	Vase	35	5
d.	Toaster	19.95	1.50

4. Calculate the percentage loss for each item (to 2 decimal places).

	Item	Cost price ($)	Selling price ($)
a.	Electric kettle	42	6
b.	Set of golf clubs	150	45
c.	Set of building blocks	16	4
d.	Five paperback books by Sydney Sheldon	60	2.50

5. Alex had a collection of 5 games, which he purchased over a period of time at $29.95 each. A friend offered to pay $70 for the whole set. Calculate the loss in dollars and express it as a percentage of the cost price.

6. **WE11** Jonny buys a pair of running shoes for $140 but they are too small. If Jonny sells the running shoes to a friend for 25% loss, determine how much the friend paid.

7. A shopkeeper at the Southbank Markets buys sheepskin boots from the wholesaler at the following prices: children's sizes — $120 per pair; adults' sizes — $170 per pair; extra-large sizes — $190 per pair. If the shopkeeper wants to make a 20% profit, determine what the sale price should be for each type.

8. P-Mart makes 30% profit on all their sales. If they pay $600 for a dining suite, calculate what price they should sell it for to make the desired profit.

Complex familiar

9. **WE12** Hans sells a restaurant for $198 000. He calculates that he has made a 10% profit on the buying price. Calculate what he paid for the restaurant originally.

10. A shopkeeper buys 20 kg of cooking chocolate for $160 and sells it in 500-g packets at $10 each. Determine the percentage profit made.

11. Michael buys a car for $12 000. It depreciates at a rate of $900 per year. If Michael wants his losses to be no more than 30% of the cost price, determine how many years from the purchase he has to sell the car.

12. By selling a collection of coins for $177, Igor makes a profit of 18%. Calculate the original cost of the collection.

13. John paid $50 for a dozen trophies. If he sells them for $14 each, calculate the percentage profit.

Complex unfamiliar

14. A retailer has purchased a particular style of jumper that is proving to be unpopular. After attempting to sell them for two consecutive seasons, the retailer decides to put them on sale at $15 each to recover part of the cost.
    Calculate the wholesale price of each jumper if the retailer suffers a 40% loss.

Fully worked solutions for this chapter are available online.

LESSON
2.6 Simple interest

SYLLABUS LINKS

- Apply percentage increase or decrease in various contexts, e.g. simple interest.
  - $I = Pin$ where I is simple interest, P is principal, i is interest rate per year and n is number of years.
- Use a spreadsheet to display examples of the above computations when multiple or repeated computations are required.

Source: General Mathematics Senior Syllabus 2024 © State of Queensland (QCAA) 2024; licensed under CC BY 4.0.

2.6.1 Calculating simple interest

When you invest money, you receive interest, and when you borrow money, you pay interest. **Simple interest** is calculated as a percentage of the amount of money invested or borrowed. It remains constant for the term of the investment or loan.

To calculate simple interest, we use the following formula.

The simple interest formula

$$\text{Simple interest} = \text{principal} \times \frac{\text{rate}}{100} \times \text{time}$$

$$I = Pin$$

- I – Simple interest: the interest on the money invested or borrowed.
- P – Principal: money to be invested or borrowed.
- i – Rate: interest rate (as a decimal) per time period.
- n – Time: number of time periods.

Note: The time period given is usually in years.

---

eles-3097

WORKED EXAMPLE 13 Calculating simple interest

Calculate the amount of simple interest earned on an investment of $4450 that returns 6.5% per annum for 3 years.

THINK	WRITE
1. Identify the components of the simple interest formula.	$P = \$4450$ $i = 6.5\%$ $\quad = 0.065$ $n = 3$
2. Substitute the values into the formula and evaluate the amount of interest.	$I = Pin$ $\quad = 4450 \times 0.065 \times 3$ $\quad = 867.75$
3. Write the answer.	The amount of simple interest earned is $867.75.

The amount of an investment is the sum of the initial investment or principal, P, and the simple interest, I.

The total amount of an investment

Amount = principal + interest
$$A = P + I$$

WORKED EXAMPLE 14 Calculating the total amount of an investment

An amount of $20 000 is invested for 6 years at 7.5% per annum simple interest. Calculate the amount of the investment after 6 years.

THINK	WRITE
1. The amount is the sum of the initial investment (the principal) and the interest (simple interest).	$A = P + I$
2. Calculate the interest using the formula $I = Pin$.	$I = Pin, P = 20\,000, i = 0.075, n = 6$ $I = 20\,000 \times 0.075 \times 6$ $\quad = 9000$
3. Calculate the amount by adding the interest to the principal.	$A = P + I$ $\quad = 20\,000 + 9000$ $\quad = 29\,000$
4. Write the answer.	The amount after 6 years is $29 000.

2.6.2 Transposing the simple interest formula

To calculate the principal, rate or time, we transpose the simple interest formula to derive the following formulas.

> ### Determining the time, interest rate or principal
>
> To determine the time: $n = \dfrac{I}{Pi}$
>
> To determine the interest rate: $i = \dfrac{I}{Pn}$
>
> To determine the principal: $P = \dfrac{I}{in}$

WORKED EXAMPLE 15 Using the transposed simple interest formula

Determine how long it will take an investment of \$2500 to earn \$1100 with a simple interest rate of 5.5% p.a.

THINK	WRITE
1. Identify the components of the simple interest formula.	$P = 2500$ $I = 1100$ $i = \dfrac{5.5}{100}$ $\quad = 0.055$
2. Substitute the values into the formula and evaluate for n.	$n = \dfrac{I}{Pi}$ $\quad = \dfrac{1100}{2500 \times 0.055}$ $\quad = 8$
3. Write the answer.	It will take 8 years for the investment to earn \$1100.

2.6.3 Simple interest with different time periods

Interest rates are usually stated in terms of a yearly rate, per annum. However, it is very common for interest to be calculated quarterly, monthly, weekly or even daily.

For example, a bank may have a yearly interest rate of 6% per annum, but if interest is calculated weekly, the rate is $\dfrac{6}{52} = 0.11538\%$ per week.

If the rate per annum is given, we can convert to different time periods as follows:

- Quarterly $i = \dfrac{\text{per annum rate}}{4}$

- Monthly $i = \dfrac{\text{per annum rate}}{12}$

- Weekly $i = \dfrac{\text{per annum rate}}{52}$

The interest rate, i, and the time period, n, must be calculated in the same time periods.

For example, 6% per annum for 2 years, with interest calculated monthly, becomes 0.5% per month $(i = \dfrac{0.06}{12} = 0.005$ per month$)$ for 24 months.

WORKED EXAMPLE 16 Calculating simple interest with a different time period

Calculate the simple interest on a loan of \$665 for 3 years at 7% p.a. calculated quarterly.

THINK	WRITE
1. Identify the information in the question.	Principal = \$665 Rate = 7% per annum Time = 3 years
2. Interest is calculated quarterly, so calculate the interest rate per quarter.	$7\% = \dfrac{7}{4} = 1.75\%$ $i = 1.75\%$ per quarter
3. Calculate the number of quarters in 3 years.	$n = 3 \times 4 = 12$ quarters
4. Use the simple interest formula to calculate interest for 3 years. *Note:* When putting 1.75% in the simple interest formula, express it as a decimal.	$I = Pin$ $= 665 \times 0.0175 \times 12$ $= 139.65$
5. Write the answer. *Note:* The simple interest after 3 years will be the same if calculated yearly, quarterly or daily.	The simple interest is \$139.65.

Exercise 2.6 Simple interest

learn on

2.6 Exercise	2.6 Exam questions on

These questions are even better in jacPLUS!
- Receive immediate feedback
- Access sample responses
- Track results and progress

Find all this and MORE in jacPLUS ▶

Simple familiar	Complex familiar	Complex unfamiliar
1, 2, 3, 4, 5, 6, 7	8, 9, 10, 11, 12	13, 14

Simple familiar

1. **WE13** Calculate the simple interest earned on an investment of:
   a. \$2575, returning 8.25% per annum for 4 years
   b. \$12 250, returning 5.15% per annum for $6\dfrac{1}{2}$ years
   c. \$43 500, returning 12.325% per annum for 8 years and 3 months
   d. \$103 995, returning 2.015% per annum for 105 months.

2. **WE14** Calculate the amount of the investment, at simple interest, on:
   a. \$500, after returning 3.55% per annum for 3 years
   b. \$2054, after returning 4.22% per annum for $7\dfrac{3}{4}$ years
   c. \$3500, after returning 11.025% per annum for 9 years and 3 months
   d. \$10 201, after returning 1.008% per annum for 63 months.

3. **WE15** Determine how long it will take an investment of:
   a. $675 to earn $216 interest with a simple interest rate of 3.2% p.a.
   b. $1000 to earn $850 interest with a simple interest rate of 4.25% p.a.
   c. $5000 to earn $2100 interest with a simple interest rate of 5.25% p.a.
   d. $2500 to earn $775 interest with a simple interest rate of 7.75% p.a.

4. a. If $2000 earns $590 in 5 years, calculate the simple interest rate.
   b. If $1800 earns $648 in 3 years, calculate the simple interest rate.
   c. If $408 is earned in 6 years with a simple interest rate of 4.25%, calculate how much was invested.
   d. If $3750 is earned in 12 years with a simple interest rate of 3.125%, calculate how much was invested.

5. **WE16** Calculate the simple interest for each of the following if interest is calculated monthly.
   a. A $8000 loan that is charged simple interest at a rate of 12.25% p.a. for 3 years
   b. A $23 000 loan that is charged simple interest at a rate of 15.35% p.a. for 6 years
   c. A $21 050 loan that is charged simple interest at a rate of 11.734% p.a. for 6.25 years
   d. A $33 224 loan that is charged simple interest at a rate of 23.105% p.a. for 54 months

6. Calculate how much simple interest is paid on each of the following if invested for 4 years.

   a. $1224 at 3.6% p.a. calculated yearly
   b. $955 at 6.024% p.a. calculated monthly
   c. $2445.50 at 4.8% p.a. calculated yearly
   d. $13 728.34 at 9.612% p.a. calculated yearly

7. A savings account with a minimum monthly balance of $800 earns $3.60 interest in a month.
   Calculate the annual rate of simple interest.

Complex familiar

8. An amount of $25 000 is invested for 5 years in an account that pays 6.36% p.a. simple interest.
   At the end of the 5 years, the money is invested for a further 2 years.
   Calculate the simple interest rate that would result in the investment amounting to $35 000 by the end of the 7 years.

9. Rohan invests $1000 at a simple interest rate of 4.5% p.a. and Ria invests $800 at a simple interest rate of 8.8%. Use a spreadsheet to calculate when Ria's investment will be greater than Rohan's investment. Give your answer correct to the nearest year.

10. A borrower has to pay 7.8% p.a. simple interest on a 6-year loan. If the total interest paid is $3744, calculate what the repayments are if they have to be made fortnightly.

11. An amount of $19 245 is invested in a fund that pays a simple interest rate of 7.8% p.a. for 42 months.
    The investor considers an alternative investment with a bank that offers a simple interest rate of 0.625% per month for the first 2.5 years and 0.665% per month after that.
    Determine which is the best investment.

12. An amount of $100 is invested in an account that earns $28 of simple interest in 8 months. Calculate the amount of interest that would have been earned in the 8 months if the annual interest rate was increased by 0.75%.

Complex unfamiliar

13. A bank offers a simple interest loan of $35 000 with monthly repayments of $545. The loan is to be paid in full in 15 years.
After 5 years of payments, the bank offers to reduce the total time of the loan to 12 years if the monthly payments are increased to $650. Calculate how much interest would be paid over the life of the loan under this arrangement.

14. A bank account pays simple interest at a rate of 7.2% p.a. calculated daily.
The account was opened with a deposit of $250 on 1 July, and regular deposits of $350 were made every month for the following 6 months. Use technology or otherwise to calculate the interest payable after 6 months.

Fully worked solutions for this chapter are available online.

LESSON
2.7 Compound interest (optional)

SYLLABUS LINKS

- (This is optional content and is covered more thoroughly in Units 3 & 4.)

Source: General Mathematics Senior Syllabus 2024 © State of Queensland (QCAA) 2024; licensed under CC BY 4.0.

2.7.1 Compound interest and the compound interest formula

Simple interest is calculated on the original investment or loan and is a constant amount for the duration of the investment or loan. However, compound interest is calculated on the balance of the account at the time (that is the initial amount plus interest from the previous time periods).

Sometimes this is expressed as earning interest on interest. Compound interest is not constant as it changes for every period of the investment or loan.

For example, the table below shows the balance after 2 years of an investment of $5000 that earns 5% compounding annually.

Time (years)	Interest	Balance
0	0	5000
1	$0.05 \times 5000 = 250$	$5000 + 250 = 5250$
2	$0.05 \times 5250 = 262.50$	$5250 + 262.50 = 5512.50$

As time progresses, the amount of interest becomes larger at each calculation. In contrast, a simple interest rate calculation on this balance would be a constant, unchanging amount of $250 each year.

To determine the final amount, A, of a compound interest investment or loan, we use the following formula.

> ## Compound interest formula
>
> $$A = P(1+i)^n$$
>
> where:
> - A is the final amount
> - P is the principal
> - i is the interest rate per compounding period
> - n is the number of compounding periods.

To calculate the compound interest, I, we subtract the initial investment or loan, P, from the final amount, A, as follows:

$$\text{Interest} = \text{amount} - \text{principal}$$
$$I = A - P$$

eles-3163

WORKED EXAMPLE 17 Calculating compound interest

Use the compound interest formula to calculate the amount of interest on an investment of $2500 at 3.5% p.a. compounded annually for 4 years, correct to the nearest cent.

THINK	WRITE
1. Identify the components of the compound interest formula, $A = P(1+i)^n$.	$P = 2500$ $i = 3.5\%$ $\quad = 0.035$ $n = 4$
2. Substitute the values into the formula and evaluate the amount of the investment.	$A = P(1+i)^n$ $\quad = 2500(1.035)^4$ $\quad = 2868.81 \text{ (to 2 decimal places)}$
3. Subtract the principal from the final amount of the investment to calculate the interest.	$I = A - P$ $\quad = 2868.81 - 2500$ $\quad = 368.81$
4. Write the answer.	The amount of compound interest is $368.81.

eles-6377

WORKED EXAMPLE 18 Calculating compound interest compounding monthly

Use the compound interest formula to calculate the amount of interest accumulated on $1735 at 7.2% p.a. for 4 years if the compounding occurs monthly. Give your answer correct to the nearest 5 cents.

THINK	WRITE
1. Identify the information in the question.	Principal = $1735 Rate = 7.2% per annum Time = 4 years

2. Interest is calculated monthly, so calculate the interest rate per month.	$i = \dfrac{0.072}{12}$ $= 0.006$
3. Determine the number of months in 4 years.	$n = 4 \times 12 = 48$ months
4. Use the compound interest formula to calculate the total amount after 4 years, using $i = 0.006$ and $n = 48$ months.	$A = P(1 + i)^n$ $A = 1735(1.006)^{48}$ $A = 2312.08$
5. Calculate the compound interest by subtracting the principal from the amount.	$I = A - P$ $I = 2312.08 - 1735$ $I = 577.08$
6. Write the answer.	The compound interest is $577.10.

2.7.2 Using compound interest to calculate inflation over several years.

Inflation is a measure of how an economy is performing over a period of time. It is an increase in the price of goods and services and a decrease in the value of our money. If wages do not increase at the same rate as inflation, then living standards decrease. Inflation is a rate that is expressed as a percentage and in Australia is called the consumer price index (CPI).

The compound interest formula is used to calculate the future cost of an item due to inflation.

WORKED EXAMPLE 19 Calculating the future value of an item due to inflation

A buyer purchased a vintage car for $300 000. If the annual inflation rate is 2.5%, determine the value of the car in 3 years' time.

THINK	WRITE
1. Recall that inflation is an application of compound interest and identify the components of the formula.	$P = 300\,000$ $i = 2.5\%$ $= 0.025$ $n = 3$
2. Substitute the values into the formula and evaluate the amount.	$A = P(1 + i)^n$ $= 300\,000(1.025)^3$ $= 323\,067.19$ (to 2 decimal places)
3. Write the answer.	The car would be valued at $323 067.19 in 3 years' time.

 Resources

✦ **Interactivity** Simple and compound interest (int-6265)

Exercise 2.7 Compound interest (optional)

2.7 Exercise	2.7 Exam questions on

Simple familiar	Complex familiar	Complex unfamiliar
1, 2, 3, 4, 5, 6	7, 8	9, 10

These questions are
even better in jacPLUS!
- Receive immediate feedback
- Access sample responses
- Track results and progress

Find all this and MORE in jacPLUS ⊙

Unless otherwise directed, where appropriate give all answers to the following questions correct to 2 decimal places or the nearest cent.

Simple familiar

1. **WE17** Use the compound interest formula to calculate the amount of compound interest (compounding annually) on an investment of:
   a. $4655 at 4.55% p.a. for 3 years
   b. $12 344 at 6.35% p.a. for 6 years
   c. $3465 at 2.015% p.a. for 8 years
   d. $365 000 at 7.65% p.a. for 20 years.

2. **WE18** Use the compound interest formula to calculate the final amount for:
   a. $675 at 2.42% p.a. for 2 years compounding weekly
   b. $4235 at 6.43% p.a. for 3 years compounding quarterly
   c. $85 276 at 8.14% p.a. for 4 years compounding fortnightly
   d. $53 412 at 4.329% p.a. for 1 year compounding daily.

3. Use the compound interest formula to calculate the principal required to yield a final amount of:
   a. $15 000 after compounding annually at a rate of 5.25% p.a. for 8 years
   b. $22 500 after compounding annually at a rate of 7.15% p.a. for 10 years
   c. $1000 after compounding annually at a rate of 1.25% p.a. for 2 years
   d. $80 000 after compounding annually at a rate of 6.18% p.a. for 15 years.

4. **WE19** An investment property was purchased for $325 000.

   If the average annual inflation rate is 2.73% p.a., determine the value of the property after 5 years.

5. An accountant earning an annual salary of $92 000 receives a pay increase in line with the yearly inflation rate of 1.8%. Determine their new annual salary.

6. An $8000 investment earns 7.8% p.a. compound interest over 3 years. Calculate how much interest is earned if the amount is compounded:
   a. annually   b. monthly   c. weekly   d. daily.

7. Shivani is given $5000 by her grandparents on the condition that she invests it for at least 3 years. Her parents help her to determine the best investment options and come up with the following choices.

   i. A local business promising a return of 3.5% compounded annually, with an additional 2% bonus on the total sum paid at the end of the 3-year period
   ii. A building society paying a fixed interest rate of 4.3% compounded monthly
   iii. A venture capitalist company guaranteeing a return of 3.9% compounded daily

   Assuming each option is equally secure, decide where Shivani should invest her money if she plans to invest for 3 years.

8. a. Use a spreadsheet to compare $1000 compounding annually with compounding quarterly at a rate of 12% p.a. for 5 years.
   b. Determine the effect of compounding at more regular intervals during the year.

9. The costs of manufacturing a smart watch decrease by 10% each year.
   The watch initially retails at $200, and the makers decrease the price in line with the manufacturing costs over the first 3 years.

   Inflation is at a steady rate of 3% over each of these years, and the price of the watch also rises with the rate of inflation. Calculate the cost of the watch for each of the 3 years according to inflation. (*Note:* Apply the inflation price increase before the manufacturing cost decrease.)

10. Francisco is a purchaser of fine art. His two favourite pieces are a sculpture he purchased 17 years ago for $12 000 and a series of prints he purchased 9 years ago for $17 000.

   a. If inflation averaged 3.3% for the period between when he bought the sculpture and when he bought the prints, determine which item cost more in real terms.
   b. The value of the sculpture has appreciated at a rate of 7.5% since it was purchased, and the value of the prints has appreciated at a rate of 6.8% since they were purchased. Calculate how much they are both worth. Round your answers correct to the nearest dollar.

Fully worked solutions for this chapter are available online.

LESSON
2.8 Exchange rates

2.8.1 Foreign currency

When travelling overseas or buying online from overseas companies, we need to exchange or convert Australian dollars (AUD or A$) to the foreign currency.

Exchange rates are the amount at which one currency can be exchanged for another currency. These rates change daily, and are published in the financial section of newspapers and reported in the news.

For example, one Australian dollar could be equivalent to:

Currency	Symbol	Exchange rate
Canadian dollar, CAD	$	0.9643
European euro, EUR	€	0.6416
Hong Kong dollar, HKD	$	5.9146
Japanese yen, JPY	¥	85.28
Malaysian ringgit, MYR	RM	3.1527
New Zealand dollar, NZD	$	1.1067
Thai baht, THB	฿	24.89
UK pound sterling, GBP	£	0.5725
US dollar, USD	$	0.7574

Converting from Australian dollars to a foreign currency

To convert from Australian dollars to a foreign currency, we use:

$$\text{foreign currency} = \text{AUD} \times \text{exchange rate}$$

WORKED EXAMPLE 20 Converting Australian dollars to foreign currency

Convert A$500 to Malaysian ringgit.

THINK	WRITE
1. Use the table to determine the exchange rate for $1 converted to RM.	A$1 = RM3.1527
2. Multiply by 500 to determine the value of A$500.	$3.1527 \times 500 = 1576.35$
3. Write the answer.	1576.35 ringgit or RM1576.35

Converting from a foreign currency to Australian dollars

$$AUD = \frac{\text{foreign currency}}{\text{exchange rate}}$$

eles-3098

WORKED EXAMPLE 21 Converting foreign currency to Australian dollars

Convert €415 to Australian dollars.

THINK	WRITE
1. Use the table to determine the exchange rate for €1 converted into AUD.	A\$1 = €0.6416
2. Divide by 0.6416.	$\dfrac{415}{0.6416} = 646.82$
3. Write the answer.	A\$646.80

Exercise 2.8 Exchange rates

learnon

2.8 Exercise	2.8 Exam questions on

Simple familiar	Complex familiar	Complex unfamiliar
1, 2, 3, 4, 5, 6, 7, 8, 9	10, 11	12

These questions are even better in jacPLUS!
- Receive immediate feedback
- Access sample responses
- Track results and progress

Find all this and MORE in jacPLUS ▶

Simple familiar

1. **WE20** Convert A\$35 to Thai baht.

2. Convert each of the following, given in Australian dollars.
   a. \$390 to US dollars
   b. \$5000 to British pounds sterling
   c. \$950 to Thai baht
   d. \$34 000 to Hong Kong dollars.

3. **WE21** Convert RM3896 to Australian dollars

4. Calculate how much you would receive in Australian dollars if you exchange:
   a. US\$4500
   b. 880 baht
   c. £3500
   d. ¥700.

5. Shae is going on an overseas trip and has budgeted to spend A\$800 in each country she visits. Calculate how much foreign currency she can spend in each of the following countries.
   a. Italy
   b. Japan
   c. Scotland
   d. USA

6. Emily buys a pair of boots online from Italy for €200 with free delivery. Calculate how much this is in Australian dollars.

7. Hiromi is on exchange with a family in Sydney and has ¥45 800 to spend. Calculate how much this is in Australian dollars.

8. Michael is going hiking in New Zealand and gets a travelcard loaded with NZ$1090. Calculate how much it costs him in Australian dollars if he has to pay A$20 for the card.

9. Determine the amount required in Australian dollars, to the nearest cent, to buy:
   a. a bottle of Coke in San Francisco for $8
   b. a bowl of noodles in Tokyo for ¥600
   c. a watch in Madrid for €540
   d. a ticket to the theatre in London for £30.

Complex familiar

10. Harry plans to visit Tokyo on business. At the start of his trip, he changes A$800 into Japanese yen. While he is in Tokyo, Harry spends 50 000 yen. Once he returns to Australia, Harry changes his leftover yen back to Australian dollars. Determine how many Australian dollars Harry has at the completion of his trip.

11. Holly travels to Germany. At the start of her trip, she changes A$660 into European euros. While she is in Germany, Holly spends 220 euros. When she returns to Australia, Holly converts her remaining euros to Australian dollars.
    Determine how many Australian dollars she will have after her trip.

Complex unfamiliar

12. During an economic crisis in 1998, Indonesia experienced severe inflation. In one week, on Monday, A$1 would have bought 9500 rupiah, whereas on Thursday, A$1 would have bought 10 900 rupiah.
    On holidays in Indonesia at this time, Joel exchanged A$120 and paid for a camera on Monday. Calculate how much he would have saved if he had waited to make the transaction on Thursday (assuming the marked price did not change).
    Give your answer in Australian dollars.

Fully worked solutions for this chapter are available online.

LESSON
2.9 Dividends

SYLLABUS LINKS

- Calculate the dividend paid on a portfolio of shares, given the dividend yield or dividend paid per share, and compare share values by calculating a price-to-earnings (P/E) ratio.

  - $\text{dividend yield} = \dfrac{\text{dividend}}{\text{share price}} \times 100$

  - $\text{P/E ratio} = \dfrac{\text{market price per share}}{\text{annual earnings per share}}$

Source: General Mathematics Senior Syllabus 2024 © State of Queensland (QCAA) 2024; licensed under CC BY 4.0.

2.9.1 Dividend paid per share

A dividend is a payment that a company makes to its shareholders for investing in their company. This means that shareholders own part of the company. When the company makes a profit, the shareholders are paid a percentage of this profit, which is called a dividend. Shares can be bought and sold on the share market or the stock exchange.

A dividend paid on a portfolio of shares can be calculated in one of the following ways:
- the dividend paid per share
- a percentage of the price of the shares, called a dividend yield.

> ### Calculating the total dividend paid based on the price per share
>
> To calculate the total dividend based on the dividend paid per share, multiply the number of shares by the dividend per share.
>
> **Total dividend = number of shares × price per share**

WORKED EXAMPLE 22 Calculating dividend paid per share

Sally owns 5000 Walidy shares. If the company pays a dividend of 24 cents per share, calculate how much Sally will receive.

THINK	WRITE
1. Convert 24 cents into dollars and multiply by the total number of shares.	Dividend = total number of shares × price per share Dividend = 0.24 × 5000 $\qquad\qquad$ = 1200
2. Write the answer.	The dividend payable will be $1200.

2.9.2 Dividend yield

Shares in different companies can vary drastically in price, from cents up to hundreds of dollars for a single share. As a company becomes more successful, the share price will rise, and as a company becomes less successful, the share price will fall.

An important factor that investors look at when deciding where to invest is the **dividend yield** (also known as **percentage dividend**) of a company. The dividend yield is calculated by dividing the dividend per share by the share price expressed as a percentage.

A growing dividend yield is a sign that a company is in a good financial position.

Calculating the dividend yield

$$\text{Dividend yield} = \frac{\text{dividend}}{\text{share price}} \times 100$$

WORKED EXAMPLE 23 Calculating the dividend yield

Calculate the dividend yield on a share that costs $13.45 with a dividend paid per share of $0.45.

THINK	WRITE
1. To determine the dividend yield, divide the dividend by the share price expressed as a decimal.	$\text{Dividend yield} = \dfrac{\text{dividend}}{\text{share price}} \times 100$ $\text{Dividend yield} = \dfrac{0.45}{13.45} \times 100$ $= 3.35$
2. Write the answer.	The dividend yield is 3.35%.

2.9.3 The price-to-earnings (P/E) ratio

The **price-to-earnings ratio** (**P/E ratio**) refers to the market price of a company's share price divided by its earnings or profit per share. It gives an indication of how much shares cost per dollar of profit earned.

The price-to-earnings ratio

$$\text{Price-to-earnings ratio} = \frac{\text{market price per share}}{\text{annual earnings per share}}$$

The higher the price-to-earnings ratio, the more you have to invest for each dollar of profit.

Exercise 2.9 Dividends

learnon

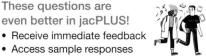

2.9 Exercise	**2.9 Exam questions** on

Simple familiar	Complex familiar	Complex unfamiliar
1, 2, 3, 4, 5, 6, 7, 8	9, 10, 11	12

These questions are even better in jacPLUS!
- Receive immediate feedback
- Access sample responses
- Track results and progress

Find all this and MORE in jacPLUS ▶

Simple familiar

1. **WE22** Nola owns 8000 shares in the company Click Dotcom. If the company pays a dividend of 18 cents per share, calculate how much will Nola receive.

2. A company paid a shareholder $3200 on their portfolio of shares. If the dividend was 33 cents per share, calculate how many shares the shareholder owned.

3. **MC** A shareholder received a dividend of $2800 on a portfolio of 6500 shares in a company. Select from the following the dividend that the company was paying per share.
   A. $0.43
   B. $43
   C. $0.043
   D. $4.30

4. Complete the following table.

	Number of shares	Dividend per share	Total dividend paid
a.		$1.04	$6500
b.	35 500	$0.55	
c.	789 500		$24 460
d.	205 500	$0.88	

5. **WE23** Calculate the dividend yield for a company with:

   a. a dividend paid per share of $0.45 and a price per share of $10.80
   b. a dividend paid per share of $0.21 and a price per share of $7.25.

6. Calculate the dividend yield paid on the following portfolios of shares.

   a. A share price of $14.60 with a total dividend paid of $2325.05 and a total of 2500 shares
   b. A share price of $22.34 with a total dividend paid of $608.32 and a total of 700 shares
   c. A share price of $45.50 with a total dividend paid of $2338.70 and a total of 1000 shares
   d. A share price of $33.41 with a total dividend paid of $5759.89 and a total of 2000 shares

7. **WE24** Calculate the price-to-earnings ratio for a company with:

   a. a current share price of $12.50 per share and an annual earning of 25 cents per share
   b. a current share price of $43.25 per share and an annual earning of $1.24 per share
   c. a current share price of $79.92 per share and an annual earning of $3.32 per share
   d. a current share price of $116.46 per share and an annual earning of $7.64 per share.

8. Calculate the earnings per share for a company with:

   a. a price-to-earnings ratio of 25.5 and a current share price of $8.75
   b. a price-to-earnings ratio of 20.3 and a current share price of $24.35
   c. a price-to-earnings ratio of 12.2 and a current share price of $10.10
   d. a price-to-earnings ratio of 26 and a current share price of $102.

Complex familiar

9. **MC** Alexandra is having trouble deciding which of the following companies to invest in. She wants to choose the company with the highest dividend yield.
   Calculate the dividend yield for each company to determine which Alexandra should choose.

   A. A clothing company with a share price of $9.45 and a dividend of 45 cents
   B. A mining company with a share price of $53.20 and a dividend of $1.55
   C. A financial company with a share price of $33.47 and a dividend of $1.22
   D. A technology company with a share price of $7.22 and a dividend of 41 cents

10. The details of two companies are shown in the following table.

Company	Share price	Net profit	Total shares
Company A	$34.50	$8 600 000	6 500 000
Company B	$1.48	$1 224 000	5 550 000

a. Calculate the price-to-earnings ratio for each company. Remember earnings are also referred to as profit.
b. If a shareholder has 500 shares in Company A and 1000 shares in Company B, calculate how much they will receive from their dividends.
c. Determine which company represents the best investment.

11. The share price of a mining company over several years is shown in the following table.

Year	2012	2013	2014	2015	2016
Share price	$44.50	$39.80	$41.20	$31.80	$29.60
Dividend per share	$1.73	$3.25	$2.74	$3.15	$3.42

a. If there are a total of 10 000 000 shares in the company and 35% of the net profit was reinvested each year, use a spreadsheet or other technology to calculate the net profit for each of the years listed.
b. Determine the price-to-earnings ratios for each of the years listed.
c. Determine which was the best year to purchase shares in the company.

Complex unfamiliar

12. You invested $2016 in shares in a company with an initial share price of $18 per share. You were paid $0.75 per share each year in a dividend. Two years later you sold that same stock at $22 per share. Evaluate your return on the investment.

Fully worked solutions for this chapter are available online.

LESSON
2.10 Review

2.10.1 Summary

doc-41471

Hey students! Now that it's time to revise this chapter, go online to:

 Access the chapter summary

 Review your results

 Practise exam questions

Find all this and MORE in jacPLUS ▶

2.10 Exercise

learn on

2.10 Exercise	**2.10 Exam questions** on

Simple familiar	Complex familiar	Complex unfamiliar
1, 2, 3, 4, 5, 6, 7, 8, 9, 10, 11, 12	13, 14, 15, 16	17, 18, 19, 20

These questions are even better in jacPLUS!
- Receive immediate feedback
- Access sample responses
- Track results and progress

Find all this and MORE in jacPLUS ▶

Simple familiar

1. **MC** The unit cost (per gram) of a 120-gram tube of toothpaste sold for $3.70 is:

   A. $32.43 B. $0.03 C. $0.44 D. $0.05

2. **MC** A 1.25-L bottle of soft drink costs $1.25. With the annual inflation rate over the next 4 years expected to be 4%, in 4 years the bottle will cost:

   A. $1.54 B. $6.00 C. $1.75 D. $1.46

3. **MC** If ServosRUs purchased petrol for 118.4 cents and sold it for 130.9 cents, the percentage profit was:

   A. 10.6% B. 9.5% C. 90% D. 1.1%

4. **MC** A basketball ring is sold for $28.50. If this represents a 24% discount from the recommended retail price (RRP), the original price was:

   A. $90.25 B. $118.75 C. $52.50 D. $37.50

5. **MC** A tradesman offers a 6.8% discount for customers who pay in cash. Calculate how much a customer would pay if they paid their bill of $244 in cash.

   A. $16.59 B. $218.48 C. $227.41 D. $261.59

6. **MC** When the simple interest formula is transposed to calculate i, determine the correct formula.

   A. $i = IPn$ B. $i = \dfrac{Pn}{PnI}$ C. $i = n$ D. $i = \dfrac{I}{Pn}$

7. **MC** David buys a skateboard online for US$190. If the exchange rate is A$1 for US$0.7574, calculate how much he pays in Australian dollars, to the nearest cent.

   A. $143.90 B. $250.85 C. $190 D. $250.86

8. **MC** The price-to-earnings ratio for a company with a share price of $2.40 and a profit of 87 cents per share is:

   A. 2.09 **B.** 2.76 **C.** 0.03 **D.** 3.27

9. **MC** Determine which of the following companies has the lowest share price.

   A. Company A with a price-to-earnings ratio of 10.4 and a profit of $1.87 per share
   B. Company B with a price-to-earnings ratio of 28.1 and a profit of 36 cents per share
   C. Company C with a price-to-earnings ratio of 14.8 and a profit of 79 cents per share
   D. Company D with a price-to-earnings ratio of 35.75 and a profit of 97 cents per share

10. **MC** The dividend yield for a company with a dividend paid per share of $0.35 and a price per share of $6.45, correct to 1 decimal place, is:

    A. 3.5% **B.** 4.5% **C.** 5.4% **D.** 6.45%

11. For each of the following, calculate the unit price for the quantity shown in brackets.

    a. 750 g of Weetbix for $4.99 (per 100 g)
    b. $16.80 for 900 g of jelly beans (per 100 g)
    c. $4.50 for 1.5 L of milk (per 100 mL)
    d. $126.95 for 15 L of paint (per L)

12. Determine the amount of GST included in the price or needing to be added to the price for the following amounts.

    a. $45.50 with GST included
    c. $448.75 with GST included
    b. $109.00 plus GST
    d. $13.25 plus GST

Complex familiar

13. Calculate the monthly repayments for a $6250 loan that is charged simple interest at a rate of 9.32% per annum for 7.25 years.

14. Adam purchased 1600 shares in a company that recently announced a dividend paid per share of $4.45.

    a. He originally purchased the shares for $2.17 each. Calculate the dividend yield of the value of the shares.
    b. Adam decides to reinvest the dividend payment in an interest bearing account at 14.4% p.a. This account adds the interest onto the investment every 6 months and calculates the interest based on the total. Calculate how much interest he will earn if he invests for 1.5 years.

15. Use a spreadsheet to calculate the price-to-earnings for each of the following companies shown in this table.

Company	Company A	Company B	Company C	Company D	Company E
Currency	Australian dollars	US dollars	European euros	Chinese yuan	Indian rupees
Share price	$23.35	$26.80	€16.20	¥133.5	₹1288
Profit per share	$1.46	$1.69	€0.94	¥8.7	₹65.5

16. Sophie purchased an investment property for $650 000, and 4 years later she sold the property for $802 190. If the higher selling price was due solely to inflation, use a spreadsheet, algebra or trial and error to determine the rate of inflation.

17. John is comparing different brands of lollies at the local supermarket. A packet of Brand A lollies costs $7.25 and weighs 250 g. A packet of Brand B lollies weighs 1.2 kg and costs $22.50.

    If the more expensive brand was to reconsider their price, calculate the price for their lollies that would match the unit price of the cheaper brand.

18. Four years ago a business was for sale at $130 000. Amanda and Callan had the money to purchase the business but missed out at the auction. Four years later the business is again for sale, but now at $185 000.

    Amanda and Callan will now need to borrow the increase in the price amount. They obtained a loan where the interest was added to the original amount at the end of each year and the interest for the next year calculated on the total amount. If there were no repayments required for 24 months, calculate the amount of interest they would owe at the end of 2 years at an annual rate of 12.75%.

19. An investment of $3655 is invested for 3 years at 6.54%. The interest made annually is added to the investment at the end of each year, and the interest calculated on the total amount. Determine the value of the investment at the end of the 3 years.

20. Tayla is planning to backpack around Australia in a year and estimates that she needs $1400 for the trip. She has already saved $1095. She decides to open a Super-Saver investment account that pays 9.55% with the interest added to her savings daily, and invest her savings.

    After one year of this investment she realises that she doesn't have enough money, so she borrows some money from her parents.

    She re-invests the original savings, the interest earned on her investment in one year and the money she borrowed for another 12 months at the same rate.

    Using a spreadsheet or otherwise, calculate how much she must borrow from her parents.

Fully worked solutions for this chapter are available online.

Answers

Chapter 2 Consumer arithmetic — managing finances

2.2 Unit cost

2.2 Exercise

1. a. per litre   b. per hour
   c. per day   d. per square metre
   e. per square metre   f. per hour

2. a. $0.59   b. $1.40

3. a. $0.80   b. $0.30

4. a. $0.96   b. $0.93

5. a. $1.46   b. $1.03

6. a. $4.08 per litre   b. $0.42 per 100 mL

7. a. $30 per kg   b. $18.45 per kg

8. Brand H

9. B

10. BBQ lamb chops: $2.54
    Porterhouse steak: $7.25
    Chicken drumsticks: $1.70

11. $20.25

12.

	Item	Size	Selling price	Unit price (per 100 g or 100 mL)
a.	Cheese	450 g	$4.78	$1.06
b.	Onions	2.5 kg	$5.75	$0.23
c.	BBQ sauce	750 g	$3.15	$0.42
d.	Milk	3.12 L	$3.12	$0.10

13. a. $4.47   b. $4.54

14. Bundle 1: $34.04/kg
    Bundle 2: $33.36/kg
    Bundle 2 is better value for money as it costs less per kilogram.

2.3 Mark-ups and discounts

2.3 Exercise

1. 178.30 cents
2. 7.69%
3. $118.80
4. 14%
5. 70.34%
6. a. 25%   b. 49.3%
7. a. $50   b. 50.25%
8. $700
9. i. a. $10.05   b. $12.95   c. $21.00   d. $7.45
   ii. a. 14.37%   b. 21.60%   c. 21.01%   d. 14.91%
10. a. 27.7%
    b. 28.8%, so **b** is the greatest discount.
11. $17.85
12. 5.04%
13. a. Normal retail price $\times 0.45$
    b. See the table at the bottom of the page.*
    c. 48.57%
14. No, as the 12.5% is calculated from different amounts. For example, a $60.00 item reduced by 12.5% ($7.50) is $52.50. A $52.50 item increased by 12.5% ($6.56) is $59.06.

2.4 Goods and services tax (GST)

2.4 Exercise

1. $3.56
2. a. $0.23   b. $6.89   c. $9.85   d. $0.13
3. a. 80c   b. 48c   c. 9c   d. 63c
4. $167.75
5. a. $126.39   b. $32.89   c. $16.17   d. $5.45
6. $98.50
7. a. $1.90   b. $0.19
8. $455.10
9. a. $33 550   b. $36 630   c. $34 705   d. $38 885

---

*13. b.

Item	Cost price	Normal retail price (255% mark-up)	Standard discount (12.5% mark-down of normal retail price)	January sale (32.25% mark-down of normal retail price)	Stocktake sale (55% mark-down of normal retail price)
Socks	$1.85	$6.57	$5.75	$4.45	$2.96
Shirts	$12.35	$43.84	$38.36	$29.70	$19.73
Trousers	$22.25	$78.99	$69.12	$53.52	$35.55
Skirts	$24.45	$86.80	$75.95	$58.81	$39.06
Jackets	$32.05	$113,78	$99.56	$77.09	$51.20
Ties	$5.65	$20.06	$17.55	$13.59	$9.03
Jumpers	$19.95	$70.82	$61.97	$47.98	$31.87

10. a. $3.18   **b.** $53.35   **c.** $5124.35 **d.** $0.21

11. $27.86

12. $0.84

13. Company A by $167

14. 15.97%

2.5 Profit and loss

2.5 Exercise

1. a. 74.94%   **b.** 40.63%   **c.** 80.08%   **d.** 50%

2. a. 55.26%   **b.** 197.5%   **c.** 45%   **d.** 63.64%

3. a. 80%   **b.** 78%   **c.** 85.71%   **d.** 92.48%

4. a. 85.71%   **b.** 70%   **c.** 75%   **d.** 95.83%

5. $79.75; 53.26%

6. $105

7. Children's price: $144, adult's price: $204, extra-large: $228

8. $780

9. $180 000

10. $150%

11. 4 years

12. $150

13. 236%

14. $25

2.6 Simple interest

2.6 Exercise

1. a. $849.75   **b.** $4100.69
 c. $44 231.34   **d.** $18 335.62

2. a. $553.25   **b.** $2725.76 **c.** $7069.34 **d.** $10 740.84

3. a. 10 years   **b.** 20 years   **c.** 8 years   **d.** 4 years

4. a. 5.9%   **b.** 12%   **c.** $1600   **d.** $10 000

5. a. $2940   **b.** $21 183.10
 c. $15 437.54   **d.** $34 543.88

6. a. $176.26   **b.** $230.12   **c.** $469.54
 d. $5278.27

7. 5.4%

8. 3.11%

9. In the 8th year

10. $75.28

11. The first investment at 7.8%

12. $28.50

13. $52 300

14. Interest payable is $41.33

2.7 Compound interest (optional)

2.7 Exercise

1. a. $664.76   **b.** $5515.98
 c. $599.58   **d.** $1 229 312.85

2. a. $708.47   **b.** $5128.17
 c. $118 035.38   **d.** $55 774.84

3. a. $9961.26   **b.** $11 278.74
 c. $975.46   **d.** $32 542.37

4. $93 656

5. Yes, a profit of $13 461.68 has been made

6. a. $2021.81   **b.** $2101.50
 c. $2107.38   **d.** $2108.90

7. The building society

8. a.

Compounding annually		Compounding quarterly	
Year	Amount	Quarter	Amount
1	$1000.00	1	$1000.00
		2	$1030.00
		3	$1060.90
		4	$1092.73
		5	$1125.51
2	$1120.00	6	$1159.27
		7	$1194.05
		8	$1229.87
		9	$1266.77
3	$1254.40	10	$1304.77
		11	$1343.92
		12	$1384.23
		13	$1425.76
4	$1404.93	14	$1468.53
		15	$1512.59
		16	$1557.97
		17	$1604.71
5	$1573.52	18	$1652.85
		19	$1702.43
		20	$1753.51

b. Compounding at more regular intervals pays more interest.

9. Year 1: $185.40, Year 2: $171.86, Year 3: $159.32

10. a. The prints ($17 000) cost more than the sculpture ($15 559.08) in real terms.

 b. The sculpture was worth $41 032. The prints were worth $30 732.

2.8 Exchange rates

2.8 Exercise

1. ฿871.15

2. a. US$295.39   **b.** £2862.50
 c. ฿23 646   **d.** HK$201 096.4

3. A$1235.77

4. a. A$5941.38   **b.** A$35.36
 c. A$6113.54   **d.** A$8.21

5. a. €513.28   **b.** ¥68 224
 c. £458   **d.** US$605.92

6. $311.72

7. $537.05

8. A$1004.91

9. a. $10.56   b. $7.04   c. $841.65   d. $52.40

10. A$213.70

11. $317.11

12. A$15.41

2.9 Dividends

2.9 Exercise

1. $1440

2. $9697

3. A

4.

a.	6250 g
b.	$19 525
c.	$0.03
d.	180 840

5. a. 4.2%   b. 2.9%

6. a. 6.37%   b. 3.89%   c. 5.14%   d. 8.62%

7. a. 50   b. 34.9   c. 24.1   d. 15.2

8. a. 34 cents/share   b. $1.20/share
   c. 83 cents/share   d. $3.92/share

9. D

10. a. Company A: $1.16; Company B: $0.19
    Company A: 29.74; Company B: 7.79

    b. $770

    c. Company B

11. a, b.

Year	Net profit	Price-to-earnings ratio
2012	$26 615 384.62	25.72
2013	$50 000 000.00	12.25
2014	$42 153 846.15	15.04
2015	$48 461 538.46	10.10
2016	$52 615 384.62	8.65

    c. 2016

12. $2464

2.10 Review

2.10 Exercise

1. B
2. D
3. A
4. D
5. C
6. D
7. D
8. B
9. B
10. C
11. a. $0.67   b. $1.87   c. $0.30   d. $8.46
12. a. $4.14   b. $10.90   c. $40.79   d. $1.33
13. $120.38
14. a. 205.07%   b. $1651.31
15. See the table at the bottom of the page.*
16. 5.4%
17. $4.70
18. $14 919.10
19. $4420.03
20. $67.79

*15

Company	Currency	Share price	Profit per share	P/E Ratio
Company A	AU dollars	$23.35	$1.46	15.99
Company B	US dollars	$26.80	$1.69	15.86
Company C	European euros	€16.20	€0.94	17.23
Company D	Chinese yuan	¥133.50	¥8.70	15.34
Company E	Indian rupees	₹1288.00	₹65.50	19.66

3 Pythagoras' theorem and mensuration

LESSON SEQUENCE

Fully worked solutions for this chapter are available online.

EXAM PREPARATION

Access exam-style questions in every lesson, available online.

on Resources

Solutions	Solutions — Chapter 3 (sol-1244)
Exam question	Exam question booklet — Chapter 3 (eqb-0260)
Digital documents	Learning matrix — Chapter 3 (doc-41473)
	Chapter summary — Chapter 3 (doc-41474)

LESSON
3.1 Overview

3.1.1 Introduction

Pythagoras' theorem is a fundamental principle in geometry that establishes a relationship between the sides of a right-angled triangle. This theorem is crucial in various fields such as physics, engineering and computer science for solving problems involving distances and measurements. It is essential in rendering 3D objects and calculating distances between points in computer-generated imagery. In astronomy, this theorem helps in calculating distances between stars and planets when considering right-angled triangle approximations.

Mensuration, on the other hand, is the branch of mathematics that deals with the measurement of geometric figures and their parameters such as length, area and volume. It involves the use of various formulas to calculate the dimensions of different shapes and solids. Mensuration is essential in surveying to determine the area and volume of land, which is crucial for agriculture, construction and property management. Have you ever wondered why tennis balls are sold in cylindrical containers? This is an example of manufacturers wanting to minimise the amount of waste in packaging.

Mensuration is essential in everyday applications such as architecture, construction and any field that requires precise measurements and spatial understanding. Together, Pythagoras' theorem and mensuration form the backbone of practical geometry, enabling us to navigate and quantify the physical world.

3.1.2 Syllabus links

Lesson	Lesson title	Syllabus key knowledge points
3.2	Pythagoras' theorem in two dimensions	○ Understand and use Pythagoras' theorem to solve practical problems in two dimensions. • $c^2 = a^2 + b^2$ where c is length of the hypotenuse and a and b are lengths of the two perpendicular sides. ○ Solve practical problems involving shape and measurement.
3.3	Pythagoras' theorem in three dimensions	○ Understand and use Pythagoras' theorem to solve practical problems and simple applications in three dimensions. • $c^2 = a^2 + b^2$ where c is length of the hypotenuse and a and b are lengths of the two perpendicular sides. ○ Solve practical problems involving shape and measurement.
3.4	Perimeter and area of polygons and circles	○ Calculate perimeters, P, of standard two-dimensional objects in practical situations, including circles, triangles, rectangles, trapeziums and parallelograms. • Circle: $C = 2\pi r$, where C is circumference and r is radius. ○ Calculate areas, A, of standard two-dimensional objects in practical situations, including circles, triangles, rectangles, parallelograms and trapeziums. • Circle: $A = \pi r^2$ where r is radius. • Triangle: $A = \frac{1}{2}bh$ where b is base length and h is perpendicular height. • Parallelogram: $A = bh$ where b is the base length and h is the perpendicular height. • Trapezium: $A = \frac{1}{2}(a+b)h$ where a and b are parallel lengths and h is perpendicular height. ○ Solve practical problems involving shape and measurement.
3.5	Perimeter and area of sectors and composite shapes	○ Calculate perimeters, P, of standard two-dimensional objects in practical situations, including sectors of circles and composites. • Sector of circle: $P = 2r + \frac{\theta}{180}\pi r$, where θ is central angle and r is radius. ○ Calculate areas, A, of standard two-dimensional objects in practical situations, including sectors of circles and composites. • Sector of circle: $A = \frac{\theta}{360}\pi r^2$ where θ is central angle and r is radius. ○ Solve practical problems involving shape and measurement.
3.6	Volume and capacity	○ Calculate volumes, V, and capacities of standard three-dimensional objects in practical situations, including rectangular prisms, cylinders, pyramids, cones, spheres and composites. • Prism: $V = Ah$ where A is base area and h is perpendicular height. • Cylinder: $V = \pi r^2 h$ where r is radius and h is perpendicular height. • Pyramid: $V = \frac{1}{3}Ah$ where A is base area and h is perpendicular height. • Cone: $V = \frac{1}{3}\pi r^2 h$ where r is radius and h is perpendicular height. • Sphere: $V = \frac{4}{3}\pi r^3$ where r is radius. ○ Solve practical problems involving shape and measurement.
3.7	Surface area of three-dimensional objects	○ Calculate surface areas, S, of standard three-dimensional objects in practical situations, including rectangular prisms, cylinders, pyramids, cones, spheres and composites. • Cylinder: $S = 2\pi rh + 2\pi r^2$ where r is radius and h is perpendicular height. • Cone: $S = \pi rs + \pi r^2$ where r is radius and s is slant height. • Sphere: $S = 4\pi r^2$ where r is radius. ○ Solve practical problems involving shape and measurement.

Source: General Mathematics Senior Syllabus 2024 © State of Queensland (QCAA) 2024; licensed under CC BY 4.0.

LESSON
3.2 Pythagoras' theorem in two dimensions

> **SYLLABUS LINKS**
>
> - Understand and use Pythagoras' theorem to solve practical problems in two dimensions.
>   - $c^2 = a^2 + b^2$ where c is length of the hypotenuse and a and b are lengths of the two perpendicular sides.
> - Solve practical problems involving shape and measurement.
>
> **Source:** General Mathematics Senior Syllabus 2024 © State of Queensland (QCAA) 2024; licensed under CC BY 4.0.

3.2.1 Pythagoras' theorem

Pythagoras' theorem allows us to calculate the length of a side of a right-angled triangle if we know the lengths of the other two sides. Consider ΔXYZ shown.

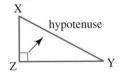

XY is the **hypotenuse** (the longest side). It is opposite the right angle.

Note that the sides of a triangle can be named in either of two ways.
1. A side can be named by the two capital letters given to the vertices at each end. This is what has been done in the figure shown to name the hypotenuse XY.
2. We can also name a side by using the lower-case letter of the opposite vertex. In the figure shown, we could have named the hypotenuse 'z'.
   Consider the right-angled triangle ABC with sides 3 cm, 4 cm and 5 cm. Squares have been constructed on each of the sides. The area of each square has been calculated ($A = S^2$) and indicated.

Note that the area of the square on the hypotenuse is equal to the sum of the areas of the squares on the other two sides.

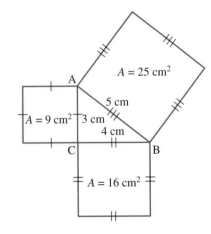

$$25 \text{ cm}^2 = 16 \text{ cm}^2 + 9 \text{ cm}^2$$
Alternatively: $(5 \text{ cm})^2 = (4 \text{ cm})^2 + (3 \text{ cm})^2$
Which means: $\text{hypotenuse}^2 = \text{base}^2 + \text{height}^2$

This result is known as **Pythagoras' theorem**.

> ## Pythagoras' theorem
>
> **The side lengths of any right-angled triangle are related according to the rule $a^2 + b^2 = c^2$, where c represents the hypotenuse (the longest side), and a and b represent the other two side lengths.**
>
>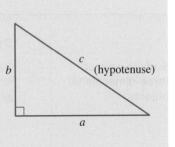

The **hypotenuse** is always the side length that is opposite the right angle. **Pythagoras' theorem** can be used to find an unknown side length of a triangle when the other two side lengths are known.

Calculate the length of the hypotenuse in the right-angled triangle shown.

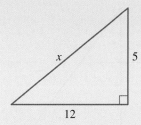

THINK	WRITE
1. Identify that the triangle is right-angled so Pythagoras' theorem can be applied.	$a^2 + b^2 = c^2$
2. Identify which side length is the hypotenuse.	x is opposite the right angle, so it is the hypotenuse; $a = 12$ and $b = 5$.
3. Substitute the known values into the theorem and simplify.	$12^2 + 5^2 = x^2$ $144 + 25 = x^2$ $169 = x^2$
4. Take the positive square root of both sides to obtain the value of x.	$\sqrt{169} = x$ $13 = x$
5. Write the answer.	The length of the hypotenuse is 13 units.

Note: We always take the positive square root when using Pythagoras' theorem because the length of a side of a triangle must be a positive number.

If the length of the hypotenuse and one other side length are known, the other side length can be found by subtracting the square of the known side length from the square of the hypotenuse.

Using Pythagoras' theorem to determine the length of a shorter side of a triangle

Rearrange Pythagoras' theorem to calculate the length of the two shorter sides.

$$a^2 + b^2 = c^2$$

Use the following equation to calculate the length of side 'a'.

$$a^2 = c^2 - b^2$$

Use the following equation to calculate the length of side 'b'.

$$b^2 = c^2 - a^2$$

Calculate the length of the unknown side in the right-angled triangle shown.

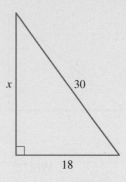

THINK	WRITE
1. Identify that the triangle is right-angled so Pythagoras' theorem can be applied.	$a^2 + b^2 = c^2$
2. Identify which side length is the hypotenuse.	30 is opposite the right angle, so it is the hypotenuse; $a = x$ and $b = 18$.
3. Substitute the known values into the theorem and simplify.	$x^2 + 18^2 = 30^2$ $\quad x^2 = 30^2 - 18^2$ $\quad\quad = 576$
4. Take the positive square root of both sides to obtain the value of x.	$x = \sqrt{576}$ $\quad = 24$
5. Write the answer.	The length of the unknown side is 24 units.

3.2.2 Pythagorean triads

Pythagorean triads (or **Pythagorean triples**) are sets of 3 numbers that satisfy Pythagoras' theorem. The first right-angled triangle we dealt with in this section had side lengths of 3 cm, 4 cm and 5 cm. This satisfied Pythagoras' theorem, so the numbers 3, 4 and 5 form a Pythagorean triad or triple.

In fact, any multiples of these numbers, for example 6, 8 and 10, or 1.5, 2 and 2.5, would also form a Pythagorean triad or triple.

Some other triads are:

$$5, \ 12, \ 13$$
$$8, \ 15, \ 17$$
$$9, \ 40, \ 41.$$

Identify whether the set of numbers 4, 6, 7 is a Pythagorean triad.

THINK	WRITE
1. Write the formula for Pythagoras' theorem.	$c^2 = a^2 + b^2$
2. Substitute the length of the longest side for the hypotenuse, c, and the lengths of the shorter sides for a and b.	$7^2 = 4^2 + 6^2$

3. Evaluate both sides of the formula.

$$7^2 = 49$$
$$4^2 + 6^2 = 16 + 36$$
$$= 52$$

4. Compare the two results.

$$7^2 \neq 4^2 + 6^2$$

5. Write the answer.

$4, 6, 7$ is *not* a Pythagorean triad.

3.2.3 Solve practical problems using Pythagoras' theorem

Pythagoras' theorem can be used to solve more practical problems. In these cases, it is necessary to draw a diagram that will help you decide the appropriate method for finding a solution. The diagram simply needs to represent the triangle; it does not need to show details of the situation described.

eles-3164

WORKED EXAMPLE 4 Solving practical problems involving Pythagoras' theorem

The fire brigade attends a blaze in a tall building. They need to rescue a person from the 6th floor of the building, which is 30 m above ground level. Their ladder is 32 m long and must be at least 10 m from the foot of the building. Assess whether the ladder can be used to reach the people needing rescue.

THINK	WRITE
1. Draw a diagram and show all given information.	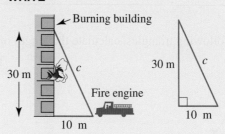
2. Write the formula after deciding if you are finding the hypotenuse or a shorter side.	$\text{Hypotenuse}^2 = \text{base}^2 + \text{height}^2$
3. Substitute the lengths of the known sides.	$c^2 = 10^2 + 30^2$
4. Evaluate the expression.	$= 100 + 900$ $= 1000$
5. Find the answer by taking the square root.	$c = \sqrt{1000}$ $= 31.62$ m
6. Write the answer.	The ladder will be long enough to make the rescue, since it is 32 m long.

on Resources

Digital documents   SkillSHEET Pythagoras' theorem (doc-29488)
SpreadSHEET Pythagoras (doc-29489)
SpreadSHEET Triads (doc-29490)
SpreadSHEET Pythagoras' theorem (doc-29491)

3.2 Exercise	3.2 Exam questions on

Simple familiar	Complex familiar	Complex unfamiliar
1, 2, 3, 4, 5, 6, 7, 8, 9, 10, 11, 12	13, 14, 15	16

These questions are even better in jacPLUS!
- Receive immediate feedback
- Access sample responses
- Track results and progress

Find all this and MORE in jacPLUS ▶

Simple familiar

1. Identify the hypotenuse in each of the following triangles.

   a.

   b.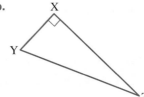

2. **WE1** Calculate the length of the hypotenuse in each of the following triangles.

   a.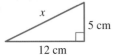

   b.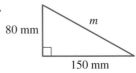

3. In each of the following triangles, calculate the length of the hypotenuse, correct to 2 decimal places.

   a.

   b.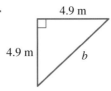

4. **WE2** Calculate the length of each unknown shorter side in the right-angled triangles shown. Round your answer to 1 decimal place.

   a.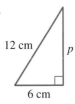

   b.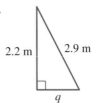

5. In each of the following right-angled triangles, calculate the length of the side marked with a pronumeral, rounded to 1 decimal place.

   a.

   b.

   c.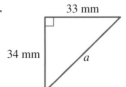

   d.

6. **MC** The hypotenuse in ΔWXY is:

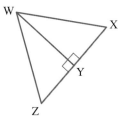

   A. WX

   B. XY

   C. YZ

   D. ZW

7. **WE3** Identify whether the following sets of numbers are Pythagorean triads.

   a. 9, 12, 15   **b.** 4, 5, 6   **c.** 30, 40, 50

   d. 3, 6, 9   **e.** 0.6, 0.8, 1.0   **f.** 7, 24, 25

8. **WE4** A cell tower is 12 m high. To support it, wires are attached to the ground 5 m from the foot of the cell tower. Calculate the length of each wire.

9. Susie needs to clean the guttering on her roof. She places her ladder 120 cm back from the edge of the guttering, which is 3 m above the ground. Calculate how long Susie's ladder will need to be (rounded to 2 decimal places). Give your answer in metres.

10. A rectangular gate is 3.5 m long and 1.3 m wide. The gate is to be strengthened by a diagonal brace as shown. Calculate how long the brace should be (correct to 2 decimal places).

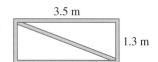

11. **MC** Decide which of the following triangles is definitely right-angled.

   A.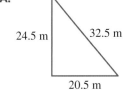

   B.

   C.

   D.

12. Complete the following Pythagorean triads.

   a. 9, ___, 15   **b.** __, 24, 25   **c.** 1.5, 2.0, __   **d.** 3, __, 5

   e. 11, 60, __   **f.** 10, __, 26   **g.** __, 40, 41   **h.** 0.7, 2.4, __

Complex familiar

13. A 250-cm ladder leans against a brick wall. The foot of the ladder is 1.2 m from the foot of the wall. Calculate how high up the wall the ladder will reach (correct to 1 decimal place). Give your answer in metres.

14. An isosceles, right-angled triangle has a hypotenuse of 10 cm. Calculate the length of the shorter sides. (*Hint:* Call both shorter sides *x*.)

15. Calculate the length of the unknown side in the diagram shown, giving your answer correct to 2 decimal places.

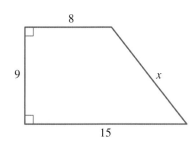

16. Consider the diagram shown.

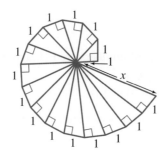

Given the side lengths marked as 1 unit, determine the length of the hypotenuse marked x.

Fully worked solutions for this chapter are available online.

LESSON
3.3 Pythagoras' theorem in three dimensions

SYLLABUS LINKS

- Understand and use Pythagoras' theorem to solve practical problems and simple applications in three dimensions.
  - $c^2 = a^2 + b^2$ where c is length of the hypotenuse and a and b are lengths of the two perpendicular sides.
- Solve practical problems involving shape and measurement.

Source: General Mathematics Senior Syllabus 2024 © State of Queensland (QCAA) 2024; licensed under CC BY 4.0.

3.3.1 Using Pythagoras' theorem in three dimensions

Many three-dimensional (3D) objects contain right-angled triangles that can be modelled with two-dimensional drawings. Using this method, we can calculate missing side length of three-dimensional objects.

3D Pythagorean theorem

The 3D Pythagorean rule for determining the longest diagonal in a rectangular prism allows for one calculation using Pythagoras' theorem instead of two separate calculations.

> **3D Pythagorean theorem**
>
> $$d^2 = x^2 + y^2 + z^2$$
>
> where x, y and z are the dimensions of the prism and d is the length of the diagonal that stretches from one corner of the box to the opposite corner.

WORKED EXAMPLE 5 Applying Pythagoras' theorem in three dimensions

Calculate the maximum length of a metal rod that would fit into a rectangular crate with dimensions $1\,\text{m} \times 1.5\,\text{m} \times 0.5\,\text{m}$:

a. using Pythagoras' theorem in two dimensions

b. using Pythagoras' theorem in three dimensions.

THINK	WRITE

a. 1. Draw a diagram of a rectangular box with a rod in it, labelling the dimensions.

a.

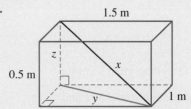

2. Draw in a right-angled triangle that has the metal rod as one of the sides, as shown in pink and labelled x.
The length of y in this right-angled triangle is not known. Draw in another right-angled triangle to calculate the length of y, as shown in green.

3. Draw in a right-angled triangle that has the metal rod as one of the sides, as shown in pink. The length of y in this right-angled triangle is not known (see step 3).
Draw in another right-angled triangle, as shown in green, to calculate the length of y (see step 4).

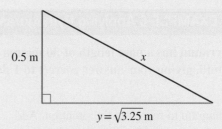

$$c^2 = a^2 + b^2$$
$$x^2 = \left(\sqrt{3.25}\right)^2 + 0.5^2$$
$$= 3.25 + 0.25$$
$$= 3.5$$
$$x = \sqrt{3.5}$$

4. Draw the right-angled triangle containing the rod and use Pythagoras' theorem to calculate the length of the rod (x).

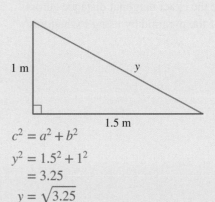

$$c^2 = a^2 + b^2$$
$$y^2 = 1.5^2 + 1^2$$
$$= 3.25$$
$$y = \sqrt{3.25}$$

5. Calculate the length of y using Pythagoras' theorem. Calculate the exact value of y.

6. Write the answer in a sentence.

b. 1. Draw a diagram of a rectangular box with a rod in it, labelling the dimensions.

The maximum length of the metal rod is $\sqrt{3.5}\,\text{m}$ ($\approx 1.87\,\text{m}$ correct to 2 decimal places).

b.

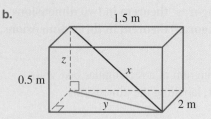

2. Write the 3D Pythagorean formula.

$d^2 = x^2 + y^2 + z^2$

3. Substitute the values for x, y, z and calculate.

$d^2 = 1.5^2 + 0.5^2 + 1^2$
$\quad = 3.5$
$d = \sqrt{3.5}$
$\quad \approx 1.87$

4. Write the answer.

The maximum length of the metal rod correct to 2 decimal places is $1.87\,\text{m}$

WORKED EXAMPLE 6 Applying Pythagoras' theorem in three dimensions

A square pyramid has a base length of 30 metres and a slant edge of 65 metres. Determine the height of the pyramid, giving your answer correct to 1 decimal place.

THINK

1. Draw a diagram to represent the situation. Add a point in the centre of the diagram below the apex of the pyramid.

WRITE

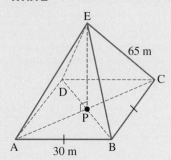

2. Determine the exact diagonal distance across the base of the pyramid by using Pythagoras' theorem.

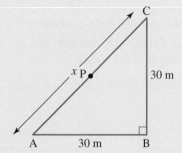

$$c^2 = a^2 + b^2$$
$$x^2 = 30^2 + 30^2$$
$$= 900 + 900$$
$$= 1800$$
$$x = \sqrt{1800}$$
$$= \sqrt{900 \times 2}$$
$$= 30\sqrt{2}$$

3. Calculate the exact distance from one of the corners on the base of the pyramid to the centre of the base of the pyramid.

$$AP = \frac{1}{2}AC$$
$$= \frac{1}{2} \times 30\sqrt{2}$$
$$= 15\sqrt{2}$$

4. Draw the triangle that contains the height of the pyramid and the distance from one of the corners on the base of the pyramid to the centre of the base of the pyramid.

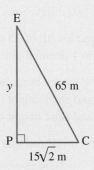

5. Use Pythagoras' theorem to calculate the height of the pyramid, rounding your answer to 1 decimal place.

$$c^2 = a^2 + b^2$$
$$65^2 = y^2 + \left(15\sqrt{2}\right)^2$$
$$4225 = y^2 + 450$$
$$y^2 = 4225 - 450$$
$$= 3775$$
$$y = \sqrt{3775}$$
$$\approx 61.4$$

6. Write the answer in a sentence.

The height of the pyramid is 61.4 metres (correct to 1 decimal place).

 Resources

⬥ **Interactivity** Pythagoras' theorem (int-6473)

Exercise 3.3 Pythagroras' theorem in three dimensions

3.3 Exercise	**3.3 Exam questions**

Simple familiar	Complex familiar	Complex unfamiliar
1, 2, 3, 4, 5, 6, 7	8, 9, 10	11, 12, 13, 14

These questions are even better in jacPLUS!
- Receive immediate feedback
- Access sample responses
- Track results and progress

Find all this and MORE in jacPLUS ▶

Simple familiar

1. **WE5** Calculate the maximum length of a metal rod that would fit into a rectangular crate with dimensions 1.2 m × 83 cm × 55 cm:

   a. using Pythagoras' theorem in two dimensions
   b. using Pythagoras' theorem in three dimensions.

2. Determine whether a metal rod of length 2.8 m would be able to fit into a rectangular crate with dimensions 2.3 m × 1.2 m × 0.8 m:

   a. using Pythagoras' theorem in two dimensions.
   b. using Pythagoras' theorem in three dimensions.

3. **WE6** A square pyramid has a base length of 25 m and a slant height of 45 m. Determine the height of the pyramid, rounding your answer to 1 decimal place.

4. Determine which of the following square pyramids has the greater height.
   Pyramid 1: base length of 18 m and slant height of 30 m
   Pyramid 2: base length of 22 m and slant height of 28 m

5. Calculate the length of the longest metal rod that can fit diagonally into the objects shown below.

   a.
   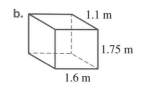

   b.
   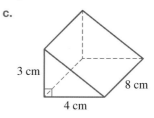

   c.

6. Calculate, correct to 2 decimal places, the height of a square pyramid with base width twice the height if the slant edge is:

   a. 20 cm   b. 48 cm   c. 5.5 cm   d. 166 cm.

7. A friend wants to pack an umbrella into her suitcase.

   a. If the suitcase measures 89 cm × 21 cm × 44 cm, determine whether her 1-m-long umbrella fits in.
   b. Give the length of the longest object that will fit into the suitcase.

Complex familiar

8. Stephano is renovating his apartment, which he accesses through two corridors. The corridors of the apartment building are 2 m wide with 2 m high ceilings, and the first corridor is at right angles to the second. Show that he can carry lengths of timber up to 6 m long to his apartment.

9. Calculate the value of x, correct to 2 decimal places, in the diagram.

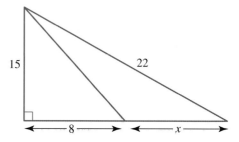

10. The Great Pyramid in Egypt is a square-based pyramid. The square base has a side length of 230.35 m, and the perpendicular height is 146.71 m.

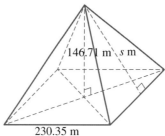

Determine the slant height, *s* m, of the great pyramid. Round your answer correct to 1 decimal place.

Complex unfamiliar

11. Angles ABD, CBD and ABC are right angles.

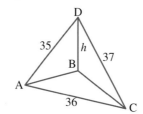

Calculate the value of *h*, correct to 3 decimal places.

12. The roof of a squash centre is constructed to allow for maximum use of sunlight.

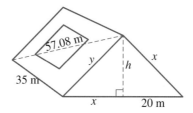

Calculate the value of *h*, giving your answer correct to 1 decimal place.

13. A semi-trailer carries a container that has the following internal dimensions: length 14.5 m, width 2.4 m and height 2.9 m.

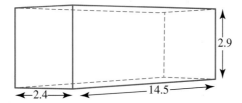

If a rectangular box with length 2.4 m, width 1.2 m and height 0.8 m is placed on the floor at one end so that it fits across the width of the container, calculate the length of the longest object that can now be placed inside if it touches the floor adjacent to the box. Give your answer correct to 2 decimal places.

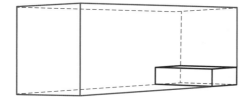

14. An ultralight aircraft is flying at an altitude of 1000 m and a horizontal distance of 10 km from its landing point. The pilot mistakenly follows a direct line to a point on the ground that is 2.5 km short of the correct landing point.

He realises his mistake when he is at an altitude of 400 m and a horizontal distance of 5.5 km from the correct landing point. He then follows a straight-line path to the correct landing point.

Calculate the total distance travelled by the aircraft from its starting point to the correct landing point, correct to the nearest metre.

Fully worked solutions for this chapter are available online.

LESSON
3.4 Perimeter and area of polygons and circles

SYLLABUS LINKS

- Calculate perimeters, P, of standard two-dimensional objects in practical situations, including circles, triangles, rectangles, trapeziums and parallelograms.
  - Circle: $C = 2\pi r$, where C is circumference and r is radius.
- Calculate areas, A, of standard two-dimensional objects in practical situations, including circles, triangles, rectangles, parallelograms and trapeziums.
  - Circle: $A = \pi r^2$ where r is radius.
  - Triangle: $A = \dfrac{1}{2}bh$ where b is base length and h is perpendicular height.
  - Parallelogram: $A = bh$ where b is the base length and h is the perpendicular height.
  - Trapezium: $A = \dfrac{1}{2}(a + b)h$ where a and b are parallel lengths and h is perpendicular height.
- Solve practical problems involving shape and measurement.

Source: General Mathematics Senior Syllabus 2024 © State of Queensland (QCAA) 2024; licensed under CC BY 4.0.

3.4.1 Units of length and area

Units of length are used to describe the distance between any two points.

The standard unit of length in the metric system is the metre. The most commonly used units of length are the millimetre (mm), centimetre (cm), metre (m) and kilometre (km).

Converting units of length

The following chart is useful when converting from one unit of length to another.

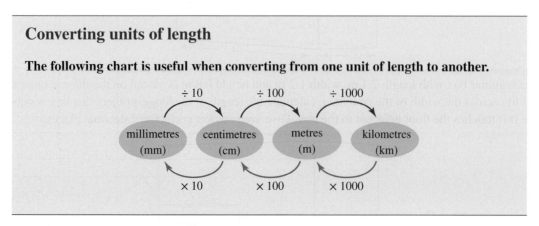

Units of **area** are named by the side length of the square that encloses that amount of space. For example, a square metre is the amount of space enclosed by a square with a side length of 1 metre.

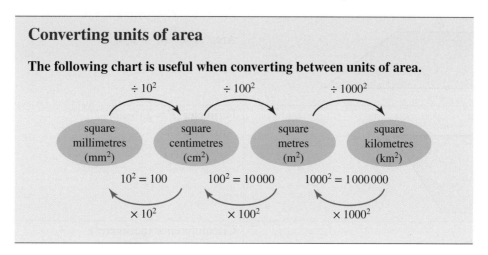

Converting units of area

The following chart is useful when converting between units of area.

3.4.2 Perimeter and area of standard shapes

You should be familiar with the methods and units of measurement used for calculating the **perimeter** (distance around an object) and area (two-dimensional space taken up by an object) of standard **polygons** and other shapes.

These are summarised in the following table.

Shape	Perimeter and area
Square	Perimeter: $P = 4l$ Area: $A = l^2$
Rectangle	Perimeter: $P = 2l + 2w$ Area: $A = lw$
Triangle	Perimeter: $P = a + b + c$ Area: $A = \dfrac{1}{2}bh$

(continued)

(continued)

Shape	Perimeter and area
Trapezium 	Perimeter: $P = a + b + c + d$ Area: $A = \dfrac{1}{2}(a + b)h$
Parallelogram	Perimeter: $P = 2a + 2b$ Area: $A = bh$
Circle	Circumference (perimeter): $C = 2\pi r$ $\quad = \pi D$ Area: $A = \pi r^2$

Note: The approximate value of π is 3.14. However, when calculating **circumference** and area, always use the π button on your calculator and make rounding off to the required number of decimal places your final step.

WORKED EXAMPLE 7 Calculating the perimeter and area of a polygon

Calculate the perimeter and area of the shape shown in the diagram.

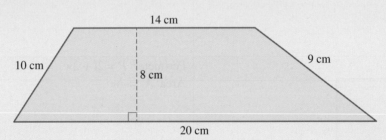

THINK

1. Identify the shape.

2. Identify the components for the perimeter formula and evaluate.

3. Write the perimeter including the units.

WRITE

Trapezium

$P = 10 + 20 + 14 + 9$
$\quad = 53$

$P = 53 \text{ cm}$

4. Identify the components for the area formula and evaluate.

$$A = \frac{1}{2}(a+b)h$$
$$= \frac{1}{2}(20+14)8$$
$$= \frac{1}{2} \times 34 \times 8$$
$$= 136$$

5. Write the answer with the correct unit.

$$A = 136 \text{ cm}^2$$

eles-6378

WORKED EXAMPLE 8 Calculating the circumference and area of a circle

Calculate the circumference and area of the shape shown in the diagram, giving your final answers correct to 2 decimal places.

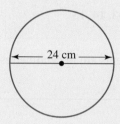

24 cm

THINK

1. Use the formula for the circumference of a circle in terms of the diameter.

2. Substitute the diameter into the equation and solve for C using the π key on your calculator.

3. Write the answer to 2 decimal places.

4. Write the formula for the area of a circle.

5. Determine the radius of the circle.

6. Substitute $r = 12$ into the formula for the area of a circle and evaluate.

7. Write the answer to 2 decimal places.

WRITE

$C = \pi \text{D}$

$C = \pi \times 24$
$= 75.3982$
$\approx 75.40 \text{ cm}$

The circumference is 75.40 cm.

$A_{\text{circle}} = \pi \times r^2$

The radius is half the length of the diameter, so:
$$r = \frac{24}{2}$$
$$= 12 \text{ cm}$$

$A_{\text{circle}} = \pi \times 12^2$
$= 144\pi$
$\approx 452.39 \text{ cm}^2$

The area is 425.39 cm^2.

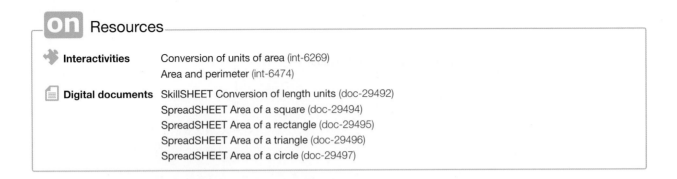
Exercise 3.4 Perimeter and area of polygons and circles

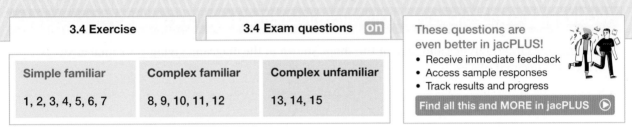

3.4 Exercise	3.4 Exam questions

Simple familiar	Complex familiar	Complex unfamiliar
1, 2, 3, 4, 5, 6, 7	8, 9, 10, 11, 12	13, 14, 15

These questions are even better in jacPLUS!
• Receive immediate feedback
• Access sample responses
• Track results and progress

Find all this and MORE in jacPLUS ▶

In the following questions, assume all measurements are in centimetres unless otherwise indicated.

Simple familiar

1. **WE7** Calculate the perimeter and area of the shape shown in the diagram.

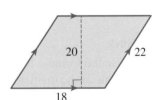

2. **WE8** Calculate the circumference and area of the shape shown in the diagram, giving your final answers correct to 2 decimal places.

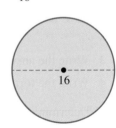

3. Calculate the perimeter and area of each of the following shapes, giving answers correct to 2 decimal places where appropriate.

a.

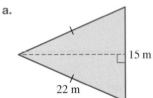

b.

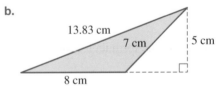

c.

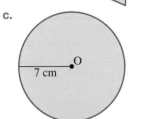

d.

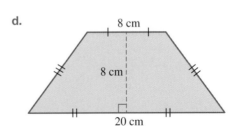

4. Correct to 2 decimal places, calculate the circumference and area of:

   a. a circle of radius 5 cm

   b. a circle of diameter 18 cm.

5. Calculate the perimeter and area of a parallelogram with side lengths of 12 cm and 22 cm, and a perpendicular distance of 16 cm between the short sides.

6. Calculate the area of a rhombus with diagonals of 11.63 cm and 5.81 cm.

7. A circle has an area of 3140 cm^2. Calculate its radius correct to 2 decimal places.

Complex familiar

8. Calculate the area of the shaded region shown in the diagram, giving your answer correct to 2 decimal places.

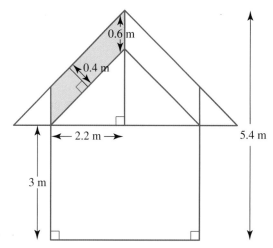

9. Calculate the perimeter of the large triangle shown in the diagram and hence calculate the shaded area.

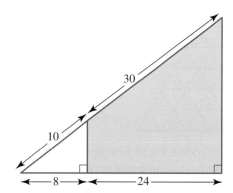

10. A window consists of a circular metal frame 2 cm wide and two straight pieces of metal that divide the inner region into four equal segments, as shown in the diagram.

    a. If the window has an inner radius of 30 cm, calculate, correct to 2 decimal places:

       i. the outer circumference of the window
       ii. the total area of the circular metal frame.

    b. If the area of the metal frame is increased by 10% by reducing the size of the inner radius, calculate the circumference of the new inner circle.

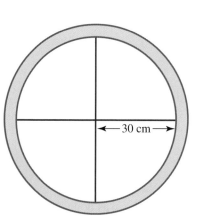

11. A rectangle has a side length that is twice as long as its width. If it has an area of 968 cm², calculate the length of its diagonal correct to 2 decimal places.

12. A semicircular section of a running track consists of eight lanes that are 1.2 m wide. The innermost line of the first lane has a total length of 100 m.

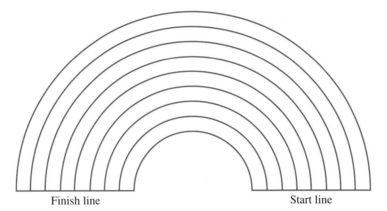

Finish line   Start line

Calculate how much further someone in lane 8 will run around the curve from the start line to the finish line.

Complex unfamiliar

13. A paved area of a garden courtyard forms an equilateral triangle with a side length of 20 m. It is paved using a series of identically sized blue and white triangular pavers as shown in the diagram.

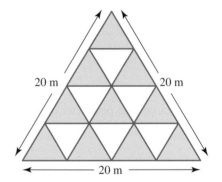

a. If the pattern is continued by adding two more rows of pavers, calculate the new perimeter and area of the paving correct to 2 decimal places.

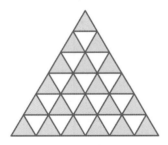

b. After the additional two rows are added, the architects decide to add two rows of rectangular pavers to each side. Each rectangular paver has a length that is twice the side length of a triangular paver, and a width that is half the side length of a triangular paver.

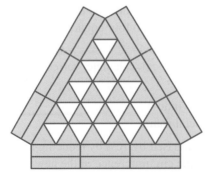

If this was done on each side of the triangular paved area, calculate the perimeter and area of the paving.

14. The London eye is Europe's largest Ferris Wheel, with a diameter of 120 m. The passenger capsules attached to the wheel travel at a speed of 15 m/minute. Calculate the time taken for one full revolution.

15. Consider the shape sketched on the Cartesian plane.

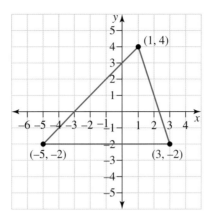

Calculate the perimeter of the shape.

Fully worked solutions for this chapter are available online.

LESSON
3.5 Perimeter and area of sectors and composite shapes

SYLLABUS LINKS

- Calculate perimeters, P, of standard two-dimensional objects in practical situations, including sectors of circles and composites.
  - Sector of circle: $P = 2r + \dfrac{\theta}{180}\pi r$, where θ is central angle and r is radius.
- Calculate areas, A, of standard two-dimensional objects in practical situations, including sectors of circles and composites.
  - Sector of circle: $A = \dfrac{\theta}{360}\pi r^2$ where θ is central angle and r is radius.
- Solve practical problems involving shape and measurement.

Source: General Mathematics Senior Syllabus 2024 © State of Queensland (QCAA) 2024; licensed under CC BY 4.0.

3.5.1 Composite shapes

A composite shape is made up of smaller, simpler shapes. Two examples are shown below.

In the second example, the two semicircles are subtracted from the square to obtain the shaded area on the left.

To calculate the areas of composite shapes, split them into standard shapes, calculate the individual areas of the standard shapes and sum the answers together.

To calculate the perimeters of composite shapes, it is often easiest to calculate each individual side length and to then calculate the total, rather than applying any specific formula.

Some composite shapes do have specific formulas.

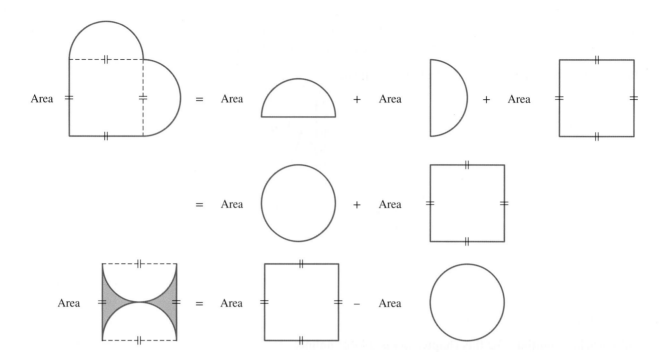

WORKED EXAMPLE 9 Calculating the area of a composite shape

Calculate the area of the object shown correct to 2 decimal places.

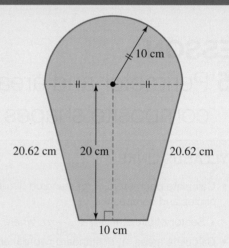

THINK

1. Identify the given information.

2. Determine the area of each component of the shape.

3. Sum the areas of the components.

4. Write the answer with the correct unit.

WRITE

The shape is a combination of a trapezium and a semicircle.

Area of trapezium: $A = \dfrac{1}{2}(a+b)h$

$\qquad = \dfrac{1}{2}(10+20)20$

$\qquad = 300 \text{ cm}^2$

Area of semicircle: $A = \dfrac{1}{2}\pi r^2$

$\qquad = \dfrac{1}{2}\pi(10)^2$

$\qquad = 50\pi \text{ cm}^2$

Total area: $300 + 50\pi \approx 457.08$

The area of the shape is 457.08 cm^2.

3.5.2 Annulus

The area between two circles with the same centre is known as an **annulus**. The area of an annulus is calculated by subtracting the area of the inner circle from the area of the outer circle.

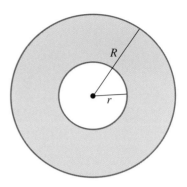

Area of an annulus

Area of annulus = area of outer circle − area of inner circle

$$A = \pi R^2 - \pi r^2$$
$$= \pi(R^2 - r^2)$$

WORKED EXAMPLE 10 Calculating the area of an annulus

Calculate the area of the annulus shown in the diagram correct to 1 decimal place.

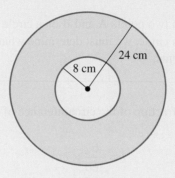

THINK	WRITE
1. Identify the given information.	The area shown is an annulus. The radius of the outer circle is 24 cm. The radius of the inner circle is 8 cm.
2. Substitute the information into the formula and simplify.	$A = \pi R^2 - \pi r^2$ $= \pi(R^2 - r^2)$ $= \pi(24^2 - 8^2)$ $= 512\pi$ ≈ 1608.5
3. Write the answer with the correct unit.	The shaded area is 1608.5 cm^2.

3.5.3 Sectors

Sectors are fractions of a circle. Because there are 360 degrees in a whole circle, the area of a sector can be found using $A = \dfrac{\theta}{360} \times \pi r^2$, where θ is the angle between the two radii that form the sector.

The perimeter of a sector is a fraction of the circumference of the related circle plus two radii:

$$P = \left(\frac{\theta}{360} \times 2\pi r \right) + 2r$$

$$= 2r \left(\frac{\theta}{360} \pi + 1 \right)$$

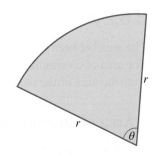

Perimeter and area of a sector

Area of a sector $= \dfrac{\theta}{360} \times \pi r^2$

Perimeter of a sector $= 2r \left(\dfrac{\theta}{360} \pi + 1 \right)$

3.5.4 Arc length

The length of the circumference between two points A and B on a circle is known as an *arc*. To calculate the arc length of a circle, we must determine what fraction of the total circumference the arc represents.

The angle $\theta°$ shown is a fraction of $360°$.

The length l of arc AB will be the same fraction of the circumference of the circle, $2\pi r$:

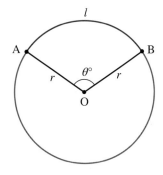

$$l = \frac{\theta}{360} \times 2\pi r$$

$$l = \frac{\pi r \theta}{180}$$

(where θ is measured in degrees).

WORKED EXAMPLE 11 Calculating the arc length, perimeter and area of a sector

Consider the diagram shown.
Calculate:
a. **the area of the shaded sector**
b. **the perimeter of the sector**
c. **the arc length, *l*.**
Give your answers correct to 2 decimal places.

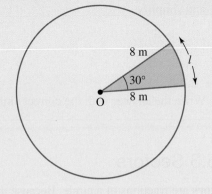

THINK	WRITE
a. 1. Write the known information from the diagram.	a. The radius is 8 m. The angle between the two radii that form the arc is 30°.

2. Substitute the values of r and θ into the area of a sector formula and simplify.

$$\text{Area of a sector} = \frac{\theta}{360} \times \pi r^2$$
$$= \frac{30}{360} \times \pi \times (8)^2$$
$$= \frac{16}{3}\pi$$
$$\approx 16.75$$

3. Write the answer with the correct unit.

The area of the sector is $16.75\,\text{m}^2$.

b. 1. Substitute the values of r and θ into the perimeter of a sector formula and simplify.

b. $\text{Perimeter of a sector} = 2r\left(\frac{\theta}{360}\pi + 1\right)$
$$= 2 \times 8 \times \left(\frac{30}{360} \times \pi + 1\right)$$
$$\approx 20.19$$

2. Write your answer with the correct unit.

The perimeter of the sector is $20.19\,\text{m}$.

c. 1. Substitute the values of r and θ into the arc length formula and simplify.

c. $\text{Arc length} = \frac{\pi r \theta}{180}$
$$= \frac{8 \times 30 \times \pi}{180}$$
$$= \frac{4\pi}{3}$$
$$\approx 4.19$$

2. Write the answer with the correct unit.

The arc length is $4.19\,\text{m}$.

3.5.5 Applications

Calculations for perimeter and area have many and varied applications, including in building and construction, painting and decorating, real estate, surveying and engineering.

When dealing with these problems, it is often useful to draw diagrams to represent the given information.

WORKED EXAMPLE 12 Solving practical problems involving perimeter and area

Calculate the total area of the triangular sails on a yacht correct to 2 decimal places if the apex of one sail is 4.3 m above its base length of 2.8 m and the apex of the other sail is 4.6 m above its base of length of 1.4 m.

THINK

1. Draw a diagram of the given information.

WRITE

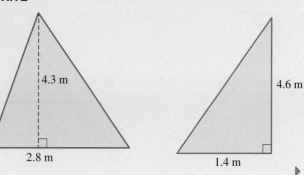

4.3 m

2.8 m

4.6 m

1.4 m

2. Identify the formulas required from the given information.	For each sail, use the formula for area of a triangle: $A = \dfrac{1}{2}bh$
3. Substitute the information into the required formulas for each area and simplify.	Sail 1: $A = \dfrac{1}{2}bh$ $= \dfrac{1}{2} \times 2.8 \times 4.3$ $= 6.02$ Sail 2: $A = \dfrac{1}{2}bh$ $= \dfrac{1}{2} \times 1.4 \times 4.6$ $= 3.22$
4. Add the areas of each of the required parts.	Area of sail $= 6.02 + 3.22$ $= 9.24$
5. Write the answer with the correct unit.	The total area of the sails is 9.24 m^2.

 Resources

Digital documents SkillSHEET Perimeter of composite shapes (doc-29498)
SkillSHEET Finding the size of a sector (doc-29499)
SkillSHEET Area of composite shapes (doc-29500)

Exercise 3.5 Perimeter and area of sectors and composite shapes

3.5 Exercise	**3.5 Exam questions** on

Simple familiar	**Complex familiar**	**Complex unfamiliar**
1, 2, 3, 4, 5, 6, 7, 8, 9, 10	11, 12, 13, 14	15, 16, 17, 18, 19, 20

These questions are even better in jacPLUS!
- Receive immediate feedback
- Access sample responses
- Track results and progress

Find all this and MORE in jacPLUS ▶

Simple familiar

1. **WE9** Calculate the area of the object shown.

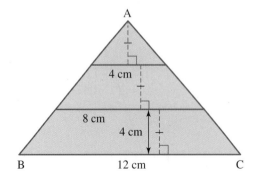

2. Calculate the perimeter and area of the object shown correct to 1 decimal place.

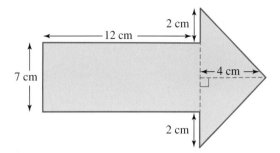

3. A circle of radius 8 cm is cut out from a square of side length 20 cm. Determine how much of the area of the square remains. Give your answer correct to 2 decimal places.

4. a. Calculate the perimeter of the shaded area inside the rectangle shown in the diagram correct to 2 decimal places.
   b. If the darker area inside the rectangle is removed, calculate the remaining area.

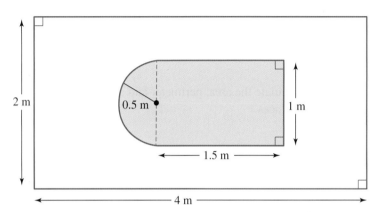

5. a. Calculate the shaded area in the diagram.

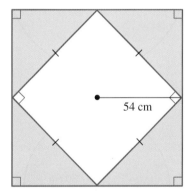

b. Calculate the unshaded area inside the square shown in the diagram, giving your answer correct to 2 decimal places.

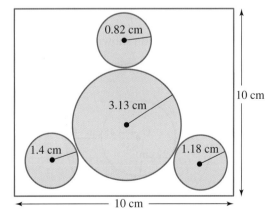

6. **WE10** Calculate the area of the annulus shown in the diagram correct to 2 decimal places.

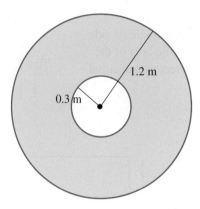

7. **WE11** Calculate the area, perimeter and arc length of the shaded sector shown in the diagram correct to 2 decimal places.

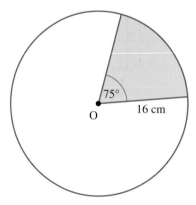

8. Calculate the area and perimeter of the shaded region shown in the diagram correct to 2 decimal places.

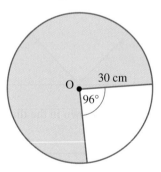

9. Calculate the area and perimeter of the shaded region shown in the diagram correct to 2 decimal places.

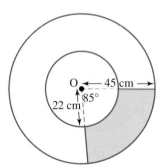

10. The area of the inner circle in the diagram shown is $\frac{1}{9}$ that of the annulus formed by the two outer circles.
Calculate the area of the inner circle correct to 2 decimal places given that the measurements are in centimetres.

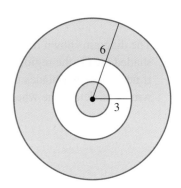

Complex familiar

11. **WE12** Calculate the area of glass in a table that consists of three glass circles. The largest circle has a diameter of 68 cm. The diameters of the other two circles are 6 cm and 10 cm less than the diameter of the largest circle.
Give your answer correct to 2 decimal places.

12. Part of the floor of an ancient Roman building was tiled in a pattern in which four identical triangles form a square with their bases.

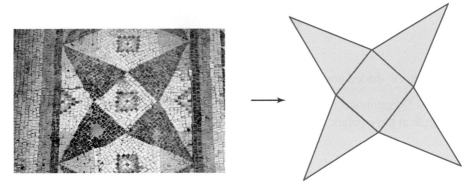

If the triangles have a base length of 12 cm and a height of 18 cm, calculate the perimeter and area they enclose, correct to 2 decimal places.
(That is, calculate the perimeter and area of the shaded region shown on the right.)

13. The vertices of an equilateral triangle of side length 2 metres touch the edge of a circle of radius 1.16 metres, as shown in the diagram.
Calculate the area of the unshaded region correct to 2 decimal places.

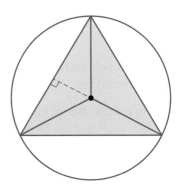

14. A trapezium is divided into five identical triangles of equal size with dimensions as shown in the diagram.

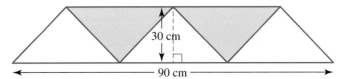

Calculate the area and perimeter of the shaded region.

15. The diagram shown represents a rear window of a car. The shaded area is the region covered by the car wiper blades. If new car wiper blades must cover at least 70% of rear windows, determine whether this car meets the regulations.

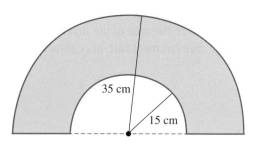

16. In the diagram, the smallest circle has a diameter of 5 cm and the others have diameters that are progressively 2 cm longer than the one immediately before.

    Calculate the area that is shaded green, correct to 2 decimal places.

17. A circle of radius 0.58 metres sits inside an equilateral triangle of side length 2 metres so that it touches the edges of the triangle at three points.

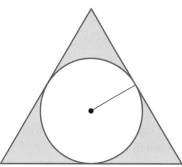

    If the circle represents an area of the triangle to be removed, determine how much area would remain once this was done.

18. An annulus has an inner radius of 20 cm and an outer radius of 35 cm. Two sectors are to be removed. If one sector has an angle at the centre of 38° and the other has an angle of 25°, calculate the remaining area. Give your answer correct to 2 decimal places.

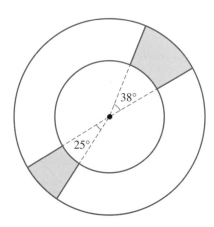

19. A new circular window installed in an arts centre has a clear glass hexagonal panel in the centre, surrounded by red glass. The radius of the window is 1.2 m.
    Calculate the area of the red glass.

20. The minute hand of a clock is 10 cm long. It is 3 cm longer than the hour hand. In one full revolution, determine how much more of the face is covered by the minute hand than by the hour hand.

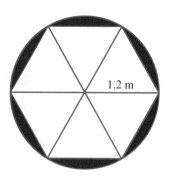

1.2 m

Fully worked solutions for this chapter are available online.

LESSON
3.6 Volume and capacity

SYLLABUS LINKS

- Calculate volumes, V, and capacities of standard three-dimensional objects in practical situations, including rectangular prisms, cylinders, pyramids, cones, spheres and composites.
    - Prism: $V = Ah$ where A is base area and h is perpendicular height.
    - Cylinder: $V = \pi r^2 h$ where r is radius and h is perpendicular height.
    - Pyramid: $V = \dfrac{1}{3}Ah$ where A is base area and h is perpendicular height.
    - Cone: $V = \dfrac{1}{3}\pi r^2 h$ where r is radius and h is perpendicular height.
    - Sphere: $V = \dfrac{4}{3}\pi r^3$ where r is radius.
- Solve practical problems involving shape and measurement.

Source: General Mathematics Senior Syllabus 2024 © State of Queensland (QCAA) 2024; licensed under CC BY 4.0.

3.6.1 Volume and capacity

The amount of space that is taken up by any solid or three-dimensional object is known as its **volume**. Volume is expressed in cubic units of measurement, such as cubic metres (m^3) or cubic centimetres (cm^3). The relationships between commonly used cubic units are shown below.

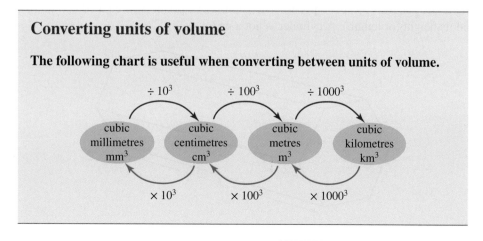

Converting units of volume

The following chart is useful when converting between units of volume.

$\div 10^3$ $\div 100^3$ $\div 1000^3$

cubic millimetres mm^3 cubic centimetres cm^3 cubic metres m^3 cubic kilometres km^3

$\times 10^3$ $\times 100^3$ $\times 1000^3$

The amount of liquid an object can hold is known as its **capacity**. The relationships between commonly used units of capacity are shown as follows.

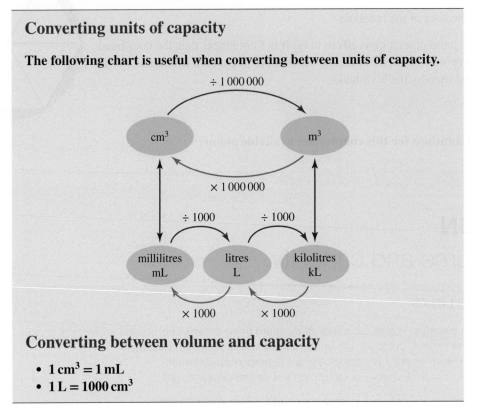

Converting units of capacity

The following chart is useful when converting between units of capacity.

$\div 1\,000\,000$

cm^3 m^3

$\times 1\,000\,000$

$\div 1000$ $\div 1000$

millilitres mL   litres L   kilolitres kL

$\times 1000$ $\times 1000$

Converting between volume and capacity

- $1\,cm^3 = 1\,mL$
- $1\,L = 1000\,cm^3$

Many standard objects have formulas that can be used to calculate their volume. If the centre point of the top of the solid is directly above the centre point of its base, the object is called a 'right solid'. If the centre point of the top is not directly above the centre point of the base, the object is an 'oblique solid'.

Note: For an oblique solid, the height, h, is the distance between the top and the base, not the length of one of the sides. (For a right solid, the distance between the top and the base equals the side length.)

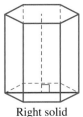

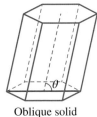

Right solid   Oblique solid

3.6.2 Volume of prisms

If a solid object has identical ends that are joined by flat surfaces, and the object's cross-section is a polygon and is the same along its length, the object is a **prism**. The volume of a prism is calculated by taking the product of the base area and its height (or length). A cylinder is not a prism.

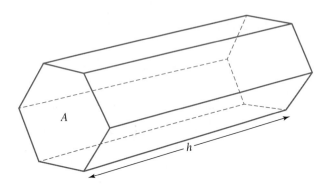

Volume of a prism

$$V = A \times h \quad (A = \text{base area})$$

Common prisms

The formulas for calculating the volume of some of the most common prisms are summarised in the following table.

Prism	Volume
Cube	$V = A \times h$
	$ = (l \times l) \times l$
	$ = l^3$
Rectangular	$V = A \times h$
	$ = (l \times w) \times h$
	$ = l \times w \times h$
Triangular	$V = A \times h$
	$ = \left(\dfrac{1}{2}bh\right) \times l$
	$ = \dfrac{1}{2}bhl$

Note: These formulas apply to both right prisms and oblique prisms, as long as you remember that the height of an oblique prism is its perpendicular height (the distance between the top and the base).

WORKED EXAMPLE 13 Calculating the volume of a triangular prism

Calculate the volume of a triangular prism with length $l = 12$ cm, triangle base length $b = 6$ cm and triangle height $h = 4$ cm.
Hence, calculate its capacity in mL.

THINK	WRITE
1. Identify the given information.	Triangular prism, $l = 12$ cm, $b = 6$ cm, $h = 4$ cm
2. Substitute the information into the appropriate formula for the solid object and evaluate.	$V = \dfrac{1}{2}bhl$ $= \dfrac{1}{2} \times 6 \times 4 \times 12$ $= 144$
3. Write the answer.	The volume is 144 cm³.
4. Convert using $1 \text{ cm}^3 = 1 \text{ mL}$.	The capacity is 144 mL.

3.6.3 Volume of cylinders

A **cylinder** is a solid object with ends that are identical circles and a cross-section that is the same along its length (like a prism). As a result, it has a curved surface along its length.

The volume of a cylinder is calculated by taking the product of the base area and the height.

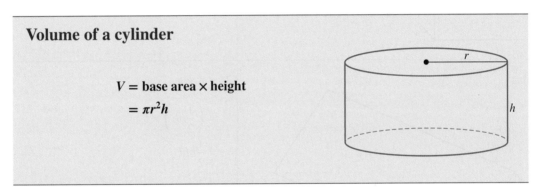

Volume of a cylinder

$V = $ base area × height
$= \pi r^2 h$

WORKED EXAMPLE 14 Calculating the volume of a cylinder

Calculate the volume of a cylinder of radius 10 cm and height 15 cm correct to the nearest cubic centimetre.
Hence, calculate its capacity in L to 2 decimal places.

THINK	WRITE
1. Identify the given information.	Cylinder, $r = 10$ cm, $h = 15$ cm
2. Substitute the information into the appropriate formula for the solid object and evaluate.	$V = \pi r^2 h$ $= \pi \times 10 \times 10 \times 15$ ≈ 4712.39

3. Write the answer with the correct unit.

The volume is 4712 cm³ correct to the nearest cubic centimetre.

4. Convert using 1000 cm³ = 1 L.

$$\text{Capacity} = \frac{4712.39}{1000}$$
$$= 4.71 \text{ L (correct to 2 decimal places)}$$
The capacity is 4.71 L.

3.6.4 Volume of cones

A **cone** is a solid object that is similar to a cylinder in that it has one end that is circular, but different in that at the other end it has a single vertex.

It can be shown that if you have a cone and a cylinder with identical circular bases and heights, the volume of the cylinder will be three times the volume of the cone. (The proof of this is beyond the scope of this course.)

The volume of a cone can therefore be calculated by using the formula for a comparable cylinder and dividing by three.

Volume of a cone

$$V = \frac{1}{3} \times \textbf{base area} \times \textbf{height}$$
$$= \frac{1}{3} \pi r^2 h$$

WORKED EXAMPLE 15 Calculating the volume of a cone

Calculate the volume of a cone of radius 20 cm and a height of 36 cm correct to 1 decimal place.

THINK	WRITE
1. Identify the given information.	Given: a cone with $r = 20$ cm and $h = 36$ cm
2. Substitute the information into the appropriate formula for the solid object and evaluate.	$V = \frac{1}{3}\pi r^2 h$ $= \frac{1}{3} \times \pi \times 20 \times 20 \times 36$ $\approx 15\,079.6$
3. Write the answer with the correct unit.	The volume is $15\,079.6 \text{ cm}^3$.

3.6.5 Volume of pyramids

A **pyramid** is a solid object whose base is a polygon and whose sides are triangles that meet at a single point. The most famous examples are the pyramids of Ancient Egypt, which were built as tombs for the pharaohs.

A pyramid is named after the shape of its base. For example, a hexagonal pyramid has a hexagon as its base polygon. The most common pyramids are square pyramids and triangular pyramids.

As with cones, the volume of a pyramid can be calculated by using the formula of a comparable prism and dividing by three.

Volume of a pyramid

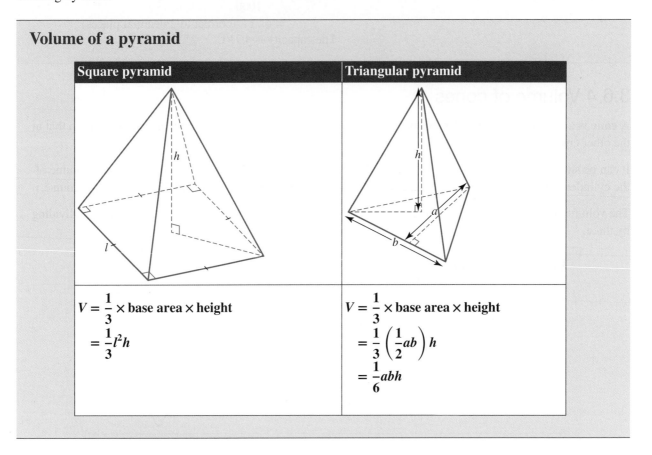

Square pyramid	Triangular pyramid
$V = \dfrac{1}{3} \times$ **base area** $\times$ **height** $= \dfrac{1}{3}l^2h$	$V = \dfrac{1}{3} \times$ **base area** $\times$ **height** $= \dfrac{1}{3}\left(\dfrac{1}{2}ab\right)h = \dfrac{1}{6}abh$

WORKED EXAMPLE 16 Calculating the volume of a pyramid

Calculate the volume of a pyramid that is 75 cm tall and has a rectangular base with dimensions 45 cm by 38 cm.

THINK	WRITE
1. Identify the given information.	Given: a pyramid with a rectangular base of 45×38 cm and a height of 75 cm
2. Substitute the information into the appropriate formula for the solid object and evaluate.	$V = \dfrac{1}{3}lwh$ $= \dfrac{1}{3} \times 45 \times 38 \times 75$ $= 42\,750$
3. Write the answer with the correct unit.	The volume is $42\,750\,\text{cm}^3$.

3.6.6 Volume of spheres

A **sphere** is a solid object that has a curved surface such that every point on the surface is the same distance (the radius of the sphere) from a central point.

The formula for calculating the volume of a sphere has been attributed to the ancient Greek mathematician Archimedes.

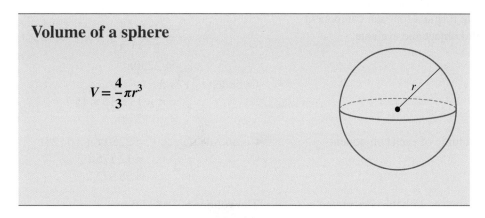

Volume of a sphere

$$V = \frac{4}{3}\pi r^3$$

WORKED EXAMPLE 17 Calculating the volume of a sphere

Calculate the volume of a sphere of radius 63 cm correct to 1 decimal place.

THINK	WRITE
1. Identify the given information.	Given: a sphere of $r = 63$ cm
2. Substitute the information into the appropriate formula for the solid object and evaluate.	$V = \dfrac{4}{3}\pi r^3$ $ = \dfrac{4}{3} \times \pi \times 63 \times 63 \times 63$ $ = 1\,047\,394.4$
3. Write the answer with the correct unit.	The volume is $1\,047\,394.4\,\text{cm}^3$.

3.6.7 Volume of composite solids

As with calculations for perimeter and area, when a solid object is composed of two or more standard shapes, we need to identify each part and add their volumes to evaluate the overall volume.

eles-6380

WORKED EXAMPLE 18 Calculating the volume of a composite solid

Calculate the volume of an object that is composed of a hemisphere (half a sphere) of radius 15 cm that sits on top of a cylinder of height 45 cm, correct to 1 decimal place.

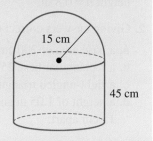

THINK	WRITE
1. Identify the given information.	Given: a hemisphere with $r = 15$ cm and a cylinder of $r = 15$ cm and $h = 45$ cm

2. Substitute the information into the appropriate formula for each component of the solid object and evaluate.

$$\text{Hemisphere: } V = \frac{1}{2}\left(\frac{4}{3}\pi r^3\right)$$
$$= \frac{1}{2}\times\left(\frac{4}{3}\times\pi\times15\times15\times15\right)$$
$$= 2250\pi$$

$$\text{Cylinder: } \quad V = \pi r^2 h$$
$$= \pi\times15\times15\times45$$
$$= 10\,125\pi$$

3. Add the volume of each component.

$$\text{Composite object: } V = 2250\pi + 10\,125\pi$$
$$= 12\,375\pi$$
$$\approx 38\,877.2$$

4. Write the answer with the correct unit.

The volume is $38\,877.2\,\text{cm}^3$.

Exercise 3.6 Volume and capacity

3.6 Exercise	3.6 Exam questions

Simple familiar	Complex familiar	Complex unfamiliar
1, 2, 3, 4, 5, 6, 7, 8, 9, 10, 11, 12	13, 14, 15, 16, 17, 18, 19, 20, 21	22, 23, 24, 25, 26

These questions are even better in jacPLUS!
• Receive immediate feedback
• Access sample responses
• Track results and progress

Find all this and MORE in jacPLUS ▶

Simple familiar

1. **WE13** Calculate the volume of a triangular prism with length $l = 2.5$ m, triangle base length $b = 0.6$ m and height $h = 0.8$ m.

2. Giving answers correct to the nearest cubic centimetre, calculate the volume of a prism that has:
   a. a base area of 200 cm² and a height of 1.025 m
   b. a rectangular base 25.25 cm by 12.65 cm and a length of 0.42 m
   c. a right-angled triangular base with one side length of 48 cm, a hypotenuse of 73 cm and a length of 96 cm
   d. a height of 1.05 m and a trapezium-shaped base with parallel sides that are 25 cm and 40 cm long and 15 cm apart.

3. The Gold Medal Pool Company sells three types of above-ground swimming pools, with base shapes that are square, rectangular or circular. Use the information in the table to list the volumes of each type from largest to smallest, giving your answers in litres.

Type	Depth	Base dimensions
Square pool	1.2 m	Length: 3 m
Rectangular pool	1.2 m	Length: 4.1 m Width: 2.25 m
Circular pool	1.2 m	Diameter: 3.3 m

4. **WE14** Giving your answer correct to the nearest cubic centimetre, calculate the volume of a cylinder of radius 22.5 cm and a height of 35.4 cm.

5. Calculate the volume of a cylinder that has:
   a. a base circumference of 314 cm and a height of 0.625 m, giving your answer correct to the nearest cubic centimetre
   b. a height of 425 cm and a radius that is three-quarters of its height, giving your answer correct to the nearest cubic metre.

6. **WE15** Calculate the volume of a cone of radius 30 cm and a height of 42 cm correct to 1 decimal place.

7. Calculate the volume of a cone that has:
   a. a base circumference of 628 cm and a height of 0.72 m, correct to the nearest whole number
   b. a height of 0.36 cm and a radius that is two-thirds of its height, correct to 3 decimal places.

8. Calculate the exact volumes of the solid objects shown in the following diagrams.

   a.
   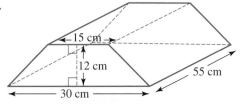

   b.
   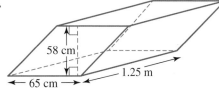

   c.
   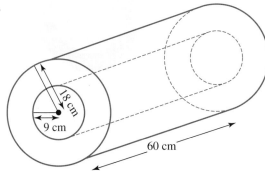

   d.
   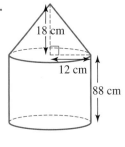

9. **WE16** Calculate the volume of a pyramid that is 2.025 m tall and has a rectangular base with dimensions 1.05 m by 0.0745 m, correct to 4 decimal places.

10. Calculate the volume of a pyramid that has:
    a. a base area of 366 cm² and a height of 1.875 m
    b. a rectangular base 18.45 cm by 26.55 cm and a height of 0.96 m
    c. a height of 3.6 m, a triangular base with one side length of 1.2 m and a perpendicular height of 0.6 m.

11. **WE17** Calculate the volume of a sphere of radius 0.27 m correct to 4 decimal places.

12. **WE18** Calculate the volume of an object that is composed of a hemisphere (half a sphere) of radius 1.5 m that sits on top of a cylinder of height 2.1 m. Give your answer correct to 2 decimal places.

Complex familiar

13. A builder uses the floor plan of the house they are building to calculate the amount of concrete they need to order for the foundations supporting the brick walls.

   a. The foundation needs to go around the perimeter of the house with a width of 600 mm and a depth of 1050 mm. Calculate how many cubic metres of concrete are required.

   b. The builder also wants to order the concrete required to pour a rectangular slab 3 m by 4 m to a depth of 600 mm. Calculate how many cubic metres of extra concrete they should order.

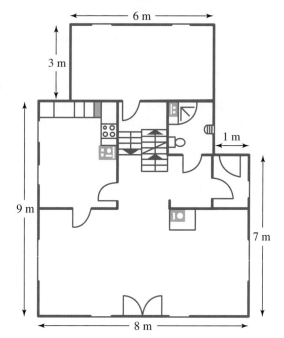

14. A company manufactures skylights in the shape of a cylinder with a hemispherical lid.

When they are fitted onto a house, three-quarters of the length of the cylinder is below the roof. If the cylinder is 1.5 m long and has a radius of 30 cm, calculate the volume of the skylight that is above the roof and the volume that is below it.

Give your answers correct to the nearest cubic centimetre.

15. The outer shape of a washing machine is a rectangular prism with a height of 850 mm, a width of 595 mm and a depth of 525 mm. Inside the machine, clothes are washed in a cylindrical stainless steel drum that has a diameter of 300 mm and a length of 490 mm.

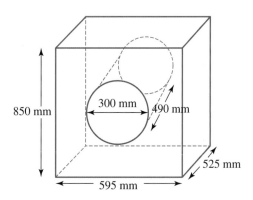

   a. Calculate the maximum volume of water, in litres, that the stainless steel drum can hold.

   b. Calculate the volume of the washing machine, in cubic metres, after subtracting the volume of the stainless steel drum.

16. The diagram shows the dimensions for a proposed house extension. Calculate the volume of insulation required in the roof if it takes up an eighth of the overall roof space.

17. Calculate the radius, correct to the nearest whole number, of a sphere that has:

   a. a volume of 248 398.88 cm^3

   b. a volume of 4.187 m^3.

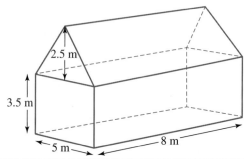

18. Calculate the volume, correct to 2 decimal places where appropriate, of an object that is composed of:

    a. a square pyramid of height 48 cm that sits on top of a cube of side length 34 cm
    b. a cone of height 75 cm that sits on top of a 60-cm-tall cylinder of radius 16 cm.

19. Tennis balls are spherical with a diameter of 6.7 cm. They are sold in packs of four in cylindrical canisters whose internal dimensions are 26.95 cm long with a diameter that is 5 mm greater than that of a ball.

    a. Calculate the volume of free space that is in a canister containing four tennis balls. Give your answer correct to 2 decimal places.
    b. The canisters are packed vertically in rectangular boxes; each box is 27 cm high and will fit exactly eight canisters along its length and exactly four along its width.
       Calculate the volume of free space that is in a rectangular box packed full of canisters. Give your answer correct to 2 decimal places.

20. An ice sculptor has a block of ice in the shape of a triangular prism.

    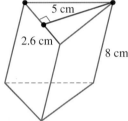

    If ice has a density of 0.91 g/mL, calculate the mass (in grams) of the block of ice.

21. A measuring jug is filled with 280 cm³ of water. The jug of water is poured into a cylindrical vase of radius 3.5 cm. Calculate the height of water in the vase.

Complex unfamiliar

22. A tank holding liquid petroleum gas (LPG) is cylindrical in shape with hemispherical ends. If the tank is 8.7 m from the top of the hemisphere at one end to the top of the hemisphere at the other, and the cylindrical part of the tank has a diameter of 1.76 m, calculate the volume of the tank to the nearest litre.

23. The glass pyramid in the courtyard of the Louvre Museum in Paris has a height of 22 m and a square base with side lengths of 35 m. A second glass pyramid at the Louvre Museum is called the Inverted Pyramid as it hangs upside down from the ceiling.

    It is one-third the volume of the larger glass pyramid but with the same square base. Evaluate the height of this second pyramid.

24. A wheat farmer needs to purchase a new grain silo and has the choice of two shapes. One is cylindrical with a conical top, and the other is cylindrical with a hemispherical top. Use the dimensions shown in the following diagrams to determine which silo has the greater volume and express this a percentage difference.

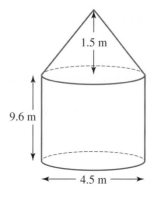

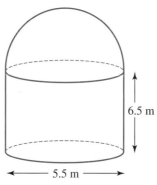

25. The volume of a hexagonal prism is 2500 cm^3. The height of the prism is 12 cm. Calculate the side length of the hexagonal base.

26. Consider the silo shown, which is full of grain.

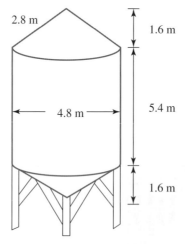

If the silo is emptied by $\frac{1}{4}$, calculate the height of the grain.

Fully worked solutions for this chapter are available online.

LESSON
3.7 Surface area of three-dimensional objects

3.7.1 Using nets to calculate surface area

The **surface area** of a solid object is equal to the combined total of the areas of each individual surface that forms it. Some objects have specific formulas for the calculation of the total surface area, whereas others require the calculation of each individual surface in turn.

Surface area is particularly important in design and construction when considering how much material is required to make a solid object.

In manufacturing, it could be important to make an object with the smallest amount of material that is capable of holding a particular volume. Surface area is also important in aerodynamics, as the greater the surface area, the greater the potential air resistance or drag.

Nets

The **net** of a solid object is a pattern or plan for its construction. Each surface of the object is included in its net. Therefore, the net can be used to calculate the total surface area of the object.

For example, the net of a cube will have six squares, whereas the net of a triangular pyramid (or tetrahedron) will have four triangles.

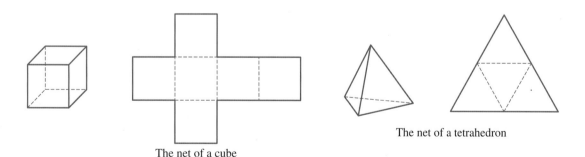

The net of a cube

The net of a tetrahedron

eles-3057

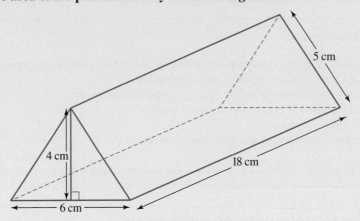

WORKED EXAMPLE 19 Using a net to calculate the total surface area

Calculate the surface area of the prism shown by first drawing its net.

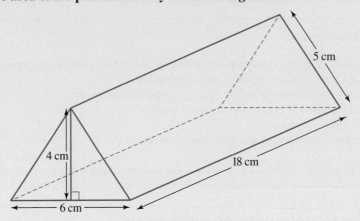

THINK

1. Identify the prism and each surface in it.

2. Redraw the given diagram as a net, making sure to check that each surface is present.

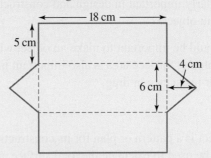

3. Calculate the area of each surface identified in the net.

4. Add the component areas and write the answer with the correct unit.

WRITE

The triangular prism consists of two identical triangular ends, two identical rectangular sides and one rectangular base.

Triangular ends:

$$A = 2 \times \left(\frac{1}{2} bh \right)$$

$$= 2 \times \left(\frac{1}{2} \times 6 \times 4 \right)$$

$$= 24$$

Rectangular sides:

$$A = 2 \times (l\,w)$$

$$= 2 \times (18 \times 5)$$

$$= 180$$

Rectangular base:

$$A = l\,w$$

$$= 18 \times 6$$

$$= 108$$

Total surface area:

$$SA = 24 + 180 + 108$$

$$= 312\,\text{cm}^2$$

The total surface area is $312\,\text{cm}^2$.

3.7.2 Surface area formulas

The surface area formulas for common solid objects are summarised in the following table.

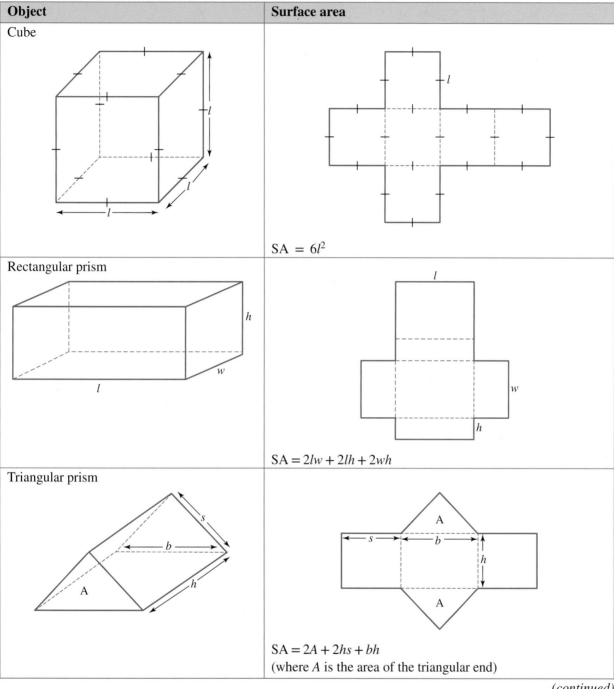

Object	Surface area
Cube	SA $= 6l^2$
Rectangular prism	SA $= 2lw + 2lh + 2wh$
Triangular prism	SA $= 2A + 2hs + bh$ (where A is the area of the triangular end)

(continued)

(continued)

Object	Surface area
Cylinder	$\text{SA} = 2\pi r^2 + 2\pi rh$ $\quad\;\; = 2\pi r\,(r + h)$
Cone	$\text{SA} = \pi rs + \pi r^2$ $\quad\;\; = \pi r\,(s + r)$ (including the circular base)
Tetrahedron	$\text{SA} = 4 \times \left(\frac{1}{2}bh\right)$

Square pyramid 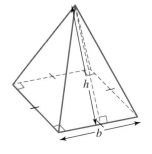	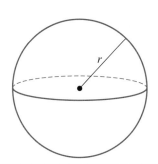
	$SA = 4 \times \left(\frac{1}{2}bh\right) + b^2$
Sphere 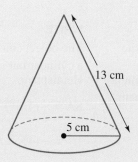	$SA = 4\pi r^2$

WORKED EXAMPLE 20 Using formulas to determine the total surface area

Calculate the surface area of the object shown by selecting an appropriate formula. Give your answer correct to 1 decimal place.

13 cm

5 cm

THINK	WRITE
1. Identify the object and the appropriate formula.	Given the object is a cone, the formula is $SA = \pi rs + \pi r^2$.
2. Substitute the given values into the formula and evaluate.	$\begin{aligned} SA &= \pi rs + \pi r^2 \\ &= \pi r(s + r) \\ &= \pi \times 5(5 + 13) \\ &= \pi \times 90 \\ &\approx 282.7 \end{aligned}$
3. Write the answer with the correct unit.	The surface area of the cone is 282.7 cm^2.

3.7.3 Surface areas of composite solids

For composite solids, be careful to include only those surfaces that form the outer part of the object.

For example, if a solid consisted of a pyramid on top of a cube, the internal surface highlighted in blue would not be included.

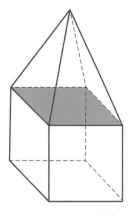

WORKED EXAMPLE 21 Calculating the total surface area of a composite solid

Calculate the surface area of the object shown correct to 2 decimal places.

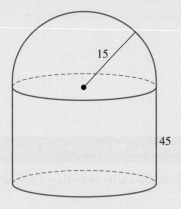

THINK	WRITE
1. Identify the components of the composite solid.	The object consists of a hemisphere that sits on top of a cylinder but not including the top of the cylinder.
2. Substitute the given values into the formula for each surface of the object and evaluate. Give exact answers, rounding the final answer only.	Hemisphere: $\text{SA} = \dfrac{1}{2}(4\pi r^2)$ $\phantom{\text{SA}} = \dfrac{1}{2}(4 \times \pi \times 15^2)$ $\phantom{\text{SA}} = 450\pi$ Cylinder (no top): $\text{SA} = \pi r^2 + 2\pi rh$ $\phantom{\text{SA}} = \pi \times 15^2 + 2 \times \pi \times 15 \times 45$ $\phantom{\text{SA}} = 1575\pi$
3. Add the area of each surface to obtain the total surface area.	Total surface area: $\text{SA} = 450\pi + 1575\pi$ $\phantom{\text{Total surface area: SA}} = 2025\pi$
4. Use a calculator to determine the value of 2025π correct to 2 decimal places and write the answer with the correct unit.	The total surface area of the object is 6361.73 cm^2.

Exercise 3.7 Surface area of three-dimensional objects

3.7 Exercise	3.7 Exam questions **on**

Simple familiar	Complex familiar	Complex unfamiliar
1, 2, 3, 4, 5, 6, 7, 8, 9, 10	11, 12, 13, 14	15, 16, 17, 18

These questions are even better in jacPLUS!
- Receive immediate feedback
- Access sample responses
- Track results and progress

Find all this and MORE in jacPLUS ⊙

Simple familiar

1. **WE19** Calculate the surface area of the prism shown by first drawing its net.

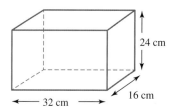

24 cm

16 cm

32 cm

2. Calculate the surface area of the tetrahedron shown by first drawing its net.

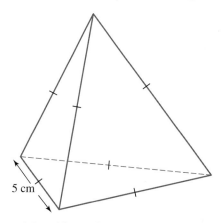

5 cm

3. **WE20** Calculate the surface areas of the objects shown by selecting appropriate formulas.

a.
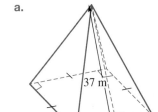

37 m

25 m

b.
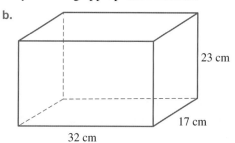

23 cm

17 cm

32 cm

4. Calculate (correct to 2 decimal places where appropriate) the surface area of:
   a. a pyramid formed by four equilateral triangles with a side length of 12 cm
   b. a sphere with a radius of 98 cm
   c. a cylinder with a radius of 15 cm and a height of 22 cm
   d. a cone with a radius of 12.5 cm and a slant height of 27.2 cm.

5. Calculate (correct to 2 decimal places where appropriate) the total surface area of:
   a. a rectangular prism with dimensions 8 cm by 12 cm by 5 cm
   b. a cylinder with a base diameter of 18 cm and a height of 20 cm
   c. a square pyramid with a base length of 15 cm and a vertical height of 18 cm
   d. a sphere of radius 10 cm.

6. A prism is 25 cm high and has a trapezoidal base whose parallel sides are 8 cm and 12 cm long respectively and are 10 cm apart.
   a. Draw the net of the prism.
   b. Calculate the total surface area of the prism.

7. **WE21** Calculate the surface area of the object shown correct to 2 decimal places.

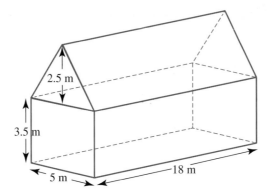

8. Calculate the surface area of the object shown correct to 2 decimal places.

9. A hemispherical glass ornament sits on a circular base that has a radius of 5 cm.

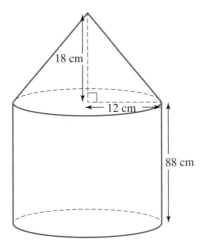

   a. Calculate its total surface area to the nearest square centimetre.
   b. If an artist attaches it to an 8-cm-tall cylindrical stand with the same circumference, calculate the new total surface area of the combined object that is created. Give your answer to the nearest square centimetre.

10. A cylindrical plastic vase is 66 cm high and has a radius of 12 cm. The centre has been hollowed out so that there is a cylindrical space with a radius of 9 cm that goes to a depth of 48 cm and ends in a hemisphere, as shown in the diagram. Giving your answers to the nearest square centimetre:

   a. calculate the area of the external surfaces of the vase
   b. calculate the area of the internal surface of the vase.

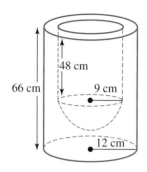

Complex familiar

11. An ice-cream shop sells two types of cones. One is 6.5 cm tall with a radius of 2.2 cm. The other is 7.5 cm tall with a radius of 1.7 cm. By first calculating the slant height of each cone correct to 2 decimal places, determine which cone (not including any ice-cream) has the greater surface area and by how much.

12. The top of a church tower is in the shape of a square pyramid that sits on top of a rectangular prism base that is 1.1 m high. The pyramid is 6 m high with a base length of 4.2 m.

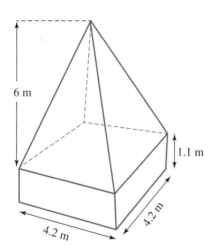

   Calculate the total external surface area of the top of the church tower if the base of the prism forms the ceiling of a balcony. Give your answer correct to 2 decimal places.

13. A dumbbell consists of a cylindrical tube that is 28 cm long with a diameter of 3 cm, and two pairs of cylindrical discs that are held in place by two locks. The larger discs have a diameter of 12 cm and a width of 2 cm, and the smaller discs are the same thickness with a diameter of 9 cm.

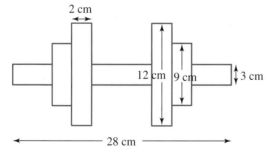

   Calculate the total area of the exposed surfaces of the discs when they are held in position as shown in the diagram. Give your answer to the nearest square centimetre.

14. A staircase has a section of red carpet down its centre strip. Each of the nine steps is 16 cm high, 25 cm deep and 120 cm wide.
    The red carpet is 80 cm wide and extends from the back of the uppermost step to a point 65 cm beyond the base of the lower step.

    a. Calculate the area of the red carpet.
    b. If all areas of the front and top of the stairs that are not covered by the carpet are to be painted white, calculate the area to be painted.

Complex unfamiliar

15. A rectangular swimming pool is 12.5 m long, 4.3 m wide and 1.5 m deep. If all internal surfaces are to be tiled at a cost of $45.50 per square meter, calculate the cost of tiling the pool.

16. A quarter-pipe skateboard ramp has a curved surface that is one-quarter of a cylinder with a radius of 1.5 m. If the surface of the ramp is 2.4 m wide, calculate the total surface area of the front, back and sides.

17. Calculate the surface area of the shape shown.

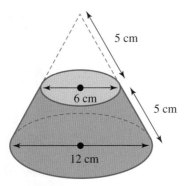

18. Calculate the surface area of the shape shown.

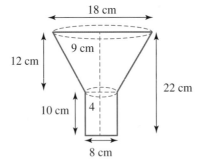

Fully worked solutions for this chapter are available online.

LESSON
3.8 Review

3.8.1 Summary

3.8 Exercise

 3.8 Exercise   **3.8 Exam questions** on

Simple familiar	Complex familiar	Complex unfamiliar
1, 2, 3, 4, 5, 6, 7, 8, 9, 10, 11, 12	13, 14, 15, 16	17, 18, 19, 20

Simple familiar

1. **MC** Select which group of three numbers in the form (a, b, c) would be the side lengths of a right-angled triangle.

   A. 6, 24, 25   **B.** 13, 14, 15   **C.** 7, 24, 25   **D.** 9, 12, 16

2. **MC** If a right-angled isosceles triangle has a hypotenuse of length 32 units, the other sides will be closest to:

   A. 21.54   **B.** 21.55   **C.** 22.62   **D.** 22.63

3. **MC** An equilateral triangle with a side length of 4 units will have a height closest to:

   A. 3.46   **B.** 4.47   **C.** 4   **D.** 3.47

4. **MC** A trapezium has a height of 8 cm and an area of 148 cm^2. Its parallel sides could be:

   A. 11 cm and 27 cm   **B.** 12 cm and 24 cm
   C. 12 cm and 26 cm   **D.** 12 cm and 25 cm

5. **MC** A circle has a circumference of 75.4 cm. Its area is closest to:

   A. 440 cm^2 **B.** 452 cm^2 **C.** 461 cm^2 **D.** 448 cm^2

6. **MC** A cylinder with a volume of 1570 cm^3 and a height of 20 cm will have a diameter that is closest to:

   A. 5 cm   **B.** 12 cm   **C.** 15 cm   **D.** 10 cm

7. **MC** A cone with a surface area of 2713 cm^2 and a diameter of 24 cm will have a slant-height that is closest to:

   A. 58 cm   **B.** 48 cm   **C.** 60 cm   **D.** 46 cm

8. **MC** A hemisphere with a radius of 22.5 cm will have a volume and total surface area respectively that are closest to:

   A. 47 689 cm^3 and 4764 cm^2 **B.** 47 689 cm^3 and 6358 cm^2
   C. 23 845 cm^3 and 6359 cm^2 **D.** 23 856 cm^3 and 4771 cm^2

9. **MC** A square pyramid with a volume of 500 cm^3 and a vertical height of 15 cm will have a surface area that is closest to:

A. 416 cm^2 B. 492 cm^2 C. 359 cm^2 D. 316 cm^2

10. **MC** An open rubbish skip is in the shape of a trapezoidal prism with the dimensions indicated in the diagram.

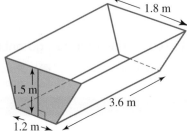

The external surface area (m^2) and volume (m^3) respectively are:

A. 8.1 and 19.8   B. 19.8 and 8.1   C. 17.75 and 8.1   D. 8.1 and 26.3

11. Calculate the length of wire required to support the mast of a yacht if the mast is 10.8 m long and the support wire is attached to the horizontal deck at a point 3.6 m from the base of the mast.

12. A surveyor is measuring a building site and wants to check that the guidelines for the foundations are square (i.e. at right angles).

The surveyor places a marker 3600 mm from a corner along one line and another marker 4800 mm from the corner along the other line.
Calculate how far apart the markers must be for the lines to be square.

Complex familiar

13. The side pieces of train carriages are made from rectangular sheets of pressed metal of length 25 metres and height 2.5 metres. Rectangular sections for the doors and windows are cut out. The dimensions of the spaces for the doors are 2 metres high by 2.5 metres wide.
The window spaces are 3025 millimetres wide by 900 millimetres high. Each sheet must have spaces cut for two doors and three windows.

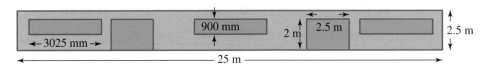

a. Calculate the total area of pressed metal that remains once the sections for the doors and windows have been removed.
b. A thin edging strip is placed around each window and around the top and sides of the door opening. Calculate the total length of edging required.

14. The supporting strut of a streetlight must be attached so that its ends are an equal distance from the top of the pole.

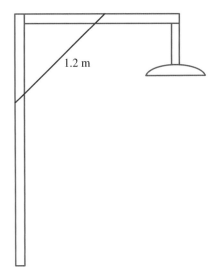

1.2 m

If the strut is 1.2 m long, determine how far the ends are from the top of the pole.

15. A semicircular arch sits on two columns as shown in the diagram. The outer edges of the columns are 7.6 m apart and the inner edges are 5.8 m apart.

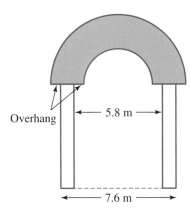

Overhang   5.8 m

7.6 m

The width of each column is three-quarters the width of the arch, and the arch overhangs the columns by one-eighth of its width on each edge. The face of the arch (the shaded area) is to be tiled.
Calculate the area the tiles will cover, correct to 2 decimal places.

16. A circular pond is placed in the middle of a rectangular garden that is 15 m by 8 m.

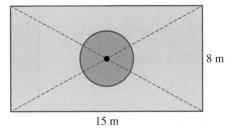

8 m

15 m

If the radius of the pond is a quarter of the distance from the centre to the corner of the garden, calculate:

a. the circumference of the pond, correct to 3 decimal places
b. the area of the garden, not including the pond, correct to 1 decimal place
c. the volume of water in the pond, correct to the nearest litre, if it is filled to a depth of 850 mm.

17. When viewed from above, a swimming pool can be seen as a rectangle with a semicircle at each end, as shown in the diagram below.

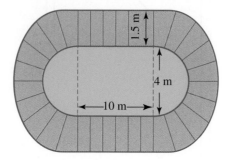

The area around the outside of the pool extending 1.5 m from the edge is to be paved.

   a. Calculate the paved area around the pool, correct to 2 decimal places.
   b. If the pool is to be filled to a depth of 900 mm in the semicircular sections and 1500 mm in the rectangular section, calculate the total volume of water in the pool to the nearest litre.

18. A rectangular piece of glass with side lengths 1000 mm and 800 mm has its corners removed for safety, as shown in the diagrams.

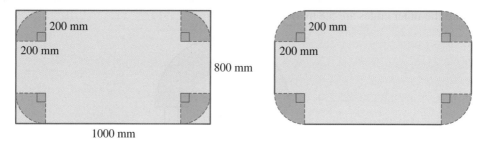

   a. Calculate the surface area of the glass after the corners have been removed, correct to 1 decimal place.
   b. Calculate the perimeter of the glass after the corners have been removed, correct to 1 decimal place.

19. A piece of timber has the dimensions 400 mm by 400 mm by 1800 mm. The top corners of the piece of timber are removed along its length. The cuts are made at an angle a distance of 90 mm from the corners, so the timber that is removed forms two triangular prisms.

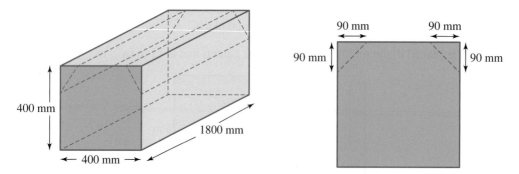

Calculate the total volume and surface area of the two smaller pieces of timber that are cut from the corners, assuming they each remain as one piece.

20. Two tunnels run under a bend in a river. One runs in a straight line from point A to point C, and the other runs in a straight line from point D to point E. Points A and D are 5 km apart, as are points C and E. Point B is 10 km from both D and E, and BD is perpendicular to BE. An access tunnel GF is to be constructed between the midpoints of AC and DE.

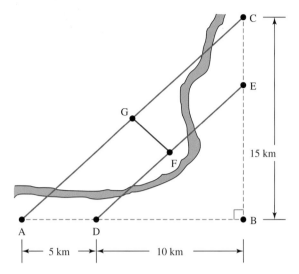

a. The inner walls of the tunnels are formed of concrete that is 3.5 metres thick. Calculate the total volume of concrete used for the tunnels, correct to 1 decimal place.

b. The inner surface of the concrete in the tunnels is sprayed with a sealant to prevent water seeping through. Calculate the total area that is sprayed with the sealant, correct to 1 decimal place.

Fully worked solutions for this chapter are available online.

Answers

Chapter 3 Pythagoras' theorem and measuration

3.2 Pythagoras' theorem in two dimensions

3.2 Exercise

1. a. PR   b. YZ
2. a. 13 cm   b. 170 mm
3. a. 10.82 cm   b. 6.93 m
4. a. 10.4 cm   b. 1.9 m
5. a. 8.9 cm   b. 22.1 cm   c. 47.4 mm
   d. 37.3 m
6. A
7. a. Yes   b. No   c. Yes
   d. No   e. Yes   f. Yes
8. 13 m
9. 3.23 m
10. 3.73 m
11. C
12. a. 9, 12, 15   b. 7, 24, 25   c. 1.5, 2.0, 2.5
    d. 3, 4, 5   e. 11, 60, 61   f. 10, 24, 26
    g. 9, 40, 41   h. 0.7, 2.4, 2.5
13. 2.2 m
14. 7.07 cm
15. 11.40
16. 4

3.3 Pythagoras' theorem in three dimensions

3.3 Exercise

1. 1.56 m
2. No, the maximum length rod that could fit would be 2.71 m long.
3. 41.4 m
4. Pyramid 1 has the greatest height.
5. a. 30.48 cm   b. 2.61 cm   c. 9.43 cm
6. a. 11.55 cm   b. 27.71 cm   c. 3.18 cm   d. 95.84 cm
7. a. Yes   b. 1.015 m
8. 6 m; a sample response can be found in the worked solutions in your online resources.
9. 8.09
10. $s = 186.5$ m
11. $h = 25.475$
12. $h = 28.5$ m
13. 13.82 m
14. 10 054 m

3.4 Perimeter and area of polygons and triangles

3.4 Exercise

1. Perimeter $= 80$ cm, area $= 360$ cm^2
2. Circumference $= 50.27$ cm, area $= 201.06$ cm^2
3. a. Perimeter $= 59$ m, area $= 155.12$ m^2
   b. Perimeter $= 28.83$ cm, area $= 20$ cm^2
   c. Perimeter $= 43.98$ cm, area $= 153.94$ cm^2
   d. Perimeter $= 48$ cm, area $= 112$ cm^2
4. a. Circumference $= 31.42$ cm, area $= 78.54$ cm^2
   b. Circumference $= 56.55$ cm, area $= 254.47$ cm^2
5. Perimeter $= 68$ cm, area $= 192$ cm^2
6. 33.79 cm^2
7. 31.61 cm
8. Area $= 1.14$ m^2
9. Perimeter $= 96$ units, area $= 360$ units^2
10. a. i. 201.06 cm   ii. 389.56 cm^2
    b. 187.19 cm
11. 49.19 cm
12. 26.39 m
13. a. Perimeter $= 90$ m, area $= 389.71$ m^2
    b. Perimeter $= 120$ m, area $= 839.7$ m^2
14. 25 minutes
15. 22.81 units

3.5 Perimeter and area of sectors and composite shapes

3.5 Exercise

1. 72 cm^2
2. Perimeter $= 48.6$ cm, area $= 106$ cm^2
3. 198.94 cm^2
4. a. 5.57 m   b. 6.11 m^2
5. a. 5831.62 cm^2 b. 56.58 cm^2
6. 4.24 m^2
7. Perimeter $= 52.94$ cm, area $= 167.55$ cm^2, arc length $= 20.94$ cm
8. Perimeter $= 198.23$ cm, area $= 2073.45$ cm^2
9. Perimeter $= 145.40$ cm, area $= 1143.06$ cm^2
10. 9.42 cm^2
11. 9292.83 cm^2
12. Perimeter $= 151.76$ cm, area $= 576$ cm^2
13. 2.50 m^2
14. Area $= 900$ cm^2, perimeter $= 194.16$ cm
15. Yes, the blue area covers 81.63% of the window.
16. 25.13 cm^2
17. 0.67 m^2
18. 2138.25 cm^2
19. 0.776 m^2
20. 160.3 cm^2

3.6 Volume and capacity

3.6 Exercise

1. 0.6 m^3

2. a. $20\,500 \text{ cm}^3$ b. $13\,415 \text{ cm}^3$
   c. $126\,720 \text{ cm}^3$ d. $51\,188 \text{ cm}^3$

3. Rectangular pool: $11\,070$ litres
   Square pool: $10\,800$ litres
   Circular pool: $10\,263.58$ litres

4. $56\,301 \text{ cm}^3$

5. a. $490\,376 \text{ cm}^3$ b. 136 m^3

6. $39\,584.1 \text{ cm}^3$

7. a. $753\,218 \text{ cm}^3$ b. 0.022 cm^3

8. a. $14\,850 \text{ cm}^3$ b. $471\,250 \text{ cm}^3$
   c. $14\,580\pi \text{ cm}^3$ d. $13\,536\pi \text{ cm}^3$

9. 0.0528 m^3

10. a. $22\,875 \text{ cm}^3$ b. $15\,675.12 \text{ cm}^3$ c. 0.432 m^3

11. 0.0824 m^3

12. 21.91 m^3

13. a. 25.2 m^3 b. 7.2 m^3

14. Volume above $= 162\,578 \text{ cm}^3$,
    Volume below $= 318\,086 \text{ cm}^3$

15. a. 34.64 litres   b. 0.231 m^3

16. 6.25 m^3

17. a. 39 cm b. 1 m

18. a. $57\,800 \text{ cm}^3$ b. $68\,361.05 \text{ cm}^3$

19. a. 467.27 cm^3 b. 9677.11 cm^3

20. 47.32 g

21. 7.28 cm

22. $19\,739$ litres

23. 7.3 m

24. The hemispherical-topped silo holds 37.35 m^3 more, which is a greater capacity of 23.25%.

25. 8.95 cm

26. 4.32 m

3.7 Surface area of three-dimensional objects

3.7 Exercise

1.

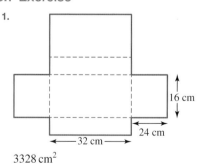

3328 cm^2

2.

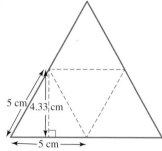

43.3cm^2

3. a. 2475 m^2 b. 3342 cm^2

4. a. 249.42 cm^2 b. $120\,687.42 \text{ cm}^2$
   c. 3487.17 cm^2 d. 1559.02 cm^2

5. a. 392 cm^2 b. 1639.91 cm^2
   c. 810 cm^2 d. 1256.64 cm^2

6. a.

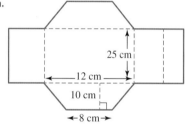

   b. 1210 cm^2

7. 390.94 m^2

8. 7902.86 cm^2

9. a. 236 cm^2 b. 487 cm^2

10. a. 5627 cm^2 b. 3223 cm^2

11. The cone with height 6.5 cm and radius 2.2 cm has the greater surface area by 6.34 cm^2.

12. 71.90 m^2

13. 688 cm^2

14. a. $34\,720 \text{ cm}^2$ b. $14\,760 \text{ cm}^2$

15. $\$4738.83$

16. 10.22 m^2

17. 226.10 cm^2

18. 1027.92 cm^2

3.8 Review

3.8 Exercise

1. C   2. D   3. A   4. D   5. B
6. D   7. C   8. D   9. A   10. B

11. 11.38 m

12. 6000 mm

13. a. 44.33 m^2 b. 36.55 m

14. 0.849 m

15. 12.63 m^2

16. **a.** 13.352 m **b.** 105.8 m^2 **c.** 12 058 L
17. **a.** 55.92 m^2 **b.** 71 310 L
18. **a.** 765 663.7 mm^2 **b.** 3256.6 mm
19. Volume = 14 580 000 mm^3
    Surface area = 1 122 405 mm^2
20. **a.** 249 cm^2 **b.** 3 664 824.9 m^2

4 Similar figures and scale factors

LESSON SEQUENCE

Fully worked solutions for this chapter are available online.

EXAM PREPARATION

Access exam-style questions in every lesson, available online.

on Resources

Solutions	Solutions — Chapter 4 (sol-1245)
Exam questions	Exam question booklet — Chapter 4 (eqb-0261)
Digital documents	Learning matrix — Chapter 4 (doc-41476)
	Chapter summary — Chapter 4 (doc-41477)

LESSON
4.1 Overview

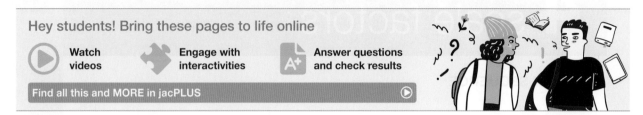
4.1.1 Introduction

To construct buildings and make sure they are functional, designers must apply the principles of mathematics. These principles involve scaled drawings in the form of plans. A plan is a drawing of the object to be built, reduced in size in a way that all the measurements correspond to the actual object.

Architects often use a different set of scales than engineers, surveyors or furniture designers. This depends on the size of what is being designed, as well as the complexity of the design.

Scale is not just used for plans; it can also be used to create a scale model of the design. A scale model is generally a physical representation of an object that maintains accurate relationships between all important aspects of the model. The scale model allows you to see some behaviour of the original object without investigating the original object itself. Scale models are used in many fields, including engineering, filmmaking, military planning, sales and hobby model building. To be considered a true scale model, all important aspects must be accurately modelled: not just the scale of the object, but also the material properties. An example could be an aerospace company wanting to test a new wing design. They could construct a scaled-down model and test it in a wind tunnel under simulated conditions.

One famous model is that of a space shuttle by Dr Maxime Faget from NASA. He needed a model to demonstrate to his colleagues that a space shuttle would be able to glide back to Earth without power — a concept unknown until then.

4.1.2 Syllabus links

Lesson	Lesson title		Syllabus links
4.2	Similarity of two-dimensional figures	○	Understand the conditions for similarity of two-dimensional figures, including similar triangles.
4.3	Linear scale factors	○	Use the scale factor for two similar figures to solve linear scaling problems.
4.4	Scale drawings — maps and plans	○	Determine measurements from scale drawings (e.g. maps and building plans) to solve problems.
4.5	Area and volume scale factors	○	Determine a scale factor and use it to solve scaling problems, e.g. calculating lengths and areas of similar figures, and calculating surface areas, volumes and capacities of similar solids.

Source: General Mathematics Senior Syllabus 2024 © State of Queensland (QCAA) 2024; licensed under CC BY 4.0.

LESSON
4.2 Similarity of two-dimensional figures

SYLLABUS LINKS

- Understand the conditions for similarity of two-dimensional figures, including similar triangles.

Source: General Mathematics Senior Syllabus 2024 © State of Queensland (QCAA) 2024; licensed under CC BY 4.0.

4.2.1 Conditions for similarity

Objects are called **similar** when they are exactly the same shape but have different sizes. Objects that are exactly the same size and shape are called **congruent**.

Similarity is an important mathematical concept that is often used for planning purposes in areas such as engineering, architecture and design. Scaled-down versions of much larger objects allow designs to be trialled and tested before their construction.

Two-dimensional objects are similar when their internal angles are the same and their side lengths are **proportional**. This means that the ratios of corresponding side lengths are always equal for similar objects. We use the symbols ~ or ||| to indicate that objects are similar.

eles-3058

> ### WORKED EXAMPLE 1 Proving two objects are similar
>
> **Demonstrate that these two shapes are similar.**
>
>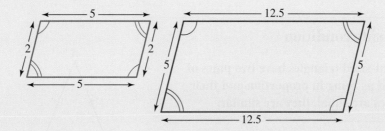
>
> | THINK | WRITE |
> |---|---|
> | 1. Confirm that the internal angles for the shapes are the same. | The diagrams indicate that all angles in both shapes match. |
> | 2. Calculate the ratio of the corresponding side lengths and simplify. | Ratios of corresponding sides: $\dfrac{12.5}{5} = 2.5$ and $\dfrac{5}{2} = 2.5$ |
> | 3. Write the answer in a sentence. | The two shapes are similar as their angles are the same and the ratios of the corresponding side lengths are equal. |

4.2.2 Similar triangles

The conditions for similarity apply to all objects, but not all of them need to be known in order to demonstrate similarity in triangles. If pairs of triangles have any of the following conditions in common, they are similar.

Note: In this chapter we will put the image first when calculating ratios of corresponding lengths. The original is blue and the image is pink.

1. Angle–angle–angle (AAA)

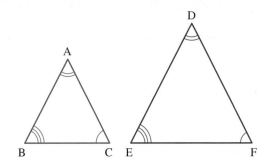

Angle-angle-angle condition

If two different-sized triangles have all three angles identified as being equal, they are similar.

$$\angle A = \angle D, \ \angle B = \angle E, \ \angle C = \angle F$$

2. Side–side–side (SSS)

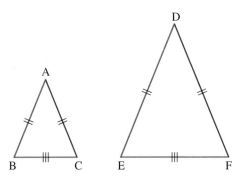

Side-side-side condition

If two different-sized triangles have all three sides identified as being in proportion, they are similar.

$$\frac{DE}{AB} = \frac{DF}{AC} = \frac{EF}{BC}$$

3. Side–angle–side (SAS)

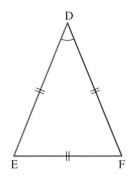

Side-angle-side condition

If two different-sized triangles have two pairs of sides identified as being in proportion and their included angles are equal, they are similar.

$$\frac{DE}{AB} = \frac{DF}{AC} \text{ and } \angle A = \angle D$$

WORKED EXAMPLE 2 Proving two triangles are similar

Demonstrate that these two triangles are similar.

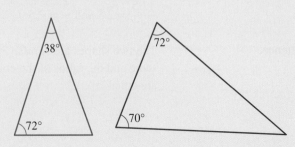

THINK	WRITE
1. Identify all possible angles and side lengths.	The angles in the blue triangle are: $38°, 72°$ and $180 - (38 + 72) = 70°$. The angles in the red triangle are: $70°, 72°$ and $180 - (70 + 72) = 38°$.
2. Use one of AAA, SSS or SAS to check for similarity.	The two triangles have all three angles identified as being equal.
3. Write the answer in a sentence.	The two triangles are similar as they satisfy the condition AAA.

WORKED EXAMPLE 3 Determining an unknown length in similar triangles

eles-6388

Calculate the value of *d* required to make the pair of triangles similar.

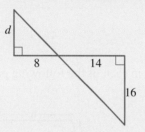

THINK

1. Identify the two separate triangles and ensure they are orientated the same way.

2. The ratio of corresponding side lengths in similar objects must be equal.
   Write down a ratio statement comparing corresponding side lengths.

3. Solve this equation for *d*. Use a calculator to assist with the arithmetic if required.

5. Write the answer.
   Note: Because *d* is a non-terminating decimal, the fractional answer must be given.

WRITE

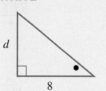

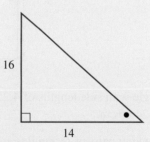

$$\frac{d}{16} = \frac{8}{14} \text{ or equivalent}$$

$$\frac{d}{16} \times 16 = \frac{8}{14} \times 16$$

$$d = \frac{64}{7}$$

For the triangles to be similar, the value of *d* must be $\frac{64}{7}$.

Exercise 4.2 Similarity of two-dimensional figures

learn on

4.2 Exercise	**4.2 Exam questions** on

Simple familiar	Complex familiar	Complex unfamiliar
1, 2, 3, 4, 5, 6, 7, 8, 9	10, 11, 12, 13, 14, 15	N/A

These questions are even better in jacPLUS!
- Receive immediate feedback
- Access sample responses
- Track results and progress

Find all this and MORE in jacPLUS ▶

Simple familiar

1. **WE1** Demonstrate that the two shapes in each of the following pairs are similar.

   a.

   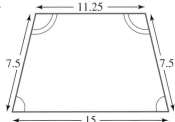
   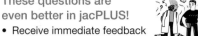

   b.

   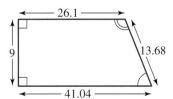

2. Demonstrate that a rectangle with side lengths of 4.25 cm and 18.35 cm will be similar to one with side lengths of 106.43 cm and 24.65 cm.

3. Identify which of the following pairs of rectangles are similar.

   a.

   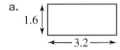

   b.

   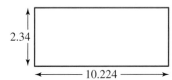

c.

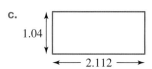

1.04

2.112

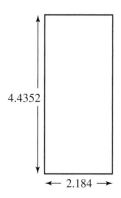

4.4352

2.184

d.

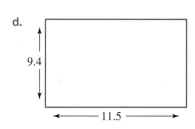

9.4

11.5

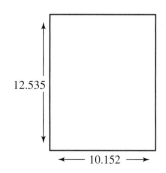

12.535

10.152

4. Identify which of the following pairs of polygons are similar.

a.

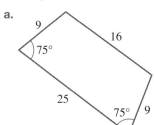

9

16

75°

25

75° 9

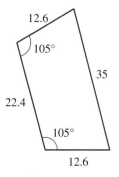

12.6

105°

35

22.4

105°

12.6

b.

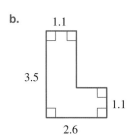

1.1

3.5

1.1

2.6

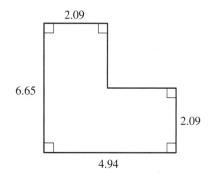

2.09

6.65

2.09

4.94

c.

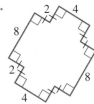

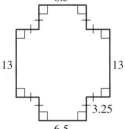

d.

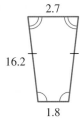

5. **WE2** Demonstrate that the two triangles in each of the following pairs are similar.

a.

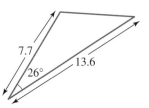

b.

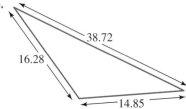

6. Identify which of the following pairs of triangles are similar.

a.

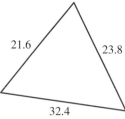

b.

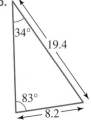

7. In each of the following groups, identify the two triangles that are similar.

a.

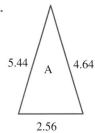
5.44   A   4.64
2.56

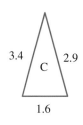

3.4   C   2.9
1.6

8.84   B   6.96
4.16

b.

4.6
A   2.8
4.375

6.3
7   B
4.2

3   C   5
4.5

c.

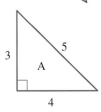
8.1
82°   A
12.6

11.2
B   23°
6.75   7.65

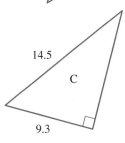
82°   9
14   C
23°

d.

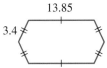
3   5
A
4

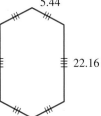
12.6
B   21

14.5
C
9.3

8. Calculate the ratios of the corresponding sides for the following pairs of objects.

a.

13.85
3.4

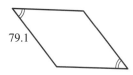
5.44
22.16

b.

35

79.1

c.

12   13

18

9. Demonstrate that the following are similar.
   a. A square of side length 8.2 cm and a square of side length 50.84 cm
   b. An equilateral triangle of side length 12.6 cm and an equilateral triangle of side length 14.34 cm

Complex familiar

10. Explain why each of the following pairs of objects must be similar.

   a.

   b.

   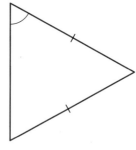

   c.

11. Evaluate the ratios of the corresponding side lengths in the following pairs of similar objects.

   a.
   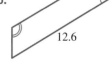
   26
   24
   7.5

   b.
   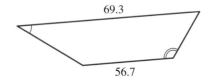
   12.6
   69.3
   56.7

12. **WE3** Calculate the value of x required to make the pair of triangles similar in each of the following diagrams.

a.

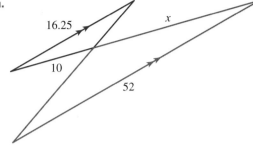

b.

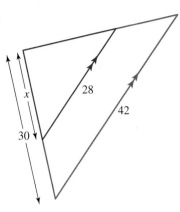

13. Evaluate the unknown side lengths in the following pairs of similar objects.

a.

b.

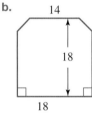

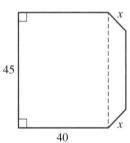

14. Calculate the values of x and y in the diagram.

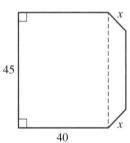

15. For the polygon shown, construct and label a similar polygon where:

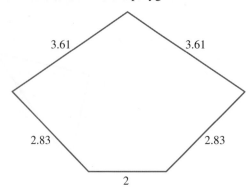

a. the corresponding sides are $\dfrac{4}{3}$ the size of those shown

b. the corresponding sides are $\dfrac{4}{5}$ the size of those shown.

Fully worked solutions for this chapter are available online.

LESSON
4.3 Linear scale factors

SYLLABUS LINKS

- Use the scale factor for two similar figures to solve linear scaling problems.

Source: General Mathematics Senior Syllabus 2024 © State of Queensland (QCAA) 2024; licensed under CC BY 4.0.

4.3.1 Calculating linear scale factors for similar triangles

Consider the pair of similar triangles shown in the diagram.

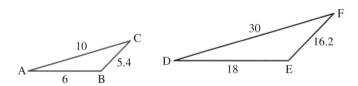

The ratios of the corresponding side lengths are:

$$DE : AB = \quad 18 : 6$$
$$= \quad 3 : 1$$

$$EF : BC = 16.2 : 5.4$$
$$= \quad 3 : 1$$

$$DF : AC = \quad 30 : 10$$
$$= \quad 3 : 1$$

Note: In this chapter we will put the image first when calculating ratios of corresponding lengths.

The side lengths of triangle DEF are all three times the lengths of triangle ABC. In this case, we would say that the **linear scale factor** is 3. The linear scale factor for similar objects can be evaluated using the ratio of the corresponding side lengths.

$\triangle ABC \sim \triangle DEF$ Linear scale factor: $\dfrac{DE}{AB} = \dfrac{EF}{BC} = \dfrac{DF}{AC} = k$

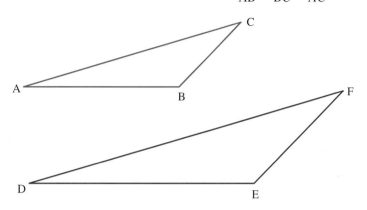

Linear scale factor

$$\text{Linear scale factor } (k) = \frac{\text{length of image}}{\text{length of object}}$$

A linear scale factor greater than 1 indicates enlargement, and a linear scale factor less than 1 indicates reduction.

eles-3059

WORKED EXAMPLE 4 Calculating a linear scale factor

Calculate the linear scale factor for the pair of similar triangles shown.

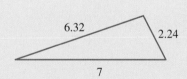   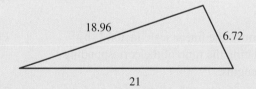

THINK

1. Identify the corresponding longest sides, the corresponding shortest sides and the corresponding third sides for both triangles.

2. Calculate the ratio of the corresponding side lengths and simplify.

3. Write the answer.

WRITE

7 and 21 are the longest corresponding sides.
2.24 and 6.72 are the shortest corresponding sides.
6.32 and 18.96 are the remaining sides.

$$k = \frac{21}{7} = \frac{18.96}{6.32} = \frac{6.72}{2.24} = 3$$

The linear scale factor is 3.

eles-6381

WORKED EXAMPLE 5 Determining an unknown length in similar triangles

For the pair of similar triangles shown, calculate the unknown side lengths.

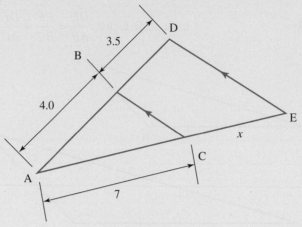

All measurements in metres

THINK

1. Identify the two separate triangles and include measurements.

2. Determine the linear scale factor, k, by calculating the ratio of the corresponding side lengths.

3. Multiplying the side length of 7 metres in the blue triangle by the linear scale factor will give the length of the corresponding side in the pink triangle, that is $(7 + x)$. Write down this mathematical statement.

4. Substitute the value of k into this equation and solve for x.

5. Write the answer.

WRITE

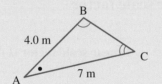

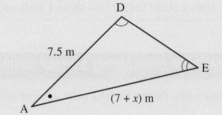

$k = \dfrac{7.5}{4} = \dfrac{15}{8} \text{ (or } 1.875)$

$7 \times k = 7 + x$

$7 \times 1.875 = 7 + x$
$13.125 = 7 + x$
$x = 13.125 - 7$
$x = 6.125$

The length of the unknown side is 6.125 metres.

Note: The above worked example could have been solved using ratios by solving $\dfrac{7+x}{7} = \dfrac{7.5}{4}$, or equivalent.

4.3.2 Applications of similar triangles

Sometimes it is not possible to measure heights of objects such as trees or tall buildings. It is, however, often possible to measure the length of a shadow cast by these objects. If we compare the length of these shadows with the length of a shadow cast by an object of known height under the same conditions, it is possible to calculate heights which are difficult to measure.

Consider the situation shown in the diagram.

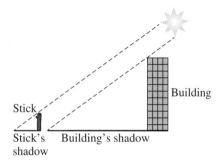

We need to determine the height of the building. Take a stick (or a person) whose height we know (or can measure) and place it, vertically, in the sunshine, near the building. The rays from the sun are parallel, so we have two **similar** right-angled triangles. (The building and the stick are at right angles to the ground.) We can measure both shadows.

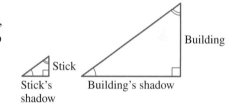

The larger right-angled triangle formed by the building and its shadow is some **scale factor** of the smaller right-angled triangle formed by the stick and the stick's shadow.

$$\text{Scale factor} = \frac{\text{length of building shadow}}{\text{length of stick shadow}}$$

We can then apply the same scale factor to the height of the stick to determine the height of the building. So,

$$\text{Height of building} = \text{height of stick} \times \text{scale factor}$$

WORKED EXAMPLE 6 Calculating the height of a tree using a shadow

At the same time as a tree cast a shadow of 14 m, a 168-cm-tall girl cast a shadow of 140 cm. Calculate the height of the tree. Give the answer to 1 decimal place.

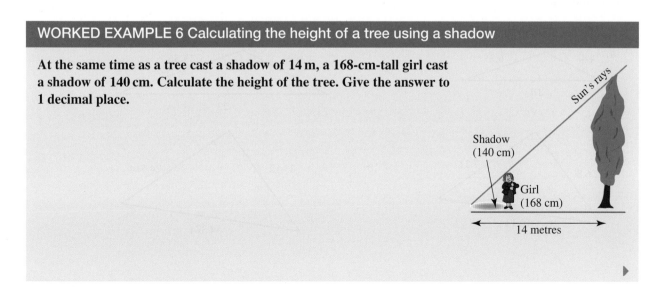

THINK	WRITE
1. Identify the two similar triangles and draw them separately. (We assume that both the girl and the tree are perpendicular to the ground.)	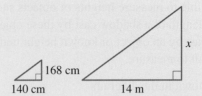 168 cm / 140 cm / 14 m / x
2. Identify the side of the triangle whose length is required.	
3. Calculate the scale factor. *Note:* Measurements must be in the same units.	Scale factor $= \dfrac{\text{length of tree shadow}}{\text{length of girl's shadow}}$ $= \dfrac{14\,\text{m}}{1.4\,\text{m}}$ $= 10$
4. Apply the scale factor to the girl's height.	Height of tree = girl's height × scale factor $= 1.68\,\text{m} \times 10$ $= 16.8\,\text{m}$
5. Write the final answer, specifying units.	The height of the tree is 16.8 m.

Exercise 4.3 Linear scale factors

learn on

4.3 Exercise	4.3 Exam questions on

Simple familiar	Complex familiar	Complex unfamiliar
1, 2, 3, 4, 5, 6, 7, 8, 9, 10	11, 12, 13, 14, 15	16, 17, 18

Simple familiar

1. **WE4** Calculate the linear scale factors for the pairs of similar triangles shown.

   a.
   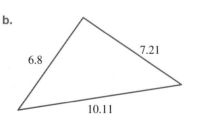
   4.4   6.3
   10

   9.68   13.86
   22

   b.
   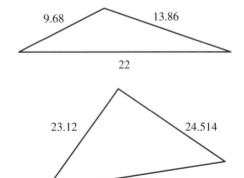
   6.8   7.21
   10.11

   23.12   24.514
   34.374

c.

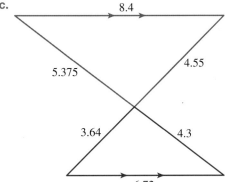

d.

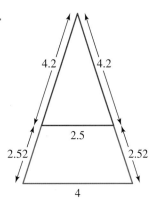

2. Calculate the linear scale factors for the pairs of similar shapes shown.

a.

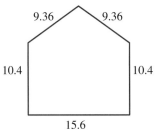

b.

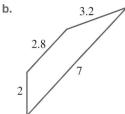

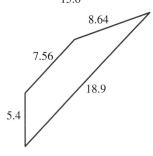

c.

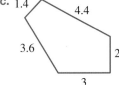

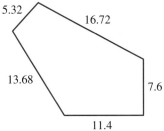

d.

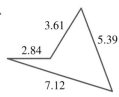

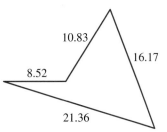

3. Calculate the linear scale factors for the following ratios of corresponding side lengths.

    a. 3 : 2   **b.** 12 : 5   **c.** 3 : 4   **d.** 85 : 68

4. Calculate the missing values for the following.

    a. $\dfrac{3}{\square} = \dfrac{\square}{12} = 6$ **b.** $\dfrac{5}{\square} = \dfrac{44}{11} = \square$ **c.** $\dfrac{\square}{7} = \dfrac{81}{9} = \square$ **d.** $\dfrac{2}{\square} = \dfrac{\square}{2} = 0.625$

5. Calculate the unknown side lengths in the pairs of similar shapes shown.

a.

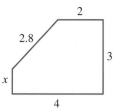

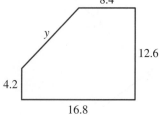

b.

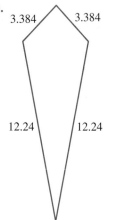

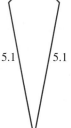

c.

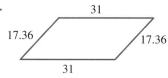

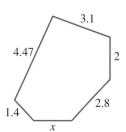

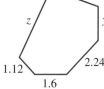

d.

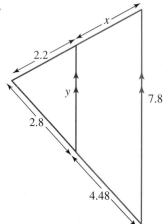

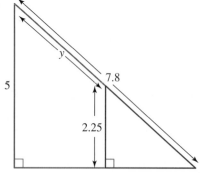

6. **WE5** Calculate the unknown side lengths in the diagrams shown.

a.

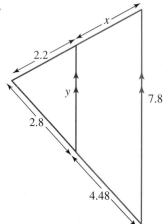

b.

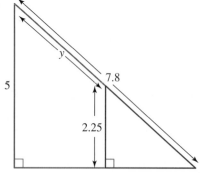

7. Calculate the length of BC in the diagram shown.

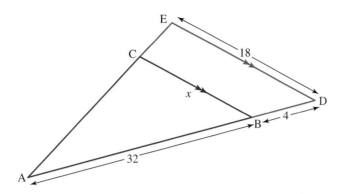

8. **WE6** At the same time as a building casts a shadow of 14.3 metres, a 2-metre stick casts a shadow of 5.3 metres. Calculate the height of the building.

9. The following diagrams represent the measurements taken from shadows of sticks and objects. Use the figures to determine the heights of the objects.

a.

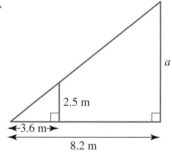

b.

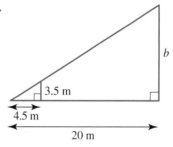

c.

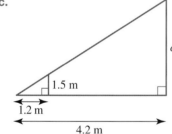

d.

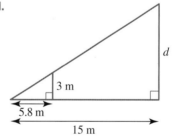

10. A boulder on the shoreline cast a shadow of 15.8 m on the beach at the same time as a 1.5-metre stick cast a shadow of 3.7 m. Determine the height of the boulder.

Complex familiar

11. The size of a flag should be proportional to the height of the flagpole. The guideline for the size of a flag is that the length and width of the flag should be 30% and 15% of the flagpole height, respectively.
If the flagpole shown is 8 m tall, calculate the dimensions of a flag appropriate for the pole. (Give your answers to the nearest cm.)

12. The side of a house casts a shadow that is 8.4 m long on horizontal ground.

  a. At the same time, an 800-mm vertical garden stake has a shadow that is 1.4 m long. Calculate the height of the house.

  b. When the house has a shadow that is 10 m long, calculate the length of the garden stake's shadow.

13. Calculate the value of x in the diagram shown.

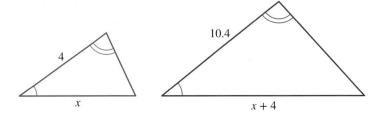

14. Calculate the value of x for the following similar shapes. Give your answer correct to 1 decimal place.

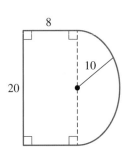
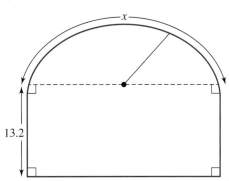

15. Calculate the value of x in the diagram.

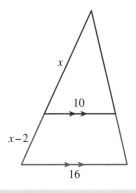

Complex unfamiliar

16. To calculate the distance across a ravine, a surveyor took a direct line of sight from point B to a fixed point A on the other side and then measured out a perpendicular distance of 18 m. From that point the surveyor measured out a smaller similar triangle as shown in the diagram.
Calculate the distance across the ravine along the line AB.

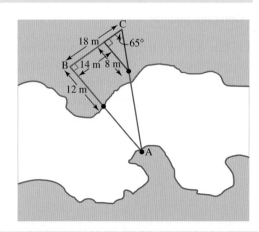

17. A section of a bridge is shown in the diagram. Calculate how high point B is above the roadway of the bridge.

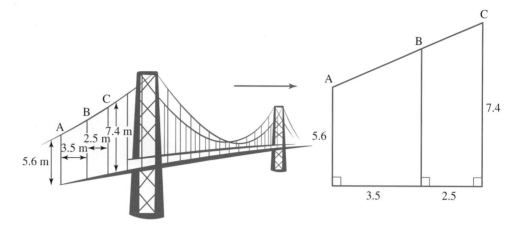

18. In a game of billiards, a ball travels in a straight line from a point one-third of the distance from the bottom of the right side and rebounds from a point three-eighths of the distance along the bottom side. The angles between the bottom side and the ball's path before and after it rebounds are equal.

a. Calculate the perpendicular distance, correct to 2 decimal places, from the bottom side after the ball has travelled a distance of 0.8 m parallel with the bottom side after rebounding.

b. If the ball has been struck with sufficient force, determine at what point on an edge of the table it will next touch. Give your answer correct to 2 decimal places.

Fully worked solutions for this chapter are available online.

LESSON
4.4 Scale drawings — maps and plans

SYLLABUS LINKS

• Determine measurements from scale drawings (e.g. maps and building plans) to solve problems.

Source: General Mathematics Senior Syllabus 2024 © State of Queensland (QCAA) 2024; licensed under CC BY 4.0.

4.4.1 Scales on maps and plans

Scales are used to draw maps and represent plans when it is not feasible to construct full-size models.

A **scale** is a ratio of the length on a drawing to the actual length.

Scale

Scale = length of drawing : actual length

The symbol : is read as 'to'.

Scales are usually written with no units. If a scale is given in two different units, the larger unit has to be converted into the smaller unit.

Sales are represented in different ways.

The following statements:

1 : 100

1 cm ⇔ 1 m

0	1	2	3	4	5 m

all represent the same scale — that is, 1 cm on the plan represents an actual distance of 1 metre.

The scale factor is expressed as a ratio.

Scale factor

Scale factor is a ratio of the same units to enlarge or reduce the size of any shape.

For example, a scale factor of $\frac{1}{2}$ or a scale (ratio) of 1 : 2 means that 1 unit on the drawing represents 2 units in actual size.

When the scale factor is a number greater than 1, the plan or drawing represents an enlargement of the original; a scale factor smaller than 1 represents a reduction of the original.

eles-6389

WORKED EXAMPLE 7 Calculating the scale of a drawing

Calculate the scale of a drawing where 2 cm on the diagram represents 1 km in reality.

THINK	WRITE
1. Convert the larger unit (km) to the smaller unit (cm).	$1 \text{ km} = 100\,000 \text{ cm}$
2. Now that both values are in the same unit, write the scale of the drawing with no units.	$2 : 100\,000$
3. Simplify the scale by dividing both sides of the scale by the highest common factor.	$\frac{2}{2} : \frac{100\,000}{2}$
Divide both sides of the scale by 2, as 2 is the highest common factor of 2 and 100 000.	$= 1 : 50\,000$
4. Write the answer.	$1 : 50\,000$

4.4.2 Calculating dimensions

The scale factor must always be included when a diagram is drawn. It is used to calculate the dimensions needed. To calculate the actual dimensions, measure the dimensions on the diagram and then divide them by the scale factor.

> ### Calculating scale factor
>
> $$\text{Scale factor} = \frac{\text{dimension on the drawing}}{\text{actual dimension}}$$
>
> **We can rearrange the scale factor formula to determine the actual dimension.**
>
> ### Calculating the actual dimensions
>
> $$\text{Actual dimension} = \frac{\text{dimension on the drawing}}{\text{scale factor}}$$

eles-6390

WORKED EXAMPLE 8 Using scale to determine actual dimensions

A scale of 1 : 200 was used for the diagram of the house shown. Calculate both the length and the width of the master bedroom, given the dimensions on the plan are length = 2.5 cm and width = 1.5 cm.

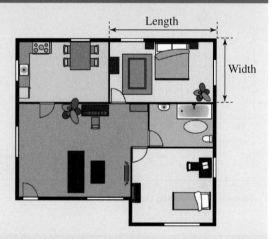

THINK	WRITE
1. State the length and the width of the room in the diagram.	Length = 2.5 cm Width = 1.5 cm
2. Write the scale of the drawing and state the scale factor.	1 : 200 This means that every 1 cm on the drawing represents 200 cm of the actual dimension. The scale factor is $\dfrac{1}{200}$.
3. Divide both dimensions by the scale factor.	Length of the bedroom $= 2.5 \div \dfrac{1}{200}$. $\qquad = 2.5 \times 200$ $\qquad = 500$ cm Width of the bedroom $= 1.5 \div \dfrac{1}{200}$ $\qquad = 1.5 \times 200$ $\qquad = 300$ cm
4. Write the answer in reasonable units.	The length of the bedroom is 5 metres and the width is 3 metres.

4.4.3 Maps and scales

Maps are always drawn using a reducing scale factor, that is at a much smaller scale. All maps have the scale written or drawn on the map.

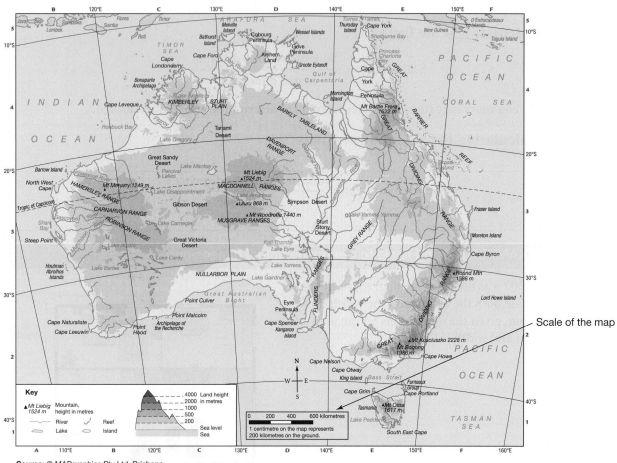

Source: © MAPgraphics Pty Ltd, Brisbane

WORKED EXAMPLE 9 Determining the scale of a map as a ratio

Determine the scale of the map as a ratio using the information in the diagram, where the length of each partition is 1 cm.

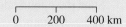

THINK	WRITE
1. State the length of each partition in the diagram.	Each partition is 1 cm.
2. State the length that each partition represents in reality.	Each partition represents 500 km.
3. Write the scale as a ratio.	1 cm : 200 km
4. Convert the larger unit into the smaller unit.	The larger unit is km. This has to be converted into cm. 200 km = 20 000 000 cm
5. Write the scale as a ratio in the same unit.	1 cm : 200 km = 1 cm : 20 000 000 cm
6. Write the scale of the map in ratio form.	The scale is 1 : 20 000 000.

4.4.4 Maps and distances

Both actual distances and distances on the map can be calculated if the scale of the map is known.

> ### Calculating actual distance from a map
>
> $$\text{Actual distance} = \frac{\text{length on the map}}{\text{scale factor}}$$

The dimensions on the map can be calculated by *multiplying* the actual dimension by the scale factor.

eles-6391

WORKED EXAMPLE 10 Calculating actual distance and distance on a map

The scale of an Australian map is 1 : 40 000 000.
a. Calculate the actual distance if the distance on the map is 3 cm.
b. Calculate the distance on the map if the actual distance is 2500 km.

THINK	WRITE
a. 1. Write the scale.	**a.** Distance on the map : actual distance $= 1 : 40\,000\,000$
2. Set up the ratios for map ratio and actual ratio.	The map ratio is $1 : 40\,000\,000$. The actual ratio is $3 : x$. (When measuring on a map in cm, this means $3\,\text{cm} : x\,\text{cm}$.)
3. Construct equivalent fractions.	$\dfrac{x}{3} = \dfrac{40\,000\,000}{1}$ $x = 3 \times 40\,000\,000$ $ = 120\,000\,000$ The distance on the map is $120\,000\,000$ cm.
4. Convert the answer into km.	$120\,000\,000\,\text{cm} = 1200\,\text{km}$
5. Write the answer.	The actual distance represented by 3 cm on the map is 1200 km.
b. 1. Write the scale.	**b.** Distance on the map: actual distance $= 1 : 40\,000\,000$
2. The actual distance is 2500 km. Convert this to cm. (When reading a map, you can choose to measure in cm or mm.)	$2500\,\text{km} = 250\,000\,000\,\text{cm}$
3. Construct equivalent fractions.	$1 : 40\,000\,000$ $x : 250\,000\,000$ $\dfrac{x}{250\,000\,000} = \dfrac{1}{40\,000\,000}$ $x = \dfrac{250\,000\,000}{40\,000\,000}$ $ = 6.25$
4. Write the answer.	The distance on the map is 6.25 cm if the actual distance is 2500 km.

4.4.5 Building plans

Detailed building plans are necessary so that builders and other tradespersons know exactly what is required to complete a project. By the time it is completed, any building will have incorporated information drawn from many facets of construction. Drawings for a domestic structure will include a survey plan, a site plan, floor plans and elevations.

Survey plan

A **survey plan** shows all boundaries of the block of land and includes the position of roadways and nearby lots, as shown in the figure.

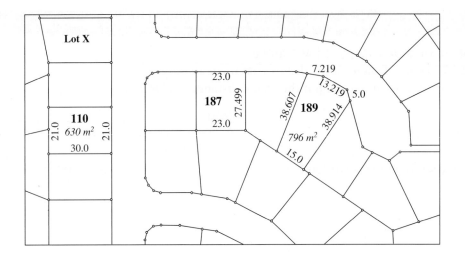

Site plan

A **site plan** shows the boundaries of the lot that is to be built on, and where the structure is to be situated on this lot. It may also show contour lines, which demonstrate the slope of the land. Contour lines are typically shown as height of the land above sea level.

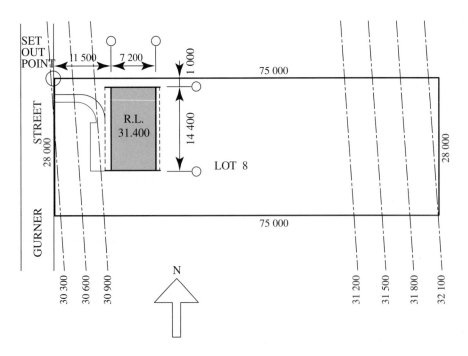

Reproduced by courtesy of Standards Australia

Floor plan

A **floor plan** shows the exact dimensions of the building, including the dimensions and names of all rooms, the size and position of doors and windows, the direction in which doors open, the thickness of walls and the location of stairs.

In simple constructions, the roof plan and the locations of all electrical and plumbing fittings are superimposed on the floor plan. For more complex constructions, plans for these services are made separately. A floor plan without electrical or plumbing fittings is shown in the following figure. If units are not indicated on a plan, it is assumed that dimensions are in millimetres.

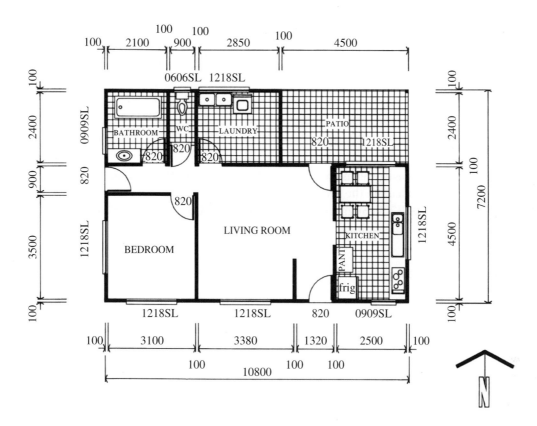

Remember

1. Scales can be represented in different ways.
2. If no units are indicated on a scale representation, any unit can be inserted, so long as the same unit is used for both the length on the drawing or map and the actual distance or length.
3. The scale factor compares the plan length with the actual length; that is, scale factor = plan length: actual length.
4. A scale factor greater than 1 represents an enlargement, and a scale factor smaller than 1 represents a reduction in size.
5. In converting from a plan length to an actual length, multiply by the scale factor. Divide by the scale factor when progressing from the actual length to the plan length.

WORKED EXAMPLE 11 Determining measurements from a survey plan

Refer to Lot X on the survey plan shown.

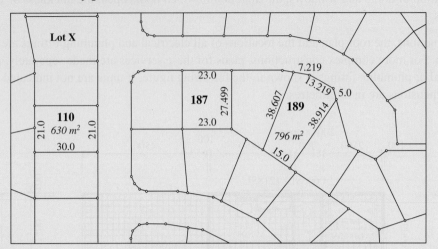

a. **State the shape of Lot X.**
b. **The two parallel sides measure 30 m and 33 m, and the front and back boundaries measure 21 m and 23 m respectively. Calculate the area of Lot X.**
c. **Lot 110 is for sale for $320 700. Suggest a sale price for Lot X by comparing its area with the area of Lot 110. Evaluate the reasonableness of your solution.**

THINK	WRITE
a. Lot X is 4-sided with one pair of parallel sides.	**a.** Lot X has the shape of a trapezium.
b. 1. Recall the formula for the area of a trapezium.	**b.** Area $= \dfrac{1}{2}(a+b)h$
2. Substitute for the variables.	$= \dfrac{1}{2}(30+33) \times 21 \, \text{m}^2$
3. Calculate the area.	$= 661.5 \, \text{m}^2$
c. 1. Calculate the price per m^2 based on Lot 110 information.	**c.** Price per m$^2 = \dfrac{\$320\,700}{630 \, \text{m}^2}$ $= \$509$ per m^2
2. Multiply by the number of square metres in Lot X.	Price $= \$509 \times 661.5$ $= \$336\,704$
3. Write the answer.	A reasonable price for Lot X would be $336 704. This is assuming that area is the most significant factor in determining the value of the land.

 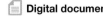 Resources

---

📄 **Digital documents** SkillSHEET Conversion of length units (doc-29492)

SkillSHEET Reading scales (How much is each interval worth?) (doc-29506)

SpreadSHEET Map scales 1 (doc-29507)

SpreadSHEET Map scales 2 (doc-29508)

Exercise 4.4 Scale drawings — maps and plans

4.4 Exercise	4.4 Exam questions on

These questions are even better in jacPLUS!
• Receive immediate feedback
• Access sample responses
• Track results and progress

Find all this and MORE in jacPLUS ▶

Simple familiar	Complex familiar	Complex unfamiliar
1, 2, 3, 4, 5, 6, 7, 8, 9, 10, 11, 12, 13, 14, 15, 16, 17, 18, 19	20, 21	N/A

Simple familiar

1. Convert to metres.

   a. 8215 mm   b. 350 cm   c. 89 km

   d. 26 mm   e. 4 cm   f. 6.4 km

2. Classify the following as enlargements or reductions.

   a. 1 : 200   b. 1 mm ⇔ 1 m

   c.
   0   1   2   3   4   5 mm
                 d.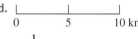
   0   5   10 km

   e. 10 : 1   f. $\dfrac{1}{10\,000}$

3. Express each of the following in the form of a simplified ratio.

   a. 2 cm ⇔ 100 m

   b. ⌞___⌞___⌞___⌞___⌞___⌟
   0   6   12 km

4. **WE7** Calculate the scale of a drawing where 100 mm on the diagram represents 2 m in reality.

5. If the scale of a diagram is 5 cm : 100 km, determine the scale using the same unit.

6. **WE8** A scale of 1 : 300 was used for the diagram of the house shown.

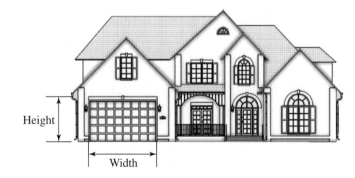

Height
Width

   Calculate both the height and width, in metres, of the garage door, given the dimensions on the plan are height = 0.8 cm and width = 2 cm.

7. **MC** A 15-m-long fence is represented by a straight line 4.5 cm long on a drawing. Calculate the scale of the drawing.

   A. 1000 : 3   B. 4.5 : 1.5

   C. 45 : 150   D. 3 : 1000

8. The floor plan shown is planned to be drawn at a scale of 1 : 200. The actual dimensions of the house are shown on the diagram.
The floor plan is not yet drawn to scale. Calculate the lengths of the dimensions shown if the floor plan was drawn to scale.

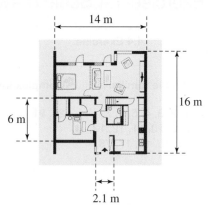

9. **WE9** Determine the scale of the map as a ratio using the information in the diagram.

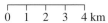

10. Determine the scale of the map as a ratio using the information in the diagram.

11. For each of the map scales shown, state the scale and determine the actual distance for the map distance given.

a. 5.1 cm

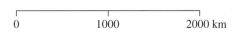

b. 27 mm

c. 38 mm

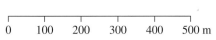

d. 9.6 cm

12. **WE10** A street map has a scale of 1 : 500 000.

a. Calculate the actual distance if the distance on the map is 2 cm.
b. Calculate the distance on the map if the actual distance is 10 km.

13. A map has a scale of 1 : 1 000 000.

a. Calculate the actual distance if the distance on the map is 1.2 cm.
b. Calculate the distance on the map if the actual distance is 160 km.

14. **MC** The distance between Perth and Adelaide is 2693 km. If this distance was drawn on a scale of 1 : 25 000 000, the distance on the map would be:

A. 10 772 mm
B. 10 772 cm
C. 0.107 72 cm
D. 10.772 cm

15. A map of Australia has a scale of 10 cm : 5000 km.

a. Write this scale in the same unit.
b. Calculate the distances, in cm, correct to 1 decimal place, on the map between:

   i. Canberra and Sydney with an actual distance of 290 km
   ii. Sydney and Brisbane with an actual distance of 925 km
   iii. Brisbane and Darwin with an actual distance of 3423 km
   iv. Darwin and Perth with an actual distance of 4042 km
   v. Perth and Adelaide with an actual distance of 2693 km
   vi. Adelaide and Melbourne with an actual distance of 727 km

Questions **16** and **17** relate to the survey plan shown.

16. `WE11` For the survey plan shown:

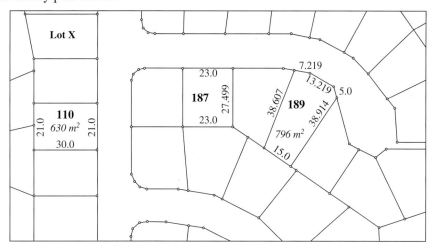

a. calculate the area in square metres of Lot 110
b. calculate the length and breadth of Lot 110
c. calculate the scale that has been used to draw the survey plan
d. redraw Lot 110 using a 1 : 500 scale
e. the area of Lot 187 is not shown. Calculate this area.

17. Lot 110 is for sale at $320 700, and Lot 189 is for sale at $399 500.

a. Identify which lot represents the better value per square metre.
b. Determine the features of a block of land that might attract a purchaser even though its dollar value per square metre may be higher than surrounding blocks. (Comparing the positions of Lots 110 and 189 can assist in your answer, but include as many other features as possible.)

18. The site plan shows Lot 8 on Gurner Street. All dimensions given are in millimetres.

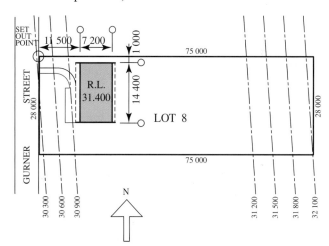

Reproduced by courtesy of Standards Australia

a. Calculate the area of Lot 8 in square metres and perches (1 perch = 25.3 m^2).
b. The shaded sketch shows the area of the dwelling proposed to be erected on this lot. Calculate the area of the proposed dwelling.
c. Calculate the scale that has been used to produce this diagram. (*Note:* This scale may not be a simple ratio.)
d. The dotted lines are contour lines (lines of height). All points along the 31 800 line are 31 800 mm above sea level.

   i. Determine whether the block is rising or falling as you walk from the Gurner Street entrance to the rear of the block.
   ii. Calculate the angle of rise or fall from the front to the rear.

19. The floor plan shows a plan of a one-bedroom dwelling. All dimensions have been given in mm. (*Note*: This scale may not be a simple ratio.)

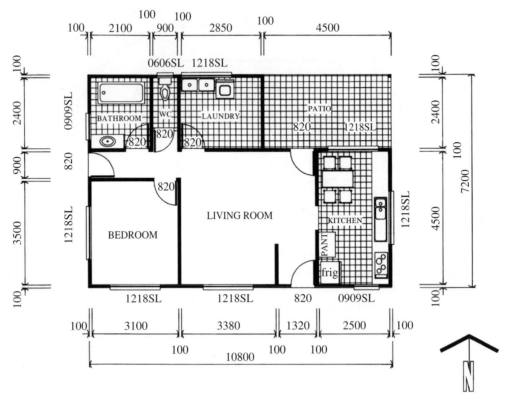

a. Calculate the area of this dwelling in square metres. (Include the patio.)
b. The patio is to be tiled using tiles costing $35 per square metre. Calculate the cost. Include an extra 5% for cutting.
c. Calculate the area of the bedroom, kitchen, laundry and bathroom.

Complex familiar

20. The figure shows a house on a block, drawn using a 1 : 250 scale.

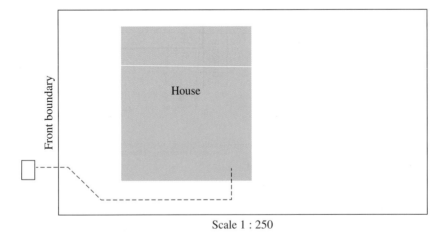

Scale 1 : 250

a. Calculate the distance between the house and the front boundary.
b. The owner intends to fence this property at a cost of $32.00 per metre, plus $250 for gates. Calculate how much it would cost to fence and install gates on this property.
c. Calculate the length of the sewer line (dashed).

21. The map shows Mooloolaba Boat Harbour.

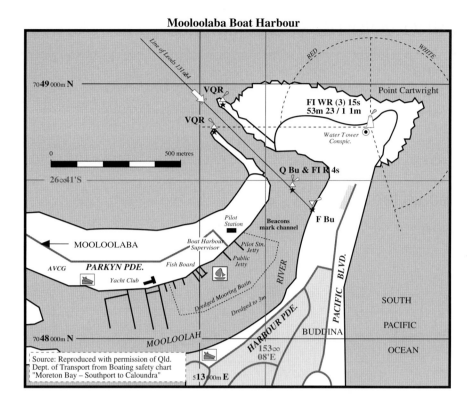

Mooloolaba Boat Harbour

a. Calculate the width of the entrance to the harbour.
b. Calculate the distance, as the seagull flies, from the yacht club to the water tower at Point Cartwright.
c. Calculate the area (in km^2) covered by this map.

Fully worked solutions for this chapter are available online.

LESSON
4.5 Area and volume scale factors

SYLLABUS LINKS

- Determine a scale factor and use it to solve scaling problems, e.g. calculating lengths and areas of similar figures, and calculating surface areas, volumes and capacities of similar solids.

Source: General Mathematics Senior Syllabus 2024 © State of Queensland (QCAA) 2024; licensed under CC BY 4.0.

4.5.1 Area scale factor

Consider three squares with side lengths of 1, 2 and 3 cm. Their areas are $1 \, \text{cm}^2$, $4 \, \text{cm}^2$ and $9 \, \text{cm}^2$ respectively.

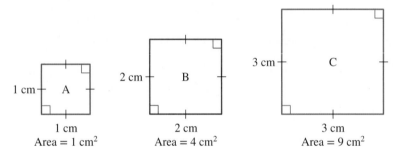

The linear scale factor between square A and square B is 2, and the linear scale factor between square A and square C is 3. When we look at the ratio of the areas of the squares, we get 4 : 1 for squares A and B, and 9 : 1 for squares A and C. In both cases, the **area scale factor** is equal to the linear scale factor raised to the power of two.

	Linear scale factor	Area scale factor
B : A	2	4
C : A	3	9

Comparing squares B and C, the ratio of the side lengths is 2 : 3, resulting in a linear scale factor of $\frac{3}{2}$ or 1.5. From the ratio of the areas we get 4 : 9, which once again indicates an area scale factor $\frac{9}{4} = 2.25$, that is, the linear scale factor to the power of two.

> **Area scale factor**
>
> **If the linear scale factor for two similar objects is k, the area scale factor will be k^2.**

This also applies to surface area for three-dimensional objects. If the dimensions of an object are enlarged by a factor of x, then the surface area will increase by a factor of x^2.

eles-3060

WORKED EXAMPLE 12 Calculating the area scale factor for a pair of similar triangles

Calculate the area scale factor for the pair of similar triangles shown in the diagram.

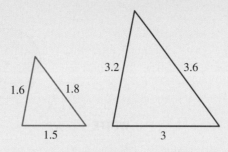

THINK

1. Calculate the linear scale factor, that is the ratio of the corresponding side lengths.

2. Square the linear scale factor to obtain the area scale factor.

3. Write the answer in a sentence.

WRITE

$$k = \frac{3}{1.5} = \frac{3.6}{1.8} = \frac{3.2}{1.6} = 2$$

$$k^2 = 2^2 = 4$$

The area scale factor is 4.

4.5.2 Volume scale factor

Three-dimensional objects of the same shape are similar when the ratios of their corresponding dimensions are equal.

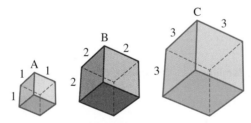

When we compare the volumes of three similar cubes, we can see that if the linear scale factor is k, the **volume scale factor** will be k^3.

	Cube B : Cube A	Scale factor
Linear	2 : 1	2
Area	4 : 1	$2^2 = 4$
Volume	8 : 1	$2^3 = 8$

	Cube C : Cube B	Scale factor
Linear	3 : 2	$\frac{3}{2}$
Area	9 : 4	$\left(\frac{3}{2}\right)^2 = \frac{9}{4}$
Volume	27 : 8	$\left(\frac{3}{2}\right)^3 = \frac{27}{8}$

	Cube C : Cube A	Scale factor
Linear	3 : 1	3
Area	9 : 1	$3^2 = 9$
Volume	27 : 1	$3^3 = 27$

Volume scale factor

If the linear scale factor for two similar objects is k, the volume scale factor will be k^3.

eles-3061

WORKED EXAMPLE 13 Calculating the volume scale factor for a pair of spheres

Calculate the volume scale factor for the pair of spheres shown in the diagram.

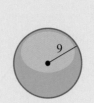
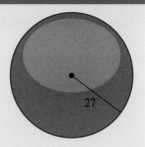

THINK	WRITE
1. Calculate the linear scale factor, that is the ratio of the corresponding dimensions.	$k = \dfrac{27}{9} = 3$
2. Cube the linear scale factor to obtain the volume scale factor.	$k^3 = 3^3 = 27$
3. Write the answer.	The volume scale factor is 27.

 Resources

Interactivities Area scale factor (int-6478)

Volume scale factor (int-6479)

Exercise 4.5 Area and volume scale factors

learn

4.5 Exercise	4.5 Exam questions on

Simple familiar	Complex familiar	Complex unfamiliar
1, 2, 3, 4, 5, 6, 7, 8, 9, 10, 11	12, 13, 14, 15	16

These questions are even better in jacPLUS!
- Receive immediate feedback
- Access sample responses
- Track results and progress

Find all this and MORE in jacPLUS

Simple familiar

1. **WE12** Calculate the area scale factor for each pair of similar triangles shown.

a.
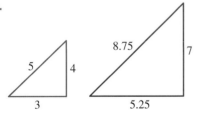

b.
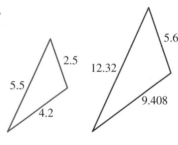

c.
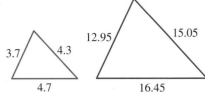

2. Calculate the area scale factor for each pair of similar objects shown.

a.

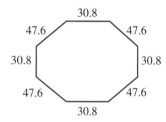

b.

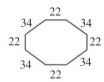

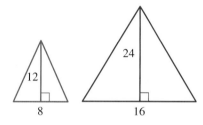

3. Consider the two similar triangles shown in the diagram.

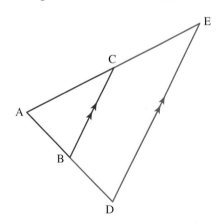

    a. Calculate the areas of the two similar triangles.

    b. State how many times larger in area the bigger triangle is.

    c. Calculate the linear scale factor.

    d. Calculate the area scale factor.

4. A rectangular swimming pool is shown on the plans for a building development with a length of 6 cm and a width of 2.5 cm. If the scale on the plans is 1 : 250:

    a. calculate the area scale factor

                                                       b. calculate the surface area of the swimming pool.

5. The area of the triangle ADE in the diagram is 100 cm², and the ratio of *DE* : *BC* is 2:1.

Calculate the area of triangle ABC.

6. The floor of a square room has an area of 12 m². Calculate the area that the room takes up in a diagram with a scale of 1 : 250.

7. **WE13** Calculate the volume scale factor for the pair of spheres shown.

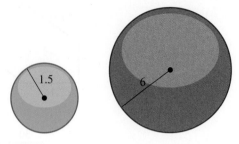

8. Calculate the volume scale factor for each pair of similar objects shown.

a.

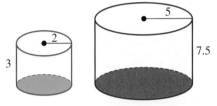

b.

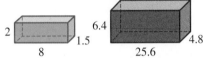

c.

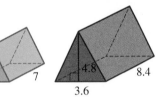

9. An architect makes a small scale model of a house out of balsa wood with the dimensions shown in the diagram.

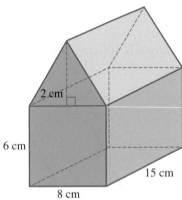

a. If the actual length of the building is 26.25 m, calculate the scale of the model.
b. Calculate the ratio of the volume of the building to the volume of the model.

10. Two similar cylinders have volumes of $400 \, \text{cm}^3$ and $50 \, \text{cm}^3$ respectively.

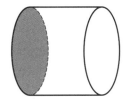

    a. Calculate the linear scale factor.

    b. If the length of the larger cylinder is 8 cm, calculate the length of the smaller one.

11. If a cube has a volume of $25 \, \text{cm}^3$ and is then enlarged by a linear scale factor of 2.5, calculate the new volume.

Complex familiar

12. A hexagon is made up of six equilateral triangles of side length 2 cm. If a similar hexagon has an area of $24\sqrt{3} \, \text{cm}^2$, calculate the linear scale factor.

13. Calculate the linear scale factor between two similar drink bottles if one has a volume of 600 mL and the other has a volume of 1.25 L.

14. If an area of $712 \, \text{m}^2$ is represented on a scale drawing by an area of $44.5 \, \text{cm}^2$, calculate the actual length that a distance of 5.3 cm on the drawing represents.

15. A model car is an exact replica of the real thing reduced by a factor of 12.

    a. If the actual surface area of the car that is spray painted is $4.32 \, \text{m}^2$, calculate the equivalent painted area on the model car.

    b. If the actual storage capacity of the car is $1.78 \, \text{m}^3$, calculate the equivalent volume for the model car.

Complex unfamiliar

16. A company sells canned fish in two sizes of similar cylindrical cans. For each size, the height is four-fifths of the diameter. If the dimensions of the larger cans are 1.5 times those of the smaller cans, derive an expression for calculating the volume of the larger cans of fish in terms of the diameter of the smaller cans.

Fully worked solutions for this chapter are available online.

LESSON
4.6 Review

4.6.1 Summary

4.6 Exercise

learn**on**

4.6 Exercise	**4.6 Exam questions** on

Simple familiar	Complex familiar	Complex unfamiliar
1, 2, 3, 4, 5, 6, 7, 8, 9, 10, 11	12, 13, 14, 15, 16, 17, 18	19, 20, 21, 22

Simple familiar

1. **MC** If a map has a scale factor of 1 : 50 000, an actual distance of 11 km would have a length on the map of:

   A. 21 cm   **B.** 22 cm   **C.** 21.5 cm   **D.** 11 cm

2. **MC** If the areas of two similar objects are 12 cm^2 and 192 cm^2 respectively, the linear scale factor will be:

   A. 3   **B.** 4   **C.** 16   **D.** 2

3. **MC** If the volumes of two similar solids are 16 cm^3 and 128 cm^3 respectively, the area scale factor will be:

   A. 2   **B.** 3   **C.** 4   **D.** 8

4. **MC** A tree casts a shadow that is 5.6 m long. At the same time a 5-m light pole casts a shadow that is 3.5 m long. The height of the tree is:

   A. 4.4 m   **B.** 7.5 m   **C.** 8.0 m   **D.** 6.0 m

5. **MC** The value of x in the diagram is:

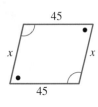
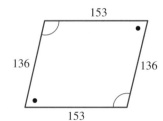

   A. 38   **B.** 41   **C.** 43   **D.** 40

6. **MC** If the scale factor of the volumes of two similar cuboids is 64 and the volume of the larger one is 1728 cm^3, the surface area of the smaller cuboid is:

   A. 64 cm^2 **B.** 27 cm^2 **C.** 54 cm^2 **D.** 9 cm^2

7. **MC** The triangle that is similar to ΔABC is:

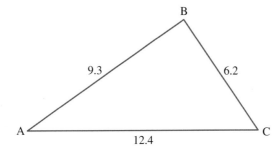

A.

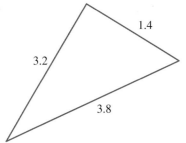

B.

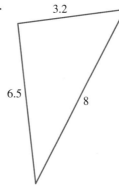

C.

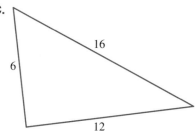

D.

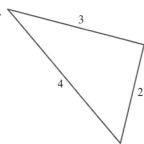

8. **MC** An enlargement diagram of a very small object is drawn at a scale of 18 : 1. If the diagram has an area of 162 cm², the equivalent area of the actual object is closest to:

   A. 0.5 cm²   **B.** 0.4 cm²   **C.** 9 cm²   **D.** 0.6 cm²

9. **MC** The plans of a house show the side of a building as 12.5 cm long. If the actual building is 15 m long, the scale of the plan is:

   A. 1 : 250   **B.** 1 : 150   **C.** 1 : 220   **D.** 1 : 120

10. **MC** The plans for a building show a concrete slab covering an area of 12.5 cm × 8.4 cm to a depth of 0.25 cm. If the plans are drawn to a scale of 1 : 225, the actual volume of concrete is closest to:

    A. 26.25 m³   **B.** 262.5 m³   **C.** 5906.25 m³   **D.** 299 m³

11. Calculate the values of the pronumerals in the following diagrams.

a.

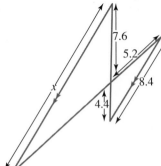

b.

c.

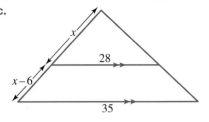

Complex familiar

12. The graphical scale on a map is shown. Determine the scale of the map as a ratio, if each partition measures 2.5 cm.

$$0 \qquad 150 \quad 300 \text{ km}$$

13. Calculate the value of x in the diagram of two similar objects shown.

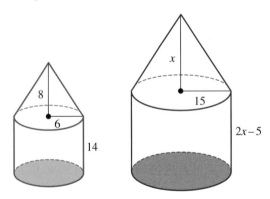

14. The volume of a solid is 1600 cm^3. If the ratio of the corresponding dimensions between this solid and a similar solid is $4:5$, calculate the volume of the similar solid.

15. Before the metric system was introduced, the area of house blocks was measured in perches (1 perch $= 25.3 \text{ m}^2$).

   a. A block of 42 perches is advertised for sale at $405 000. Convert the area to square metres and calculate the price per square metre.
   b. One lot is 850 m^2 and another is 28 perches. Identify which lot is larger.

16. A triangle has side lengths of $32 \text{ mm}, 45 \text{ mm}$ and 58 mm.

   a. Calculate the side lengths of a larger similar triangle using a corresponding side ratio of $2:3$.
   b. Calculate the side lengths of the larger triangle in a drawing with a scale of $5:2$.

17. A farmer divides paddock ABCD into two separate paddocks along the line FE as shown in the diagram. All distances are shown in metres.

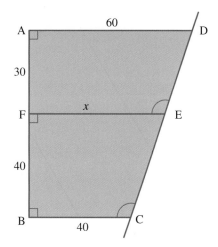

a. Derive the ratio of the corresponding sides of the original paddock, ABCD, with paddock BCEF in its simplest form.

b. Calculate the length of fencing required to separate the two paddocks along the line FE. Give your answer correct to the nearest centimetre.

18. On a map that is drawn to a scale of 1 : 225 000, the distance between two points is 88 cm.

a. Calculate the actual distance between the two points.

A boat sets out to travel from one point to the other, but the navigator makes an error. After travelling 100 km, the crew realise they are directly south of a point that they should have reached after travelling 90 km in a direct line to their destination.

b. If they continue in their current direction, calculate how much further they have to travel to be directly south of the intended destination.

c. Calculate how far away from the intended destination the boat will be when it reaches the point on its course that is directly to the south.

d. When the boat is at the point directly to the south of the intended destination, determine how far away it will be on the map.

Complex unfamiliar

19. The top third of an inverted right cone is removed as shown in the diagram.

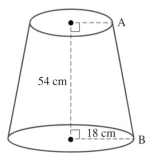

Calculate the distance along the edge of the remaining part of the cone from A to B.

20. A rectangular box with dimensions $94\,\text{cm} \times 31\,\text{cm}$ leans against a wall as shown in the diagram.

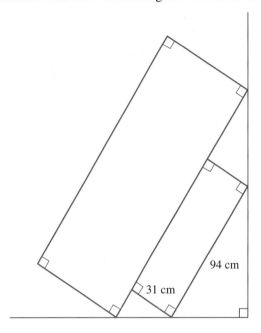

A larger box with corresponding side lengths in a ratio of $4:7$ leans against the first box. If the smaller box touches the floor at a point that is $52\,\text{cm}$ from the base of the wall, determine how far up the wall the larger box reaches.

21. A swimmer is observed in the water from the top of a vertical cliff that is $5\,\text{m}$ tall. A line of sight is taken from a point $2\,\text{m}$ back from the edge of the cliff.
The swimmer, the edge of the cliff and a point $1.8\,\text{m}$ above the cliff are in line.
If the swimmer moves a further $2.45\,\text{m}$ away from the cliff, determine from how far above the cliff the new line of sight should be taken.

22. A caterer sells takeaway coffee in three different cup sizes.
The ratio of corresponding dimensions between the small cup and the large cup is $4:5$.
If the caterer charges \$5.00 for a small cup, \$5.50 for a medium cup and \$6.00 for a large cup, determine which is the better value for the customer.

200 mL   300 mL

Fully worked solutions for this chapter are available online.

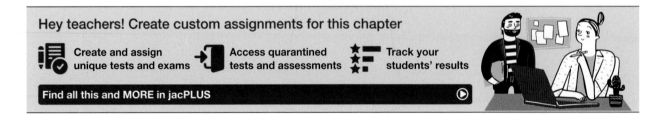

Hey teachers! Create custom assignments for this chapter

Create and assign unique tests and exams

Access quarantined tests and assessments

Track your students' results

Find all this and MORE in jacPLUS

Answers

Chapter 4 Similar figures and scale factors

4.2 Similarity of two-dimensional figures

4.2 Exercise

1. a. $\dfrac{15}{8} = \dfrac{11.25}{6} = \dfrac{7.5}{4} = 1.875$, and all angles are equal.

   b. $\dfrac{41.04}{11.4} = \dfrac{26.1}{7.25} = \dfrac{9}{2.5} = \dfrac{13.68}{3.8} = 3.6$, and all angles are equal.

2. $\dfrac{106.43}{18.35} = \dfrac{24.65}{4.25} = 5.8$, and all angles are equal.

3. a. $\dfrac{4.8}{3.2} = 1.5, \dfrac{2.24}{1.6} = 1.4$; not similar

   b. $\dfrac{10.224}{2.84} = 3.6, \dfrac{2.34}{0.65} = 3.6$; similar

   c. $\dfrac{4.4352}{2.112} = 2.1, \dfrac{2.184}{1.04} = 2.1$; similar

   d. $\dfrac{12.535}{11.5} = 1.09, \dfrac{10.152}{9.4} = 1.08$; not similar

4. a. $\dfrac{35}{25} = 1.4, \dfrac{22.4}{16} = 1.4, \dfrac{12.6}{9} = 1.4$ and all angles are equal; similar

   b. $\dfrac{4.94}{2.6} = 1.9, \dfrac{6.65}{3.5} = 1.9, \dfrac{2.09}{1.1} = 1.9$ and all angles are equal; similar

   c. $\dfrac{13}{8} = 1.625, \dfrac{6.5}{4} = 1.625, \dfrac{3.25}{2} = 1.625$ and all angles are equal; similar

   d. $\dfrac{3.24}{2.7} = 1.2, \dfrac{19.44}{16.2} = 1.2, \dfrac{2.16}{1.8} = 1.2$ and all angles are equal; similar

5. a. $\dfrac{44.275}{7.7} = \dfrac{78.2}{13.6} = 5.75$, SAS

   b. $\dfrac{38.72}{17.6} = \dfrac{16.28}{7.4} = \dfrac{14.85}{6.75} = 2.2$, SSS

6. a. $\dfrac{32.4}{18} = \dfrac{21.6}{12} = 1.8, \dfrac{23.8}{14} = 1.7$; not similar

   b. $\dfrac{54.32}{19.4} = \dfrac{22.96}{8.2} = 2.8$ and all angles are equal; similar

7. a. A and C   b. B and C   c. A and C   d. A and B

8. a. 1.6 : 1   b. 2.26 : 1   c. 3.6 : 1

9. a. $\dfrac{50.84}{8.2} = 6.2$, all side lengths are in proportion and all angles are equal.

   b. $\dfrac{14.34}{12.6} = 1.138$, all side lengths are in proportion and all angles are equal.

10. a. All angles are equal and side lengths are in proportion.

    b. All measurements (radius and circumference) are in proportion.

    c. All angles are equal and side lengths are in proportion.

11. a. 4 : 3   b. 5.5 : 1

12. a. 32   b. 20

13. a. 4.8   b. 7.07

14. $x = 5.6, y = 7$

15. a.

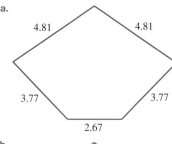

b.

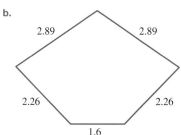

4.3 Linear scale factors

4.3 Exercise

1. a. 2.2   b. 3.4   c. 1.25   d. 1.6

2. a. 5.2   b. 2.7   c. 3.8   d. 3

3. a. 1.5   b. 2.4   c. 0.75   d. 1.25

4. a. $\dfrac{3}{\boxed{0.5}} = \dfrac{\boxed{72}}{12} = 6$

   b. $\dfrac{5}{\boxed{1.25}} = \dfrac{44}{11} = \boxed{4}$

   c. $\dfrac{\boxed{63}}{7} = \dfrac{81}{9} = \boxed{9}$

   d. $\dfrac{2}{\boxed{3.2}} = \dfrac{\boxed{1.25}}{2} = 0.625$

5. a. $x = 1, y = 11.76$ b. 1.41

   c. 2.8   d. $x = 2, y = 1.6, z = 3.576$

6. a. $x = 3.52, y = 3$ b. $y = 4.29$

7. 16

8. 5.4 m

9. a. 5.7 m   b. 15.6 m

   c. 5.3 m   d. 7.8 m

10. 6.4 m

11. 240 cm and 120 cm

12. a. 4.8 m   b. 1.67 m

13. 2.5

14. 51.8

15. 5

16. 24 m

17. 6.65 m

18. a. 0.36 m

   b. 0.67 m from the bottom of the left side

4.4 Scale drawings — maps and plans

4.4 Exercise

1. a. 8.215 m   **b.** 3.5 m   **c.** 89 000 m
   d. 0.026 m   **e.** 0.04 m   **f.** 6400 m

2. a. Reduction   **b.** Reduction   **c.** Enlargement
   d. Reduction   **e.** Enlargement   **f.** Reduction

3. a. 1 : 5000   **b.** 1 : 200 000

4. 1 : 20

5. 1 : 2 000 000

6. Height $=$ 2.4 m
   Width $=$ 6 cm

7. D

8. 7 cm, 8 cm, 3 cm, 1.05 cm

9. 1 : 200 000

10. 1 : 50 000

11. a. Scale 5 : 200 000 000, 2040 km

   b. Scale 1 : 500 000, 13 500 m

   c. Scale 1 : 10 000, 380 m

   d. Scale 1 : 500 000, 48 km

12. a. 10 km   **b.** 2 cm

13. a. 12 km   **b.** 16 cm

14. D

15. a. 1 : 50 000 000

   b. i. 0.6 cm

     ii. 1.9 cm

    iii. 6.8 cm

    iv. 8.1 cm

     v. 5.4 cm

    vi. 1.5 cm

16. a. 630 m^2

   b. 30 m, 21 m

   c. Approx. 1 : 1500

   d. Rectangle 6 cm by 4.2 cm

   e. 632 m^2

17. a. Lot 189

   b. Does it front a main road? Is it low lying? Slope of land, views, aspect.

18. a. 2100 m^2, 83 perches

   b. 104 m^2

   c. Approx. 1 : 800

   d. i. Rising

     ii. 1.4°

19. a. 77.8 m^2

   b. Approx. $400

   c. In order 10.85 m^2, 11.25 m^2, 6.84 m^2, 5.04 m^2

20. a. 4.75 m   **b.** $3034   **c.** 18.75 m

21. a. Approx. 75 m

   b. Approx. 1000 m

   c. Approx. 2 km^2

4.5 Area and volume scale factors

4.5 Exercise

1. a. $\dfrac{49}{16} = 3.0625$ **b.** $\dfrac{3136}{625} = 5.0176$

   c. $\dfrac{49}{4} = 12.25$

2. a. $\dfrac{64}{25} = 2.56$ **b.** $\dfrac{49}{25} = 1.96$

3. a. 48 and 192 square units

   b. 4

   c. 2

   d. 4

4. a. 62 500   **b.** 93.75 m^2

5. 25 cm^2

6. 1.92 cm^2

7. 64

8. a. 15.625   **b.** 32.768   **c.** 1.728

9. a. 1 : 175   **b.** 5 359 375 : 1

10. a. 2   **b.** 4 cm

11. 390.625 cm^3

12. 2

13. 1.28

14. 21.2 m

15. a. 300 cm^2 **b.** 1030 cm^3

16. $V = \dfrac{27\pi D^3}{40}$

4.6 Review

4.6 Exercise

1. B   **2.** B
3. C   **4.** C
5. D   **6.** C
7. D   **8.** A
9. D   **10.** D

11. a. 14.51

   b. $x = 20, y = 33.6$

   c. 8

12. 1 : 6 000 000

13. 20

14. 3125 cm^3

15. a. 1063 m^2, $381/m^2 **b.** 850 m^2 is larger.

16. a. 48 mm, 67.5 mm, 87 mm

   b. 120 mm, 168.75 mm, 217.5 mm

17. a. 7 : 4   **b.** 5143 cm

18. a. 198 km   **b.** 120 km   **c.** 95.90 km   **d.** 42.62 cm

19. 55.32 cm

20. 137.04 cm

21. 1.24 m

22. Large cup

5 Linear and non-linear relationships

LESSON SEQUENCE

Fully worked solutions for this chapter are available online.

EXAM PREPARATION

Access exam-style questions in every lesson, available online.

on Resources

Solutions	Solutions — Chapter 5 (sol-1246)
Exam questions	Exam question booklet — Chapter 5 (eqb-0262)
Digital documents	Learning matrix — Chapter 5 (doc-41478)
	Chapter summary — Chapter 5 (doc-41480)

LESSON
5.1 Overview

5.1.1 Introduction

Formulas are used every day, for example to calculate areas, volumes and lengths. These are the types of formulas that you may be aware of from previous studies. There are many very famous formulas that you may not have heard of directly, but which have had a huge impact on the development of our understanding of the universe.

Here are three of the most famous formulas in history:

1. **Isaac Newton's Law of Universal Gravitation:**

$$F = G\frac{m_1 m_2}{r^2}$$

This formula explains why the planets move the way they do and how gravity works, both on Earth and in the wider universe. This formula was first published in 1687.

2. **The mass–energy equivalence equation:**

$$E = mc^2$$

This formula was developed as part of Einstein's theory of relativity, which looks at the relationship between space and time. This theory was first proposed in 1905. The theory of relativity rocked the world of physics and deepened our knowledge of the universe.

3. **The Pythagorean theorem:**

$$c^2 = a^2 + b^2$$

5.1.2 Syllabus links

Lesson	Lesson title	Syllabus links
5.2	**Substitution**	○ Substitute numerical values into linear and simple non-linear algebraic expressions, and evaluate.
5.3	**Formulas**	○ Find the value of a pronumeral in linear and simple non-linear equations given the values of the other pronumerals, transposing equations where necessary.
		○ Use a spreadsheet or an equivalent technology to construct a table of values from a formula, including two-by-two tables for formulas with two variable quantities.
5.4	**Transposition**	○ Find the value of a pronumeral in linear and simple non-linear equations given the values of the other pronumerals, transposing equations where necessary.

Source: General Mathematics Senior Syllabus 2024 © State of Queensland (QCAA) 2024; licensed under CC BY 4.0.

LESSON
5.2 Substitution

SYLLABUS LINKS

- Substitute numerical values into linear and simple non-linear algebraic expressions and evaluate.

Source: General Mathematics Senior Syllabus 2024 © State of Queensland (QCAA) 2024; licensed under CC BY 4.0.

5.2.1 Identifying linear and non-linear relationships

A **linear relationship** is a relationship between two **variables** that when plotted gives a straight line. Many real-life situations can be described by linear relations, such as water being added to a tank at a constant rate, or the same amount of money being deposited into a bank account in regular intervals.

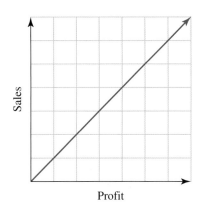

A **non-linear relationship** between two variables when plotted does **not** give a straight line. Real-life situations described by non-linear relationships include the growth of a plant over a number of weeks, and the relationship between the side length of a square and its area.

When written as algebraic relationships, linear equations can be identified if the power of both variables is 1, and non-linear equations can be identified if the power of at least one variable is greater than or less than 1.

> **Identifying linear and non-linear relationships**
>
> **When a linear relationship is expressed as an equation, the highest power of both variables in the equation is 1. Remember x can be written as x^1.**
>
> **For example,**
> $$y = 2x$$
>
> **When a non-linear relationship is expressed as an equation, the power of at least one of the variables is less than or greater than 1.**
>
> **For example,**
> $$y = 4x^2$$

eles-3039

WORKED EXAMPLE 1 Identifying linear equations

Identify which of the following equations are linear and which are not linear.

a. $y = 4x + 1$

b. $b = c^2 - 5c + 6$

c. $y = \sqrt{x}$

d. $m^2 = 6(n - 10)$

e. $d = \dfrac{3t + 8}{7}$

f. $y = 5^x$

THINK	WRITE
a. 1. Identify the variables.	a. y and x
2. Write the power of each variable.	y has a power of 1.
	x has a power of 1.
3. Check if the equation is linear.	Since both variables have a power of 1, this is a linear equation.
b. 1. Identify the two variables.	b. b and c
2. Write the power of each variable.	b has a power of 1.
	c has a power of 2.
3. Check if the equation is linear.	c has a power of 2, so this is not a linear equation.
c. 1. Identify the two variables.	c. y and x
2. Write the power of each variable.	y has a power of 1.
Note: A square root is a power of $\dfrac{1}{2}$.	x has a power of $\dfrac{1}{2}$.
3 Check if the equation is linear.	x has a power of $\dfrac{1}{2}$, so this is not a linear equation.
d. 1. Identify the two variables.	d. m and n
2. Write the power of each variable.	m has a power of 2.
	n has a power of 1.
3. Check if the equation is linear.	m has a power of 2, so this is not a linear equation.
e. 1. Identify the two variables.	e. d and t
2. Write the power of each variable.	d has a power of 1.
	t has a power of 1.
3. Check if the equation is linear.	Since both variables have a power of 1, this is a linear equation.
f. 1. Identify the two variables.	f. y and x
2. Write the power of each variable.	y has a power of 1.
	x is the power.
3. Check if the equation is linear.	Since x is the power, this is not a linear equation.

5.2.2 Substituting numerical values into equations

A **formula** or **rule** is an expression or equation that expresses the relationship between certain quantities.

For example, $y = x + 3$.

When a variable in a formula is replaced by a number, we say that the number is **substituted** into the formula.

WORKED EXAMPLE 2 Substituting a numerical value into a linear equation

If $y = 5x - 4$, substitute the given value of x into the formula to calculate the value of y in each case.

a. $x = 2$

b. $x = -6$

THINK

a. 1. Write the formula.

2. Substitute 2 for x.

3. Equate for y and write the answer.

b. 1. Write the formula.

2. Substitute -6 for x.

3. Equate for y and write the answer.

This worked example could also be set up as a table like the one shown. Complete the table given the rule $y = 5x - 4$.

WRITE

a. $y = 5x - 4$

$y = 5 \times 2 - 4$

$y = 10 - 4$

$y = 6$

b. $y = 5x - 4$

$y = 5 \times -6 - 4$

$y = -30 - 4$

$y = -34$

x	$y = 5x - 4$
2	6
-6	-34

5.2.3 Substituting numerical values into expressions

An algebraic expression contains a letter that represents an unknown number. This letter is known as a pronumeral (or variable).

If the value of the pronumerals (or variables) are known, it is possible to **evaluate** (work out the value of) an algebraic expression by using **substitution**. This is done by replacing the pronumeral with its corresponding value.

When evaluating algebraic expressions, the rules for **BIDMAS (Brackets, Indices, Division, Multiplication, Addition and Subtraction)** must still apply and be remembered; for example, $3x$ means $3 \times x$ and ab means $a \times b$.

WORKED EXAMPLE 3 Substituting positive numerical values into expressions

Determine the values of the following expressions if $a = 4$ and $b = 3$.

a. $5b$

b. $\dfrac{8a}{4}$

c. $4(b^3 - 1)$

THINK

a. 1. Substitute $b = 3$ into the expression.

2. Evaluate and write the answer.

b. 1. Substitute $a = 4$ into the expression.

2. Evaluate and write the answer.

c. 1. Substitute $b = 3$ into the expression.

2. Evaluate and write the answer.

WRITE

a. $5b = 5 \times 3$

$= 15$

b. $\dfrac{8a}{4} = \dfrac{8 \times 4}{4}$

$= \dfrac{32}{4}$

$= 8$

c. $4(b^3 - 1) = 4(3^3 - 1)$

$= 4(26)$

$= 104$

Both positive and negative numbers can be substituted into an equation. When substituting negative numbers into an equation, be careful of the signs.

WORKED EXAMPLE 4 Substituting positive and negative numbers into expressions

Determine the values of the following expressions if $p = -2$ and $q = 5$.

a. $-2p + 2$

b. $\dfrac{3}{4q}$

c. $7(5 - 3p^2)$

THINK	WRITE
a. 1. Substitute $p = -2$ into the expression.	a. $-2p + 2 = -2 \times -2 + 2$
2. Evaluate and write the answer.	$= 4 + 2$
	$= 6$
b. 1. Substitute $q = 5$ into the expression.	b. $\dfrac{3}{4q} = \dfrac{3}{4 \times 5}$
2. Evaluate and write the answer.	$= \dfrac{3}{20}$
c. 1. Substitute $p = -2$ into the expression.	c. $7(5 - 3p^2) = 7(5 - 3 \times (-2)^2)$
2. Evaluate and write the answer.	$= 7(5 - 3 \times 4)$
	$= 7(5 - 12)$
	$= 7 \times -7$
	$= -49$

Resources

Interactivities Input and output tables (int-4001)
Finding a formula (int-4002)
Substitution (int-4003)

Exercise 5.2 Substitution

learn**on**

5.2 Exercise	5.2 Exam questions on

Simple familiar	Complex familiar	Complex unfamiliar
1, 2, 3, 4, 5, 6, 7, 8, 9, 10	11, 12, 13, 14, 15	N/A

These questions are even better in jacPLUS!
- Receive immediate feedback
- Access sample responses
- Track results and progress

Find all this and MORE in jacPLUS

Simple familiar

1. **WE1** Identify which of the following equations are linear and which are not linear.

a. $y^2 = 7x + 1$

b. $t = 7x^3 - 6x$

c. $y = 3(x + 2)$

d. $m = 2^{x+1}$

e. $4x + 5y - 9 = 0$

f. $x = \dfrac{6 - y}{4}$

2. **WE2** Evaluate the following equations to find the value of y in each case if $x = 3$ and $z = 8$.

   a. $y = 5 + x$ b. $y = x + 14$ c. $y = z + 2$ d. $y = 17 + z$

   e. $y = x - 2$ f. $y = 21 - x$ g. $y = z - 3$ h. $y = 21 - z$

3. **WE3** Determine the value of the following expressions if $a = 3$ and $b = 6$.

   a. $2a$ b. $-5b$ c. $7a$ d. $15b$

   e. $12a$ f. $-3 \times 2b$ g. $5a \times 2$ h. $-8a$

4. Evaluate the following expressions if $m = 8$ and $n = 3$.

   a. $\dfrac{m}{2}$ b. $\dfrac{n}{3}$ c. $\dfrac{24}{m}$ d. $\dfrac{15}{n}$

   e. $-\dfrac{88}{m}$ f. $-\dfrac{36}{n}$ g. $\dfrac{4m}{5}$ h. $\dfrac{20n}{15}$

5. Evaluate the following expressions if $j = 1$ and $k = 6$.

   a. $3(j + 5)$ b. $2(k + 5)$ c. $7(11 - j)$ d. $12(17 - k)$

   e. $5j(j + 4)$ f. $2k(k - 2)$ g. $6j(j - 3)$ h. $5k(k - 2)$

6. **WE4** Determine the values of the following expressions if $p = -3$ and $q = 12$.

   a. $5q$ b. $-3p$ c. $5 + q$ d. $13 - p$

   e. $-\dfrac{39}{p}$ f. $\dfrac{2q}{8}$ g. $5(2 - p)$ h. $q(q - 8)$

7. Determine the values of the following expressions if $u = -2$ and $v = -10$.

   a. $-7u$ b. $-3v$ c. $15 + v$ d. $27 - u$

   e. $-\dfrac{56}{u}$ f. $\dfrac{6v}{4}$ g. $9(2 - u)$ h. $v(v - 8)$

8. Determine the values of the following expressions if $m = -6$ and $n = 0$.

   a. $13 - n$ b. $-2(3 + m)$ c. m^2

   d. $7m + 50$ e. $9n(10 - n)$ f. $\dfrac{7n}{(2 - n)}$

9. Evaluate the following tables, given the rule:

   a. $y = 5x - 4$

x	y
0	
1	
2	
3	

   b. $y = -2x + 16$

x	y
3	
6	
9	
12	

10. Evaluate the following tables, given the rule:

    a. $y = 4x + 10$

x	y
-4	
-2	
0	
2	

    b. $y = -15x + 35$

x	y
-8	
-4	
0	
4	

11. Calculate the value of the algebraic expression $\dfrac{x^2(5x-1)}{2x}$ when $x=4$.

12. Determine the value of x that makes the algebraic expression $6x(3x-1)$ equal 60.

13. **MC** The value of x that makes the algebraic expression $\dfrac{3x^2(7x+3)}{2x}$ equal 33 is:

    A. $x=1$ **B.** $x=2$ **C.** $x=-1$ **D.** $x=-2$

14. The area of a circle is calculated by using the formula $A=\pi r^2$. Calculate the radii of the following circles to 2 decimal places:

    a. A circular garden bed of area $78\,\text{m}^2$
    b. A circular dinner plate of area $250\,\text{cm}^2$
    c. A dartboard of area $1598\,\text{cm}^2$

15. The time in seconds, T, for a pendulum to complete one swing is given by the formula $T=2\pi\sqrt{\dfrac{l}{g}}$, where $g=9.8\,\text{m/s}^2$ and l is the length of the pendulum in metres.
    Determine the length of a pendulum that will complete one swing in 2.5 s.
    Give your answer to the nearest cm.

Fully worked solutions for this chapter are available online.

LESSON
5.3 Formulas

SYLLABUS LINKS

- Find the value of a pronumeral in linear and simple non-linear equations given the values of the other pronumerals, transposing equations where necessary.
- Use a spreadsheet or an equivalent technology to construct a table of values from a formula, including two-by-two tables for formulas with two variable quantities.

Source: General Mathematics Senior Syllabus 2024 © State of Queensland (QCAA) 2024; licensed under CC BY 4.0.

5.3.1 Evaluating formulas

A formula is a special equation or rule that describes the relationship between different quantities.

To be able to use a formula, you need to know:
- what the pronumerals represent
- the correct units for the pronumerals
- what the formula is used for and the relationship between the variables. For example, in directly proportional expressions, as one variable increases, so does the other; in inversely proportional expressions, as one variable increases, the other decreases; and in polynomial expressions, as one variable increases, the other increases at a much greater rate.

Three examples are shown in the table.

Formula	What it is used for	What the pronumerals represent	Unit requirements
$A = lw$ (directly proportional)	Calculating the area of a rectangle	A represents the area of the rectangle. l represents the length of the rectangle. w represents the width of the rectangle.	l and w must be in the same units.
$s = \dfrac{d}{t}$ (inversely proportional)	Calculating speed given distance and time	s represents speed. d represents distance. t represents time.	s must reflect the units for d and t, e.g. km/h.
$E = mc^2$ (a polynomial)	Calculating the energy contained in a given mass	E represents the amount of energy. m represents the mass. c represents the speed of light.	m must be in kg. c must be in m/s.

Often equations are related to mathematical formulas and require substitution to calculate the required answer.

eles-3102

WORKED EXAMPLE 5 Substituting values for two variables into a formula

The surface area of a cylinder is calculated by using the formula
$SA = 2\pi r^2 + 2\pi rh$, where r is the radius of the circular base and h is
the height of the cylinder.
(*Note:* This formula is both directly proportional and a polynomial.)
a. If a small paint can is a cylinder with a height of 8 cm and a radius
of 5 cm, calculate its surface area to 2 decimal places.
b. If the height of the cylinder is doubled to 16 cm, calculate the effect
on the surface area to 2 decimal points.
Comment on your results using mathematical reasoning.

THINK	WRITE
a. 1. Identify the formula required to answer the question.	a. $SA = 2\pi r^2 + 2\pi rh$
2. Identify the variables given and make sure the units are consistent.	$h = 8$ cm and $r = 5$ cm
3. Substitute the variable value into the formula.	$SA = 2\pi r^2 + 2\pi rh$ $SA = 2\pi(5)^2 + 2\pi(5)(8)$ $SA = 408.41$
4. Write the answer with correct units.	Surface area $= 408.41$ cm^2
b. 1. Substitute the variable value into the formula.	b. $SA = 2\pi(5)^2 + 2\pi(5)(16)$ $SA = 659.73$
2. Write the answer with correct units.	Surface area $= 659.73$ cm^2 The height doubling has directly and proportionally increased the surface area.

5.3.2 Using technology to evaluate formulas

Often multiple calculations are to be done using the same formula. This can be done quickly with a spreadsheet. The following example looks at calculating the body mass index (BMI) of people with different weights and heights. BMI is an indicator of whether a person is underweight, overweight or a healthy weight.

BMI is calculated using a person's weight and height. Overweight is defined as a BMI of 25–29.9; obesity is defined as a BMI equal to or greater than 30. Your BMI is calculated by dividing your weight (in kilograms) by your height squared (in metres)

WORKED EXAMPLE 6 Using technology to construct a table of values from a formula

Use a spreadsheet to calculate the BMI of people with weights of 50 kg to 60 kg with increments of 2 kg and with heights of 1.60 m to 1.70 m with increments of 0.01 m.

THINK

1. Identify the formula required to calculate the BMI.

2. Identify the variables given and make sure the units used are correct.

3. Set up the spreadsheet with the specified increments. In cell B3 input the formula $(=\$B\$2/(A3)\wedge 2)$. This calculates the BMI of a person with weight 50 kg and height. It also uses the $ to lock the cell for copying the formula.

WRITE

$$BMI = \frac{weight}{(height)^2}$$

height (m) and weight (kg)

	A	B	C	D	E	F	G
1					Weight (kg)		
2	Height (m)	50	52	54	56	58	60
3	1.6	19.53125					
4	1.61						
5	1.62						
6	1.63						
7	1.64						
8	1.65						
9	1.66						
10	1.67						
11	1.68						
12	1.69						
13	1.7						

4. Copy this cell down vertically to calculate the BMI for all the heights for a 50-kg person. Do this by clicking on the B3 cell and moving the cursor to the bottom right-hand corner; the cross will go black. Then drag this down to cell B13.

	A	B	C	D	E	F	G
1					Weight (kg)		
2	Height (m)	50	52	54	56	58	60
3	1.6	19.53125					
4	1.61	19.28938					
5	1.62	19.05197					
6	1.63	18.81892					
7	1.64	18.59012					
8	1.65	18.36547					
9	1.66	18.14487					
10	1.67	17.92822					
11	1.68	17.71542					
12	1.69	17.50639					
13	1.7	17.30104					

5. Do the same thing for cell C3, but adjust to use the 52 kg weight. In cell C3 type (=C2/(A3)^2). Then copy this vertically the same way as in the previous section.

	A	B	C	D	E	F	G
1				Weight (kg)			
2	Height (m)	50	52	54	56	58	60
3	1.6	19.53125	20.3125				
4	1.61	19.28938	20.06095				
5	1.62	19.05197	19.81405				
6	1.63	18.81892	19.57168				
7	1.64	18.59012	19.33373				
8	1.65	18.36547	19.10009				
9	1.66	18.14487	18.87066				
10	1.67	17.92822	18.64534				
11	1.68	17.71542	18.42404				
12	1.69	17.50639	18.20665				
13	1.7	17.30104	17.99308				

6. Repeat this for each of the columns to complete the table of BMI values.

	A	B	C	D	E	F	G
1				Weight (kg)			
2	Height (m)	50	52	54	56	58	60
3	1.6	19.53125	20.3125	21.09375	21.875	22.65625	23.4375
4	1.61	19.28938	20.06095	20.83253	21.6041	22.37568	23.14726
5	1.62	19.05197	19.81405	20.57613	21.33821	22.10029	22.86237
6	1.63	18.81892	19.57168	20.32444	21.0772	21.82995	22.58271
7	1.64	18.59012	19.33373	20.07733	20.82094	21.56454	22.30815
8	1.65	18.36547	19.10009	19.83471	20.56933	21.30395	22.03857
9	1.66	18.14487	18.87066	19.59646	20.32225	21.04805	21.77384
10	1.67	17.92822	18.64534	19.36247	20.0796	20.79673	21.51386
11	1.68	17.71542	18.42404	19.13265	19.84127	20.54989	21.2585
12	1.69	17.50639	18.20665	18.9069	19.60716	20.30741	21.00767
13	1.7	17.30104	17.99308	18.68512	19.37716	20.0692	20.76125

Exercise 5.3 Formulas

learn on

5.3 Exercise	**5.3 Exam questions** on

Simple familiar	Complex familiar	Complex unfamiliar
1, 2, 3, 4, 5, 6	7, 8, 9, 10, 11, 12	13, 14

These questions are even better in jacPLUS!
- Receive immediate feedback
- Access sample responses
- Track results and progress

Find all this and MORE in jacPLUS ▶

Simple familiar

1. Evaluate the following formulas for C if $a = -5$ and $b = 0$.

   a. $C = a + b$
   b. $C = 2a$
   c. $C = b - a$
   d. $C = -2a(3 + b)$
   e. $C = 12b$
   f. $C = 2ab$
   g. $C = 4b(b - a)$
   h. $C = \dfrac{7b}{(2 - a)}$

2. Evaluate the following formulas for R if $p = -2$ and $q = -5$.

   a. $R = \dfrac{(p - q)}{-3p}$
   b. $R = \dfrac{4q^2}{-10p}$
   c. $R = p^2(3p - 2q)$
   d. $R = p^2 - q^2$

3. **WE5** The surface area of a cylinder is calculated by using the formula $SA = 2\pi r^2 + 2\pi rh$, where r is the radius of the circular base and h is the height of the cylinder.

   a. Given a cylinder has a height of 13 cm and a radius of 4 cm, calculate its surface area to 2 decimal places.
   b. Calculate its surface area to 2 decimal places if the height is doubled to 26 cm.

4. The area of a triangle is $A = \frac{1}{2}bh$, where b is the length of the base and h is the vertical height.
   Calculate the area of the triangle shown.

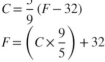

12 m

18 m

5. The surface area of a cylinder and a sphere are found by using the following formulas, where r stands for the radius and h represents the height.

   $$SA_{\text{cylinder}} = 2\pi r^2 + 2\pi rh \quad SA_{\text{sphere}} = 4\pi r^2$$

   Calculate the surface areas of the following to 2 decimal places.

   a. Cylinder with a radius of 3 cm and height of 8 cm
   b. Cylinder with a radius of 15 cm and height of 9 cm
   c. Cylinder with a radius of 12.5 m and height of 18 m
   d. Cylinder with a radius of 2 300 cm and height of 35.8 m
   e. Sphere with a radius of 5 cm
   f. Sphere with a radius of 12 m
   g. Sphere with a radius of 7.25 cm
   h. Sphere with a diameter of 25 m

6. **WE6** Use a spreadsheet to calculate the BMI of people with weights of 60 kg to 70 kg with increments of 2 kg and with heights of 1.70 m to 1.80 m with increments of 0.01 m.

Complex familiar

7. Use a spreadsheet to calculate the net force (F) for masses (m) of 100 kg to 200 kg, with increments of 10 kg and with accelerations (a) of 0.5 m/s² to 1.5 m/s², with increments of 0.1 m/s², given Newton's Second Law is $F = m \times a$.

8. Use a spreadsheet to calculate the kinetic energy (KE) of objects with masses (m) of 100 kg to 200 kg with increments of 10 kg and with velocities (v) 15 m/s to 25 m/s with increments of 1 m/s, given $KE = \frac{1}{2}mv^2$.

9. In Australia we measure our daily temperature in degrees Celsius, °C, whereas in the United States for example, they measure temperature in degrees Fahrenheit, °F.
   To convert between the two, the following formulas can be used and rounded to 2 decimal places:

   $$C = \frac{5}{9}(F - 32)$$

   $$F = \left(C \times \frac{9}{5}\right) + 32$$

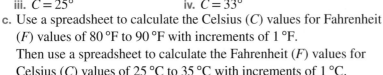

   a. Convert the following Fahrenheit temperatures to Celsius:
      i. $F = 100°$ ii. $F = 50°$
      iii. $F = 78°$ iv. $F = 25°$
   b. Convert the following Celsius temperatures to Fahrenheit:
      i. $C = 45°$ ii. $C = 0°$
      iii. $C = 25°$ iv. $C = 33°$
   c. Use a spreadsheet to calculate the Celsius (C) values for Fahrenheit (F) values of 80 °F to 90 °F with increments of 1 °F.
      Then use a spreadsheet to calculate the Fahrenheit (F) values for Celsius (C) values of 25 °C to 35 °C with increments of 1 °C.

10. Determine the volumes of the following shapes after identifying the appropriate formula to use.

a.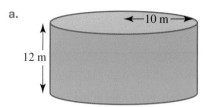

b.

15 cm

8 cm

6 cm

11. Determine the volumes of the following shapes after investigating the appropriate formula to use.

a.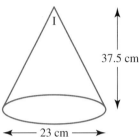

37.5 cm

23 cm

b.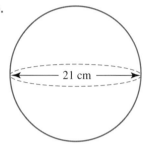

21 cm

12. A local supermarket hires out glasses when a customer purchases their party supplies from them. They use the formula $C = 0.025n + 29$, where C represents the fixed cost and n represents the number of glasses hired. Calculate the following fees.

a. The total fee when 66 glasses were hired
b. The total fee when 116 glasses were hired
c. The total half-yearly fee for glass hire when 34, 48, 21, 103, 87 and 77 glasses were hired over the half-year
d. The yearly fee when on average 58 glasses were hired each month

Complex unfamiliar

13. The volume of a rectangular prism is given by $V = L \times W \times H$. Identify the dimensions of two possible rectangular prisms that have a volume of $120\,\text{m}^3$.

14. Two electricity bills are calculated by using the formulas $C_A = 0.16K + 75$ and $C_B = 0.12K + 90$, where C is the cost of the bill and K is the amount of kilowatt-hours (kWh) of electricity used. Calculate the amount of kilowatt-hours used when the two bills are the same.

Fully worked solutions for this chapter are available online.

LESSON
5.4 Transposition

SYLLABUS LINKS

• Find the value of a pronumeral in linear and simple non-linear equations given the values of the other pronumerals, transposing equations where necessary.

Source: General Mathematics Senior Syllabus 2024 © State of Queensland (QCAA) 2024; licensed under CC BY 4.0.

5.4.1 Transposing linear equations

If we are given a **linear equation** between two variables, we are able to **transpose** this relationship. That is, we can change the equation so that the variable on the right-hand side of the equation becomes the stand-alone variable on the left-hand side of the equation.

WORKED EXAMPLE 7 Transposing a linear equation

Transpose the linear equation $y = 4x + 7$ to make x the subject of the equation.

THINK	WRITE
1. Isolate the variable on the right-hand side of the equation (by subtracting 7 from both sides).	$y - 7 = 4x + 7 - 7$ $y - 7 = 4x$
2. Divide both sides of the equation by the coefficient of the variable, x (in this case 4).	$\dfrac{y-7}{4} = \dfrac{4x}{4}$ $\dfrac{y-7}{4} = x$
3. Transpose the relation by interchanging the left-hand side and the right-hand side.	$x = \dfrac{y-7}{4}$

Inverse operations

Inverse operations are operations that UNDO each other

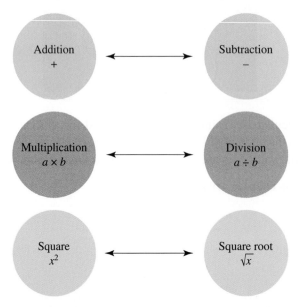

5.4.2 Transposing formulas

A formula is an equation showing the relationship between two or more quantities.

For example, as was mentioned earlier in the chapter, body mass index (B) is calculated using a person's mass (M) in kilograms and their height (H) in metres.

$$B = \frac{M}{H^2}$$

Sometimes formulas are rearranged so that a relationship is more obvious, or to make the formula easier to use. This is often called making a different variable the subject. This is when the new variable is to the left of the equal sign.

For example, making the mass the subject of the body mass index formula gives us:

$$M = B \times H^2$$

To make a different variable the subject requires you to rearrange or transpose the formula. This is done by using inverse operations performed on both sides of the equation.

WORKED EXAMPLE 8 Transposing a formula in one step

Transpose the formula $B = \dfrac{M}{H^2}$ to make M the subject of the formula.

THINK	WRITE
1. Write down the formula.	$B = \dfrac{M}{H^2}$
2. To make M the subject, use inverse operations to leave M on the right-hand side by itself. Undo the division by H^2 by multiplying both sides by H^2, as shown in red.	$H^2 \times B = \dfrac{M}{H^2} \times H^2$ $H^2 \times B = M$
3. Rewrite the equation with the subject first and the variables alphabetically.	$M = H^2 \times B$ $\quad = BH^2$

WORKED EXAMPLE 9 Transposing a formula in two steps

Transpose the formula $a = \dfrac{v - u}{t}$ to make v the subject.

THINK	WRITE
1. Write the formula.	$a = \dfrac{v - u}{t}$
2. Undo the steps used to build the formula around v, starting with the last step first. Multiply both sides by t, as shown in red.	$a \times t = \dfrac{v - u}{t} \times t$
3. Then add u to both sides, as shown in red.	$a \times t + u = v - u + u$
4. Rewrite the formula with the subject on the left-hand side.	$v = at + u$

eles-6383

eles-3103

WORKED EXAMPLE 10 Transposing a formula using the opposite operation of squared

Transpose the formula $B = \dfrac{M}{H^2}$ to make H the subject.

THINK	WRITE
1. Write the formula.	$B = \dfrac{M}{H^2}$
2. Undo the steps used to build the formula around H, starting with multiplying both sides by H^2.	$B \times H^2 = \dfrac{M}{\cancel{H^2}} \times \cancel{H^2}$ $B \times H^2 = M$
3. Then divide both sides by B.	$\dfrac{B \times H^2}{B} = \dfrac{M}{B}$ $H^2 = \dfrac{M}{B}$
4. Then take the square root of both sides.	$H^2 = \dfrac{M}{B}$ $\sqrt{H^2} = \sqrt{\dfrac{M}{B}}$ (positive square root as height is positive) $H = \sqrt{\dfrac{M}{B}}$
5. Rewrite the formula.	$H = \sqrt{\dfrac{M}{B}}$

---

on Resources

Interactivities   Backtracking (int-4045)
                  Inverse operations (int-4043)
                  Rearranging formulas (int-6040)

Exercise 5.4 Transposition

learn on

5.4 Exercise	**5.4 Exam questions** on

These questions are even better in jacPLUS!
- Receive immediate feedback
- Access sample responses
- Track results and progress

Find all this and MORE in jacPLUS ▶

Simple familiar	Complex familiar	Complex unfamiliar
1, 2, 3, 4, 5, 6	7, 8, 9, 10	11, 12, 13

Simple familiar

1. **WE7** Transpose the linear equation $y = 6x - 3$ to make x the subject of the equation.

2. Transpose the linear equation $6y = 3x + 1$ to make x the subject of the equation.

3. Transpose the following linear equations to make x the subject.

   a. $y = 2x + 5$

   b. $3y = 6x + 8$

4. **WE8** Transpose each of the following formulas to make the variable shown in the brackets the subject of the formula.

   a. $C = AB + D$ (A)

   b. $v = u + at$ (t)

   c. $V = \dfrac{Ah}{3}$ (h)

   d. $A = \dfrac{h(a + b)}{2}$ (b)

5. **WE9** Transpose each of the following formulas to make the variable shown in the brackets the subject of the formula.

   a. $A = 2\pi r(r + h)$ (h)

   b. $S = \dfrac{a}{1 - r}$ (r)

   c. $I = \dfrac{PRT}{100}$ (R)

   d. $A = PR^n$ (P)

6. **WE10** Transpose each of the following formulas to make the variable shown in the brackets the subject of the formula.

   a. $A = x^2$ (x)

   b. $V = \dfrac{\pi r^2 h}{3}$ (r)

   c. $v^2 = u^2 + 2as$ (u)

   d. $c = \sqrt{a^2 + b^2}$ (a)

Complex familiar

7. The formula $F = \dfrac{9C}{5} + 32$ converts degrees Celsius, °C, to degrees Fahrenheit, °F. While travelling in the USA you discover that the weather reports give the weather forecast in degrees Fahrenheit.

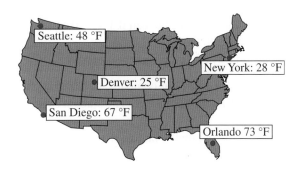

   a. Transpose the formula to make C the subject.
   b. Use your transposed formula from part **a** to convert the temperatures shown to degrees Celsius.

8. A local courier company uses the formula $C = 3.5h + 5$, where h is the number of kilometres and C is the cost of the delivery in dollars, to calculate the total delivery cost.

   a. Transpose the formula to make h the subject of the formula.
   b. If the cost of a delivery is \$43.50, calculate how many kilometres the delivery was from the courier company base.

9. Transpose the equation $T = \dfrac{x-3}{x-1}$ to make x the subject.

10. Transpose the following equations to make the variable in brackets the subject.

   a. $z = \dfrac{a(x-w)^2 + 3}{e} - p$ (w)

   b. $w = \dfrac{8w - x}{3(q-r)} + 10$ (r)

   c. $a = b\sqrt{cd + e}$ (d)

   d. $m = \dfrac{e+3}{z} - tr$ (z)

Complex unfamiliar

11. Consider the can shown below.

   A teacher asks students to calculate the height of the can if $V = 210\,\text{cm}^3$ and $r = 3\,\text{cm}$. One of the students calculated the height of the can to be 22.28 cm.
   Explain using mathematical reasoning whether the student is correct.

12. A student was discussing the formula for a hire car company, $C = 0.95D + 30$, where C is the total cost of the car hire and D is the distance travelled in kilometres. The student says that to hire a car from this rental company costs 95 cents a kilometre.
   Explain whether the student is correct using mathematical reasoning.

13. Consider the waffle cone shown below.

   Calculate the radius of the waffle cone when $V = 105\,\text{cm}^3$ and $h = 8\,\text{cm}$.

Fully worked solutions for this chapter are available online.

LESSON
5.5 Review

5.5.1 Summary

doc-41480

Hey students! Now that it's time to revise this chapter, go online to:

Access the chapter summary

Review your results

Practise exam questions

Find all this and MORE in jacPLUS

5.5 Exercise

learn on

5.5 Exercise	**5.5 Exam questions** on

Simple familiar	Complex familiar	Complex unfamiliar
1, 2, 3, 4, 5, 6, 7, 8, 9, 10, 11, 12	13, 14, 15, 16	17, 18, 19, 20

These questions are even better in jacPLUS!
- Receive immediate feedback
- Access sample responses
- Track results and progress

Find all this and MORE in jacPLUS

Simple familiar

1. **MC** The value of the expression $3(6x - 8)$ when $x = 9$ is:
   - **A.** 138
   - **B.** 186
   - **C.** 124
   - **D.** 46

2. **MC** Kinetic energy is calculated using the formula $KE = \frac{1}{2}mv^2$, where m is the mass of the object in kg, v is its velocity in m/s and KE is in joules (J). If an object of mass 12 kg is moving at a velocity of 10 m/s, it has a kinetic energy of:
   - **A.** 1200 J
   - **B.** 120 J
   - **C.** 1000 J
   - **D.** 600 J

3. **MC** The algebraic expression $3p(5q - 7)$, when $p = -2$ and $q = -3$, is:
   - **A.** 144
   - **B.** −144
   - **C.** 132
   - **D.** −132

4. **MC** If Qua is getting paid $12.50 per hour at his part-time job and over the weekend he works 6 hours of normal time and 4 hours of double time, the amount he earns over the weekend is:
   - **A.** $75
   - **B.** $100
   - **C.** $125
   - **D.** $175

5. **MC** There are 1.093 613 3 yards for every metre. The number of yards Usain Bolt ran to win the 100 m sprint at the Olympics, to 2 decimal places, was:
   - **A.** 109.63
   - **B.** 91.44
   - **C.** 190.36
   - **D.** 109.36

6. **MC** The surface area of a cone is shown. If the cone has a sloping height of 12 m and a base radius of 6 m, the surface area of the cone is closest to:
   - **A.** 339.29 m^2
   - **B.** 120 m^2
   - **C.** 303.29 m^2
   - **D.** 297.87 m^2

   $SA = \pi rs + \pi r^2$

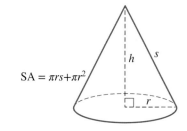

7. **MC** A formula for calculating velocity is $v = u + at$, where v is the final velocity, u the initial velocity, a is the acceleration and t represents time. If an object initially starts from rest and accelerates at 2 m/s² for 12 seconds, it reaches a velocity of:

   A. 12 m/s   **B.** 24 m/s   **C.** 36 m/s   **D.** 10 m/s

8. Transpose the circular motion formula to make the velocity, v, the subject.

$$a = \frac{v^2}{r}$$

9. Transpose the circular motion formula to make the radius, r, the subject.

$$a = \frac{v^2}{r}$$

10. To calculate the displacement of an object, the formula used is $x = \frac{1}{2}(u + v)t$. Transpose the formula to make the velocity, v, the subject of the formula.

11. To calculate the displacement of an object, the formula used is $x = \frac{1}{2}(u + v)t$. Transpose the formula to make the time, t, the subject of the formula.

12. The volume of a cone is calculated using the formula $V = \frac{\pi r^2 h}{3}$. Calculate the radius, r, to 2 decimal places when the cone has a volume of 100 cm³ and $h = 12$ cm.

Complex familiar

13. Given the values $a = 3$ and $b = -3$, determine the values of the following expressions.

   a. $\dfrac{a^2 - 4}{b + 2}$ **b.** $\left(a + \dfrac{1}{b}\right)^2$ **c.** $a^3 \times \dfrac{1}{b^3}$ **d.** $\sqrt[3]{\dfrac{a}{b}}$

14. Calculate the radius to 2 decimal places of a circle with:

   a. a circumference of 91 m
   b. a circumference of 100 cm
   c. a circumference of 57 m
   d. an area of 28 cm².

15. The formula for the volume of a cone is $V = \frac{1}{3}\pi r^2 h$.

   If a cone has a volume of 62 cm³, determine the radius of the cone if it has a height of:

   a. 5 cm
   b. 30 mm

16. Transpose each of the following formulas.

   a. $a^2 + b^2 = c^2$, making b the subject
   b. $E = mc^2$, making c the subject
   c. $V = \frac{4}{3}\pi r^3$, making r the subject

17. A new game has been created by students for the school fair. To win the game you need to hit the target with 5 darts in the shaded region.

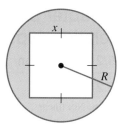

a. If $A = 80 \, \text{cm}^2$ and $x = 3 \, \text{cm}$, calculate the value of R.

b. The students found that the best size for the game board is when $R = 10$ and $x = 5$. Determine the percentage of the total board that is shaded, to the nearest whole number.

18. For the expressions given, explain the following using mathematical reasoning.

a. Why the expression $15 - 5x$ will always be positive if x is a negative number

b. Why the expression $-4 - x^2$ will always be negative

c. What the two x values are that make the expression $36 - x^2$ equal to zero

19. You are investigating getting your business card printed for your new game store. A local printing company charges $250 for the cardboard used and an hourly rate for labour of $40.

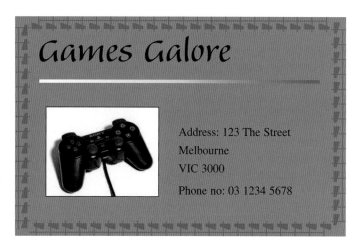

a. You have budgeted $1000 for the printing job. Calculate how many hours of labour you can afford. Give your answer to the nearest minute.

b. An alternative to printing is photocopying. The company charges 15 cents per side for the first 10 000 cards and then 10 cents per side for the remaining cards. Compare and determine which is the cheaper option for 18 750 single-sided cards using mathematical reasoning and by how much.

20. You have inherited $2000, and you decided to invest this money into an interest-bearing account paying $r\%$ interest compounded annually. The amount, $\$A$, to which your investment will grow is given by the formula $A = P\left(1 + \dfrac{r}{100}\right)^n$, where P is the principal amount and n is the number of years.

a. Calculate r when your $2000 grows to $2500 over 2 years.

b. If the interest rate is 7%, use trial and error to determine the time it takes (to the nearest year) for your initial inheritance of $2000 to double in value.

Fully worked solutions for this chapter are available online.

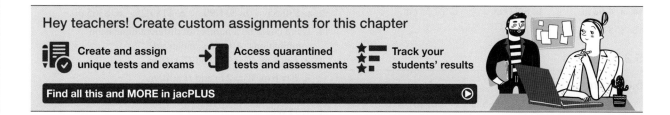

Hey teachers! Create custom assignments for this chapter

Create and assign unique tests and exams

Access quarantined tests and assessments

Track your students' results

Find all this and MORE in jacPLUS

Answers

Chapter 5 Linear and non-linear relationships

5.2 Substitution

5.2 Exercise

1. a. Not linear   b. Not linear   c. Linear
   d. Not linear   e. Linear   f. Linear

2. a. 8   b. 17   c. 10   d. 25
   e. 1   f. 18   g. 5   h. 13

3. a. 6   b. -30 c. 21   d. 90
   e. 36   f. -36 g. 30   h. -24

4. a. 4   b. 1   c. 3   d. 5
   e. -11 f. -12 g. $\dfrac{32}{5}$ h. 4

5. a. 18   b. 22   c. 70   d. 132
   e. 25   f. 48   g. -12 h. 120

6. a. 60   b. 9   c. 17   d. 16
   e. 13   f. 3   g. 25   h. 48

7. a. 14   b. 30   c. 5   d. 29
   e. 28   f. -15 g. 36   h. 180

8. a. 13   b. 6   c. 36   d. 8
   e. 0   f. 0

9. a.

x	y
0	-4
1	1
2	6
3	11

b.

x	y
3	10
6	4
9	-2
12	-8

10. a.

x	y
-4	-6
-2	2
0	10
2	18

b.

x	y
-8	155
-4	95
0	35
4	-25

11. 38

12. $x = 2$; using technology $x = 2$, $x = \dfrac{-5}{3}$.

13. D

14. a. 4.98 m   b. 8.92 cm   c. 22.55 cm

15. 155 cm

5.3 Formulas

5.3 Exercise

1. a. -5 b. -10 c. 5   d. 30
   e. 0   f. 0   g. 0   h. 0

2. a. $\dfrac{1}{2}$ b. 5   c. 16   d. -21

3. a. 427.26 cm²   b. 753.98 cm²

4. 108 m²

5. a. 207.35 cm²   b. 2 261.95 cm²   c. 2 395.46 m²
   d. 8 497.38 m²   e. 314.16 cm²   f. 1 809.56 m²
   g. 660.52 cm²   h. 1 963.50 m²

6.

	A	B	C	D	E	F	G
1				Weight (kg)			
2	Height (m)	60	62	64	66	68	70
3	1.7	20.76125	21.45329	22.14533	22.83737	23.52941	24.22145
4	1.71	20.51913	21.20311	21.88708	22.57105	23.25502	23.93899
5	1.72	20.28123	20.95727	21.63332	22.30936	22.9854	23.66144
6	1.73	20.04745	20.71569	21.38394	22.05219	22.72044	23.38869
7	1.74	19.81768	20.47827	21.13886	21.79945	22.46003	23.12062
8	1.75	19.59184	20.2449	20.89796	21.55102	22.20408	22.85714
9	1.76	19.36983	20.0155	20.66116	21.30682	21.95248	22.59814
10	1.77	19.15158	19.78997	20.42836	21.06674	21.70513	22.34352
11	1.78	18.937	19.56824	20.19947	20.8307	21.46194	22.09317
12	1.79	18.72601	19.35021	19.97441	20.59861	21.22281	21.84701
13	1.8	18.51852	19.1358	19.75309	20.37037	20.98765	21.60494

7.

	A	B	C	D	E	F	G	H	I	J	K	L
1					Mass (kg)							
2	Acceleration	100	110	120	130	140	150	160	170	180	190	200
3	0.5	50	55	60	65	70	75	80	85	90	95	100
4	0.6	60	66	72	78	84	90	96	102	108	114	120
5	0.7	70	77	84	91	98	105	112	119	126	133	140
6	0.8	80	88	96	104	112	120	128	136	144	152	160
7	0.9	90	99	108	117	126	135	144	153	162	171	180
8	1	100	110	120	130	140	150	160	170	180	190	200
9	1.1	110	121	132	143	154	165	176	187	198	209	220
10	1.2	120	132	144	156	168	180	192	204	216	228	240
11	1.3	130	143	156	169	182	195	208	221	234	247	260
12	1.4	140	154	168	182	196	210	224	238	252	266	280
13	1.5	150	165	180	195	210	225	240	255	270	285	300

8.

	A	B	C	D	E	F	G	H	I	J	K	L
1					Mass (kg)							
2	Velocity	100	110	120	130	140	150	160	170	180	190	200
3	15	11250	12375	13500	14625	15750	16875	18000	19125	20250	21375	22500
4	16	12800	14080	15360	16640	17920	19200	20480	21760	23040	24320	25600
5	17	14450	15895	17340	18785	20230	21675	23120	24565	26010	27455	28900
6	18	16200	17820	19440	21060	22680	24300	25920	27540	29160	30780	32400
7	19	18050	19855	21660	23465	25270	27075	28880	30685	32490	34295	36100
8	20	20000	22000	24000	26000	28000	30000	32000	34000	36000	38000	40000
9	21	22050	24255	26460	28665	30870	33075	35280	37485	39690	41895	44100
10	22	24200	26620	29040	31460	33880	36300	38720	41140	43560	45980	48400
11	23	26450	29095	31740	34385	37030	39675	42320	44965	47610	50255	52900
12	24	28800	31680	34560	37440	40320	43200	46080	48960	51840	54720	57600
13	25	31250	34375	37500	40625	43750	46875	50000	53125	56250	59375	62500

9. a. i. 37.78 °C   ii. 10.00 °C
   iii. 25.56 °C   iv. -3.89 °C

   b. i. 113.00 °F   ii. 32.00 °F
   iii. 77.00 °F   iv. 91.40 °F

   c.

	A	B	C	D	E
1	Fahrenheit	Celsius		Celsius	Fahrenheit
2	80	26.66667		25	77
3	81	27.22222		26	78.8
4	82	27.77778		27	80.6
5	83	28.33333		28	82.4
6	84	28.88889		29	84.2
7	85	29.44444		30	86
8	86	30		31	87.8
9	87	30.55556		32	89.6
10	88	31.11111		33	91.4
11	89	31.66667		34	93.2
12	90	32.22222		35	95

10. a. 3 769.91 m³   b. 720 cm³

11. a. 5 193.45 cm³   b. 4 849.05 cm³

12. a. $30.65   b. $31.90   c. $183.25   d. $365.40

13. There could be a number of different dimension combinations. A few examples are:
    i. $L = 4$ m, $W = 3$ m and $H = 10$ m
    ii. $L = 6$ m, $W = 2$ m and $H = 10$ m
    iii. $L = 6$ m, $W = 4$ m and $H = 5$ m

14. 375 kWh

5.4 Transposition

5.4 Exercise

1. $x = \dfrac{y+3}{6}$

2. $x = 2y - \dfrac{1}{3}$

3. a. $x = \dfrac{y-5}{2}$ b. $x = \dfrac{3y-8}{6}$

4. a. $A = \dfrac{C-D}{B}$ b. $t = \dfrac{v-u}{a}$

   c. $h = \dfrac{3V}{A}$ d. $b = \dfrac{2A}{h} - a$

5. a. $h = \dfrac{A}{2\pi r} - r$ b. $r = 1 - \dfrac{a}{S}$

   c. $R = \dfrac{100I}{PT}$ d. $P = \dfrac{A}{R^n}$

6. a. $x = \sqrt{A}$ b. $r = \sqrt{\dfrac{3V}{\pi h}}$

   c. $u = \sqrt{v^2 - 2as}$ d. $a = \sqrt{c^2 - b^2}$

7. a. $C = \dfrac{5}{9}(F - 32)$

   b. Seattle: 8.9 °C
   San Diego: 19.4 °C
   Denver: −3.9 °C
   New York: −2.2 °C
   Orlando: 22.8 °C

8. a. $h = \dfrac{C-5}{3.5}$ b. 11 km

9. $x = \dfrac{T-3}{T-1}$

10. a. $w = x - \sqrt{\dfrac{e(z+p)-3}{a}}$

   b. $r = q - \dfrac{8w-x}{3(w-10)}$

   c. $d = \dfrac{\left(\frac{a}{b}\right)^2 - e}{c}$

   d. $z = \dfrac{e+3}{m+tr}$

11. Please see the worked solutions.

12. The student is partially correct. It does cost 95 cents per kilometre; however, there is also an initial charge of $30 for the car hire.

13. 3.54 cm

5.5 Review

5.5 Exercise

1. A
2. D
3. C
4. D
5. D
6. A
7. B
8. $v = \sqrt{ar}$
9. $r = \dfrac{v^2}{a}$
10. $v = \dfrac{2x}{t} - u$
11. $t = \dfrac{2x}{u+v}$
12. 2.82 cm
13. a. −5   b. $\dfrac{64}{9}$ c. −1   d. −1
14. a. 14.48 m   b. 15.92 cm   c. 9.07 m   d. 2.99 cm
15. a. 3.44 cm   b. 4.44 cm
16. a. $b = \sqrt{c^2 - b^2}$

   b. $c = \sqrt{\dfrac{E}{m}}$

   c. $r = \sqrt[3]{\dfrac{3V}{4\pi}}$

17. a. 5.32 cm
   b. 92%
18. a. Two multiplied negatives means we are always adding a positive number to 15.

   b. A number squared is always positive, so when we subtract it from −4 the result will always be negative.

   c. We need to subtract 36 from 36 to get zero. So $x = 6$ or $x = -6$, because they both equal 36 when squared.

19. a. 18 hours, 45 minutes
   b. The printing is cheaper by $1375.
20. a. $3007.31   b. $4969.69   c. 11.8%   d. 10 years

6 Linear equations

LESSON SEQUENCE

Fully worked solutions for this chapter are available online.

EXAM PREPARATION

Access exam-style questions in every lesson, available online.

on Resources

 Solutions   Solutions — Chapter 6 (sol-1247)

 Exam questions   Exam question booklet — Chapter 6 (eqb-0263)

 Digital documents Learning matrix — Chapter 6 (doc-41481)
                       Chapter summary — Chapter 6 (doc-41483)

LESSON
6.1 Overview

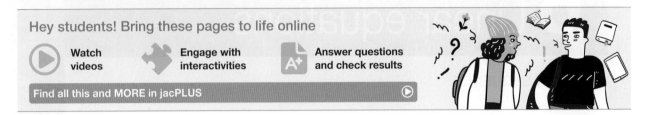
6.1.1 Introduction

Linear equations have been around for more than 4000 years. However, it wasn't until the seventeenth century that progress was made in linear algebra by the founder of calculus, Leibnitz. This was followed by work by Cramer and was adapted further by Gauss. Linear equations themselves were invented in 1843 by Irish mathematician Sir William Rowan Hamilton. He made important contributions to mathematics, and his work was also used with quantum mechanics.

Today, linear equations are more relevant than ever, underpinning many modern technologies and industries. They are crucial in computer science for algorithms and machine learning, where they help in optimising and predicting outcomes.

In economics and finance, linear equations model relationships between variables such as supply and demand, or the pricing of financial instruments. In technology, they are integral to developing software, particularly in data analysis and artificial intelligence. For everyday applications, linear equations simplify decision-making processes. For instance, they can model the cost of a ride-share trip, helping users compare prices across different services. They are also used in environmental science to track and predict changes in climate patterns by analysing data trends. From planning budgets to optimising routes for delivery services, linear equations continue to be a versatile and indispensable tool in our modern world.

6.1.2 Syllabus links

Lesson	Lesson title	Syllabus links
6.2	**Solving linear equations**	○ Solve linear equations, including equations with variables on both sides and equations with rational solutions.
6.3	**Developing linear equations and solving practical problems**	○ Develop a linear equations from a description in words.
		○ Solve practical problems involving linear equations.

Source: General Mathematics Senior Syllabus 2024 © State of Queensland (QCAA) 2024; licensed under CC BY 4.0.

LESSON
6.2 Solving linear equations

SYLLABUS LINKS

- Solve linear equations, including equations with variables on both sides and equations with rational solutions.

Source: General Mathematics Senior Syllabus 2024 © State of Queensland (QCAA) 2024; licensed under CC BY 4.0.

6.2.1 Equations with variables on one side

Equations are mathematical statements that show two equal expressions. This means that the left-hand side and the right-hand side of an equation are equal.

A linear equation has variables whose highest power is 1; for example $y = 2x + 5$.

Linear equations can be solved using inverse operations. When solving equations, the last operation performed on the variable when building the equation is the first operation undone by applying inverse operations to both sides of the equation.

eles-3030

WORKED EXAMPLE 1 Solving linear equations with variables on one side

Solve the following linear equations to determine the unknowns.

a. $5x = 12$ b. $8t + 11 = 20$ c. $12 = 4(n - 3)$ d. $\dfrac{4x - 2}{3} = 5$

THINK	WRITE
a. 1. Identify the operations performed on the unknown.	**a.** $5x = 5 \times x$ So the operation is $\times 5$.
2. Write the opposite operation.	The opposite operation is $\div 5$.
3. Perform the opposite operation on both sides of the equation.	Step 1 ($\div 5$): $5x = 12$ $\dfrac{5x}{5} = \dfrac{12}{5}$ $x = \dfrac{12}{5}$
4. Write the answer in its simplest form.	$x = \dfrac{12}{5}$
b. 1. Identify the operations performed in order on the unknown.	**b.** $8t + 11$ The operations are $\times 8$, $+ 11$.
2. Write the opposite operations.	$\div 8$, $- 11$
3. Perform the opposite operations in reverse order on both sides of the equation, one operation at a time.	Step 1 ($- 11$): $8t + 11 = 20$ $8t + 11 - 11 = 20 - 11$ $8t = 9$

Step 2 ($\div 8$):
$$8t = 9$$
$$\frac{8t}{8} = \frac{9}{8}$$
$$t = \frac{9}{8}$$

4. Write the answer in its simplest form. $\quad t = \dfrac{9}{8}$

c. 1. Identify the operations performed in order on the unknown. (Remember operations in brackets are performed first.)

c. $4(n - 3)$

The operations are $-3, \times 4$.

2. Write the opposite operations. $\quad +3, \div 4$

3. Perform the opposite operations on both sides of the equation in reverse order, one operation at a time.

Step 1 ($\div 4$):
$$12 = 4(n - 3)$$
$$\frac{12}{4} = \frac{4(n - 3)}{4}$$
$$3 = n - 3$$

Step 2 ($+3$):
$$3 = n - 3$$
$$3 + 3 = n - 3 + 3$$
$$6 = n$$

4. Write the answer in its simplest form. $\quad n = 6$

d. 1. Identify the operations performed in order on the unknown.

d. $\dfrac{4x - 2}{3}$

The operations are $\times 4, -2, \div 3$.

2. Write the opposite operations. $\quad \div 4, +2, \times 3$

3. Perform the opposite operations on both sides of the equation in reverse order, one operation at a time.

Step 1 ($\times 3$):
$$\frac{4x - 2}{3} = 5$$
$$3 \times \frac{4x - 2}{3} = 5 \times 3$$
$$4x - 2 = 15$$

Step 2 ($+2$):
$$4x - 2 = 15$$
$$4x - 2 + 2 = 15 + 2$$
$$4x = 17$$

Step 3 ($\div 4$):
$$4x = 17$$
$$\frac{4x}{4} = \frac{17}{4}$$
$$x = \frac{17}{4}$$

4. Write the answer in its simplest form. $\quad x = \dfrac{17}{4}$

6.2.2 Equations with variables on both sides

When a variable appears on both sides of an equation, inverse operations are used to collect the variables into a single term.

eles-3166

WORKED EXAMPLE 2 Solving a linear equation with the variable on both sides

Solve $3x - 12 = 5x + 4$.

THINK	WRITE
1. Move all the x values to one side by subtracting $3x$ from both sides.	$3x - 12 = 5x + 4$ $3x - 3x - 12 = 5x - 3x + 4$ $-12 = 2x + 4$
2. To get the x values by themselves, subtract 4 from both sides.	$-12 = 2x + 4$ $-12 - 4 = 2x + 4 - 4$ $-16 = 2x$
3. Divide both sides by 2, the coefficient of x.	$-16 = 2x$ $\dfrac{-16}{2} = \dfrac{2x}{2}$ $-8 = x$
4. Write the answer.	$x = -8$

To solve linear equations containing fractions, multiply both sides of the equations by the lowest common denominator first to remove the fraction.

eles-6384

WORKED EXAMPLE 3 Solving a linear equation in fraction form

Solve $\dfrac{3a - 2}{2} = \dfrac{a + 11}{3}$.

THINK	WRITE
1. The lowest common multiple of 2 and 3 is 6. Rewrite each fraction so that the numerator and denominator are converted into equivalent fractions, as shown in blue.	$\dfrac{3a-2}{2} \times \dfrac{3}{3} = \dfrac{a+11}{3} \times \dfrac{2}{2}$ $\dfrac{3(3a-2)}{6} = \dfrac{2(a+11)}{6}$
2. To remove the fractions, multiply each side by 6, as shown in red, and simplify.	$\dfrac{3(3a-2)}{{}^1\cancel{6}} \times \dfrac{\cancel{6}}{1} = \dfrac{2(a+11)}{{}^1\cancel{6}} \times \dfrac{\cancel{6}}{1}$ $3(3a-2) = 2(a+11)$
3. Expand the brackets then collect the pronumeral terms by subtracting $2a$ from both sides, as shown in purple, and simplify.	$9a - 6 = 2a + 22$ $9a - 6 - 2a = 2a + 22 - 2a$ $7a - 6 = 22$
4. Add 6 to both sides, as shown in pink, and simplify.	$7a - 6 + 6 = 22 + 6$ $7a = 28$
5. Divide both sides by 7, as shown in pink, and simplify.	$\dfrac{{}^1\cancel{7}a}{{}^1\cancel{7}} = \dfrac{{}^4\cancel{28}}{{}^1\cancel{7}}$
6. Write the answer.	$a = 4$

Exercise 6.2 Solving linear equations

learn on

6.2 Exercise	**6.2 Exam questions** on

Simple familiar	Complex familiar	Complex unfamiliar
1, 2, 3, 4, 5, 6, 7, 8, 9, 10, 11, 12, 13, 14, 15	16, 17	18, 19

Simple familiar

1. **WE1** Solve the following equations.

    a. $x - 7 = 12$
    b. $m + 7 = 15$
    c. $p + 12 = 25$
    d. $q - 23 = 27$
    e. $y + 12 = 8$
    f. $k + 21 = 10$

2. Solve the following equations.

    a. $5x = 30$
    b. $4y = 24$
    c. $7z = 56$
    d. $9b = 36$
    e. $3g = -39$
    f. $-8p = 88$

3. Solve the following equations.

    a. $2m - 5 = 7$
    b. $3d - 12 = 24$
    c. $5x - 16 = 14$
    d. $9x - 29 = 25$

4. Solve the following equations.

    a. $2.8 + 0.2b = 18$
    b. $4 - 2x = 7$
    c. $2 - 4g = -6$
    d. $5 = 2x + 13$

5. Solve the following equations.

    a. $3(x - 5) = 24$
    b. $5(2g + 3) = 7$
    c. $8 = 3(x - 2)$
    d. $7 + 2(2x - 9) = 0$

6. **WE2** Solve the following equations.

    a. $3m - 4 = m + 8$
    b. $6g - 18 = g + 52$
    c. $4t + 14 = 10t + 80$
    d. $3w - 27 = 7w - 75$

7. Solve the following equations.

    a. $-7(-2 + 5x) - 6x = 13$
    b. $8(2 - x) + 1 = 9$
    c. $-(3 - 4x) + 3 = 4$
    d. $2(x + 1) + 3(x + 2) = 6$

8. **WE3** Solve the following equations.

    a. $\dfrac{x}{2} + 3 = 5$
    b. $\dfrac{2s}{3} - 7 = 2$
    c. $3 + \dfrac{3x}{2} = 12$
    d. $-11 - \dfrac{x}{5} = -5$
    e. $\dfrac{a - 5}{3} = -2$
    f. $8 = \dfrac{a + 4}{11}$

9. Solve the following equations.

    a. $3j - 22 = -4 - 3j$
    b. $26x - 125 = x$
    c. $2(3x + 1) = 5(2x + 3)$
    d. $-2(3a + 1) = 5(1 - a)$

10. Solve the following linear equations to determine the unknowns.

    a. $2(x+1) = 8$

    b. $n - 12 = -2$

    c. $4d - 7 = 11$

    d. $\dfrac{x+1}{2} = 9$

11. Determine the exact values of the unknowns in the following linear equations.

    a. $14 = 5 - x$

    b. $\dfrac{2(3-x)}{3} = 5$

12. Solve the following linear equations for the pronumerals given in brackets.

    a. $v = u + at \quad (a)$

    b. $\dfrac{x}{p} - r = s \quad (x)$

13. The equation $w = 10t + 120$ represents the amount of water in a tank, w (in litres), at any time, t (in minutes).

    Determine the time, in minutes, that it takes for the tank to have the following amounts of water.

    a. 450 litres

    b. 1200 litres

14. Solve the following equations.

    a. $\dfrac{b+4}{3} = \dfrac{b-8}{5}$

    b. $\dfrac{3y-2}{7} = \dfrac{y+6}{4}$

    c. $\dfrac{4-3y}{2} = \dfrac{1-5y}{3}$

    d. $4 + t = \dfrac{4t+1}{3}$

15. Solve the following equations to determine the unknowns. Express your answers in exact form.

    a. $\dfrac{2-5x}{8} = \dfrac{3}{5}$

    b. $\dfrac{6(3y-2)}{11} = \dfrac{5}{9}$

    c. $\left(\dfrac{4x}{5} - \dfrac{3}{7}\right) + 8 = 2$

    d. $\dfrac{7x+6}{9} + \dfrac{3x}{10} = \dfrac{4}{5}$

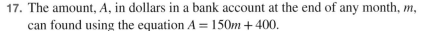

Complex familiar

16. The distance travelled, d (in kilometres), at any time t (in hours) can be found using the equation $d = 95t$.

    Calculate the time in hours that it takes to travel the following distances. Give your answers correct to the nearest minute.

    a. 190 km

    b. 250 km

    c. 65 km

    d. 356.5 km

17. The amount, A, in dollars in a bank account at the end of any month, m, can found using the equation $A = 150m + 400$.

    a. Determine how many months it would take to have the following amounts of money in the bank account.

       i. $1750

       ii. $3200

    b. Determine how many years it would take to have $10 000 in the bank account. Give your answer correct to the nearest month.

18. The temperature, C, in degrees Celsius can be found using the equation $C = \dfrac{5(F-32)}{9}$, where F is the temperature in degrees Fahrenheit. Nora needs to set her oven at 190°C, but her oven's temperature is measured in Fahrenheit.
    Determine the temperature in Fahrenheit that Nora should set her oven to.

19. The height of a plant can be found using the equation
    $h = \dfrac{2(3t+15)}{3}$, where h is the height in cm and t is time in weeks.
    When the plant reaches 60 cm it is given additional plant food. The plant's growth each week for the next 4 weeks is found using the equation $g = t + 2$, where g is the growth each week in cm and t is the time in weeks since additional plant food was given.
    Determine the height of the plant in cm for the next 4 weeks.

Fully worked solutions for this chapter are available online.

LESSON
6.3 Developing linear equations and solving practical problems

SYLLABUS LINKS

- Develop a linear equations from a description in words.
- Solve practical problems involving linear equations.

Source: General Mathematics Senior Syllabus 2024 © State of Queensland (QCAA) 2024; licensed under CC BY 4.0.

6.3.1 Developing linear equations from word descriptions

In practical problems, solving linear equations can answer everyday questions such as the time required to have a certain amount in the bank, the time taken to travel a certain distance, or the number of participants needed to raise a certain amount of money for charity.

Thus, in practical situations it can be useful to create linear equations to solve a problem. The first step is to identify the unknown and choose a variable to represent it.

The linear equation can then be solved as before, and we can use the result to answer the original question.

eles-3031

WORKED EXAMPLE 4 Developing a linear equation from a description in words

Cans of soft drinks are sold at SupaSave in packs of 12 costing $5.40. Form and solve a linear equation to determine the price of 1 can of soft drink.

THINK	WRITE
1. Identify the unknown and choose a pronumeral to represent it.	Let S be the price of a can of soft drink; therefore, $S =$ price of a can of soft drink.
2. Use the given information to write an equation in terms of the pronumeral. *Note:* $5.40 is written as 5.4 in the equation.	$12S = 5.4$
3. To solve for S, divide both sides by 12.	$\dfrac{12S}{12} = \dfrac{5.4}{12}$ $S = 0.45$
4. Interpret the solution in terms of the original problem.	The price of 1 can of soft drink is $0.45 or 45 cents.

General strategy to develop a linear equation from a description in words

- **Read the problem carefully and identify the unknown quantity.**
- **Choose a variable to represent this quantity.**
- **If there are two unknown quantities, find a way to write the second unknown in terms of the first.**
- **Formulate a linear equation for the chosen variable, interpreting the words as mathematical operations.**

You can then solve the equation using one of the methods shown in Lesson 6.2.

TIP

When considering a word problem, you might find highlighting key words useful to help you formulate an equation.

The key words in this case are the words that indicate a relationship between numbers, such as the sum, decrease, double, or less than.

Exercise 6.3 Developing linear equations and solving practical problems

learnon

6.3 Exercise	6.3 Exam questions on

Simple familiar	Complex familiar	Complex unfamiliar
1, 2, 3, 4, 5, 6, 7, 8	9, 10, 11, 12	13

These questions are
even better in jacPLUS!
- Receive immediate feedback
- Access sample responses
- Track results and progress

Find all this and MORE in jacPLUS ▶

Simple familiar

1. **WE4** Artists' pencils at the local art supply store sell in packets of 8 for $17.92. Form and solve a linear equation to determine the price of 1 artists' pencil.

2. Natasha is trying to determine which type of cupcake is the best value for money. The three options Natasha is considering are:
   - 4 red velvet cupcakes for $9.36
   - 3 chocolate delight cupcakes for $7.41
   - 5 caramel surprise cupcakes for $11.80.
   Form and solve linear equations for each type of cupcake to determine which has the cheapest price per cupcake.

3. Three is added to a number and the result is then divided by four, giving an answer of nine. Determine the number.

4. One pair of parallel sides in a parallelogram is 3 times the length of the other pair. Determine the side lengths if the perimeter of the parallelogram is 84 cm.

5. Six times the sum of four plus a number is equal to one hundred and twenty-six. Calculate the number.

6. Tommy is saving for a remote-controlled car that is priced at $49. He has $20 in his piggy bank. Tommy saves $3 of his pocket money every week and puts it in his piggy bank.
   The amount of money in dollars, M, in his piggy bank after w weeks can be found using the rule $M = 3w + 20$. Determine how many weeks it will take for Tommy to have saved enough money to purchase the remote-controlled car.

7. Fred is saving for a holiday and decides to deposit $40 in his bank account each week. At the start of his saving scheme he has $150 in his account.
   a. Calculate how much money Fred will have in his account at the end of the fourth week.
   b. The holiday Fred wants to go on will cost $720 dollars. Determine a rule to describe Fred's savings. Hence, calculate how many weeks it will take Fred to save up enough money to pay for his holiday.

8. Sabrina is a landscape gardener and has been commissioned to work on a rectangular piece of garden. The length of the garden is 6 metres longer than the width, and the perimeter of the garden is 64 metres. Determine the parameters of the garden.

9. One week Jordan bought a bag of his favourite fruit and nut mix at the local market. The next week he saw that the bag was on sale for 20% off the previously marked price. Jordan purchased two more bags at the reduced price. Jordan spent $20.54 in total for the three bags. Calculate the original price of a bag of fruit and nut mix.

10. Yuri is doing his weekly grocery shop and is buying both carrots and potatoes. He calculates that the average weight of a carrot is 60 g and the average weight of a potato is 125 g. Furthermore, he calculates that the average weight of the carrots and potatoes that he purchases is 86 g.
If Yuri's shopping weighed 1.29 kg in total, determine how many of each vegetable he purchased.

11. Ho has a water tank in his back garden that can hold up to 750 L in water. At the start of a rainy day (at 0:00) there is 165 L in the tank, and after a heavy day's rain (at 24:00) there is 201 L in the tank.
Assuming that the rain fell consistently during the 24-hour period, set up a linear equation to represent the amount of rain in the tank at any point during the day to determine the time of day at which the amount of water in the tank reached 192 L.

12. A large fish tank is being filled with water. After 1 minute the height of the water is 2 cm and after 4 minutes the height of the water is 6 cm.
The height of the water, h, in cm after t minutes can be modelled by a linear equation.

a. Determine whether the fish tank was empty of water before being filled. Justify your answer by using calculations.

b. Construct an equation between the height of water in the fish tank and the time to determine the height of the water in the fish tank after five minutes.

13. Michelle and Lydia live 325 km apart. On a Sunday they decide to drive to each other's respective towns. They pass each other after 2.5 hours. If Michelle drives an average of 10 km/h faster than Lydia, calculate the speed at which they are both travelling.

Fully worked solutions for this chapter are available online.

LESSON
6.4 Review

6.4.1 Summary

6.4 Exercise

learn on

6.4 Exercise	6.4 Exam questions on

Simple familiar	Complex familiar	Complex unfamiliar
1, 2, 3, 4, 5, 6, 7, 8, 9, 10, 11, 12	13, 14, 15, 16, 17, 18	19, 20

Simple familiar

1. **MC** The value of x in the linear equation $3(2x + 5) = 12$ is:

   A. -1.5 B. -1 C. -0.5 D. 2

2. **MC** The equation $v = u + at$ is an equation for motion, given an initial velocity, a rate of acceleration and a period of time.
   The correct solution to this equation for a is:

   A. $a = \dfrac{v - u}{t}$ B. $a = \dfrac{u - v}{t}$ C. $a = v - tu$ D. $a = u - tv$

3. **MC** The value of x in the linear equation $4(x - 1) = 12$ is:

   A. 3 B. 4 C. 2 D. 5

4. **MC** The solution to the equation $5 + \dfrac{x}{2} = 12$ is:

   A. $x = 1$ B. $x = 19$ C. $x = 16$ D. $x = 14$

5. **MC** Five is added to 3 times a number and the result is 17. The number is:

   A. 5 B. 12 C. 3 D. 4

6. **MC** The solution to the equation $-8x = -96$ is:

   A. -12 B. 11 C. 8 D. 12

7. **MC** Identify the equation for which equation $x = -5$ is a solution.

   A. $x - 25 = -20$ B. $\dfrac{x}{25} = 5$ C. $-6x = 30$ D. $\dfrac{40}{x} = 8$

8. **MC** The solution to the equation $2(r+3) = 3(r-2)$ is:

   A. -12 **B.** 12 **C.** -2 **D.** 2

9. **MC** The solution to the equation $\dfrac{2(5-x)}{7} = 1$ is:

   A. $-\dfrac{2}{3}$ **B.** $\dfrac{3}{2}$ **C.** $-\dfrac{3}{2}$ **D.** $\dfrac{3}{2}$

10. **MC** The solution to the equation $mp + x = s$ for p is:

    A. $p = \dfrac{s-x}{m}$ **B.** $p = \dfrac{x-s}{m}$ **C.** $p = \dfrac{m-x}{s}$ **D.** $p = \dfrac{x-m}{s}$

11. Paulie wants to make a group booking of tickets to a show. Each adult ticket cost $69.50 and each child's ticket costs $34.70. There are 15 people in the group.
    Determine how many adult tickets and how many children's tickets Paulie bought, if she spent a total of $729.30.

12. Solve each of the following equations for the unknown.

    a. $3x + 15 = 14$

    b. $\dfrac{5 - 2m}{3} = -2$

Complex familiar

13. Solve the following equations.

    a. $\dfrac{3x - 4}{6} = \dfrac{x + 2}{5}$

    b. $\dfrac{3 - y}{2} = \dfrac{1 + 4y}{3}$

14. Petra is doing a survey of how humans and pets are in her extended family. She gives the following information to her friend Juliana:
    - There are 33 humans and pets (combined).
    - The combined number of legs (pets and owners) is 94.
    - Each human has 2 legs and each pet has 4 legs.

    a. If h is the number of humans, express the number of pets in terms of h.
    b. Write expressions for the total number of human legs and the total number of pet legs in terms of h.

15. Solve each of the following equations for the unknown.

    a. $\dfrac{3s}{4} + 6 = 10$

    b. $\dfrac{-2(3t + 1)}{5} + 3 = 9$

16. Consider the linear equation $p = 5(2n + 4)$.
    Determine the value of n when p is 310.

17. The following numbers are the first 5 terms in a number sequence.
$$3, \frac{11}{3}, \frac{13}{3}, \frac{15}{3}, \frac{17}{3}$$
    Write an equation that determines the value of the terms for this number sequence.

18. If $x + y = 30$, $y + z = 46$ and $x + z = 45$, calculate $x + y + z$.

19. A study of the homework habits of a group of students showed that the amount of weekly homework, in hours, completed by the students had an effect on their performance on the weekly assessment tasks. The table shown represents the number of weekly homework hours spent by the group of students and the average percentage mark they achieved on their weekly assessment tasks.

Hours of homework, h	1	2	3	4	5	6	7
Average percentage mark, m	22	31	40	49	58	67	76

The teacher used this study to demonstrate to their students the effect of the amount of weekly homework on the students' performance.
Freda argued that this study only worked up to a certain number of hours of homework.
Using this study, justify whether Freda has a valid argument.

20. A local supermarket bought milk for $2.45 per litre and stored it in two refrigerators. During the night, one refrigerator broke down and ruined 22 litres. The remaining milk was sold for $4.20 per litre. Determine how many litres the supermarket sold if they made a profit of $91.35.

Fully worked solutions for this chapter are available online.

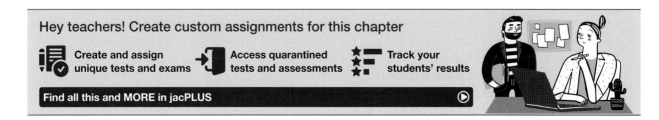

Hey teachers! Create custom assignments for this chapter

Create and assign unique tests and exams

Access quarantined tests and assessments

Track your students' results

Find all this and MORE in jacPLUS

Answers

Chapter 6 Linear equations

6.2 Solving linear equations

6.2 Exercise

1. a. $x = 19$ b. $m = 8$ c. $p = 13$
   d. $q = 50$ e. $y = -4$ f. $k = -11$

2. a. $x = 6$ b. $y = 6$ c. $z = 8$
   d. $b = 4$ e. $g = -13$ f. $p = -11$

3. a. $m = 6$ b. $d = 12$
   c. $x = 6$ d. $x = 6$

4. a. $b = 76$ b. $x = \dfrac{-3}{2}$
   c. $g = 2$ d. $x = -4$

5. a. $x = 13$ b. $g = \dfrac{-4}{5}$
   c. $x = \dfrac{14}{3}$ d. $x = \dfrac{11}{4}$

6. a. $m = 6$ b. $g = 14$
   c. $t = -11$ d. $w = 12$

7. a. $x = \dfrac{1}{41}$ b. $x = 1$
   c. $x = 1$ d. $x = \dfrac{-2}{5}$

8. a. $x = 4$ b. $s = \dfrac{27}{2}$ c. $x = 6$
   d. $x = -30$ e. $a = -1$ f. $a = 84$

9. a. $j = 3$ b. $x = 5$
   c. $x = \dfrac{-13}{4}$ d. $a = -7$

10. a. $x = 3$ b. $n = 10$
    c. $d = 4.5$ d. $x = 17$

11. a. $x = -9$ b. $x = -4.5$

12. a. $a = \dfrac{v - u}{t}$ b. $x = p(r + s)$

13. a. 33 minutes b. 108 minutes

14. a. $b = -22$ b. $y = 10$
    c. $y = -10$ d. $t = 11$

15. a. $x = \dfrac{-14}{25}$ b. $y = \dfrac{163}{162}$
    c. $x = \dfrac{-195}{28}$ d. $x = \dfrac{12}{97}$

16. a. 2 hours b. 2 hours 38 minutes
    c. 41 minutes d. 3 hours 45 minutes

17. a. i. 9 months
    ii. 19 months ($= 18.67$ months)
    b. 5 years, 4 months

18. $374\,°F$

19. 63 cm, 67 cm, 72 cm, 78 cm

6.3 Developing linear equations and solving practical problems

6.3 Exercise

1. $2.24

2. The red velvet cupcakes are the cheapest per cupcake — $2.34, compared to $2.47 for chocolate delight and $2.36 for caramel surprise.

3. 33

4. 10.5 cm and 31.5 cm

5. 17

6. 10 weeks

7. a. $310
   b. $A\,(\$) = 150 + 40w$; 15 weeks

8. 13 m by 19 m

9. $7.90

10. 9 carrots and 6 potatoes

11. 6:00 pm

12. a. At $t = 0$, $h = \dfrac{2}{3}$ cm. Hence, the fish tank was not empty of water before being filled.
    b. At $t = 5$, $h = 7\dfrac{1}{3}$ cm.

13. Michelle: 70 km/h, Lydia: 60 km/h

6.4 Review

6.4 Exercise

1. C 2. A
3. B 4. D
5. D 6. D
7. C 8. B
9. B 10. A

11. Paulie purchased 6 adult tickets and 9 children's tickets.

12. a. $x = 3$ b. $m = 5.5$

13. a. $x = \dfrac{32}{9}$ b. $y = \dfrac{7}{11}$

14. a. $33 - h$
    b. Total number of human legs $= 2h$; total number of pet legs $4(33 - h)$

15. a. $s = 5\dfrac{1}{3}$ b. $t = -5\dfrac{1}{3}$

16. $n = 19$

17. $t = \dfrac{2}{3}n + \dfrac{7}{3}$

18. $\dfrac{121}{2} = 60.5$

19. When $h = 10$, the average percentage mark is 103%. As this is not possible, Freda has a valid argument that the study does not work for 10 hours or more.

20. 105 litres

7 Straight-line graphs

LESSON SEQUENCE

Fully worked solutions for this chapter are available online.

EXAM PREPARATION

Access exam-style questions in every lesson, available online.

 Resources

 Solutions   Solutions — Chapter 7 (sol-1248)

 Exam question   Exam question booklet — Chapter 7 (eqb-0264)

📄 **Digital documents** Learning matrix — Chapter 7 (doc-41485)
                        Chapter summary — Chapter 7 (doc-41486)

LESSON
7.1 Overview

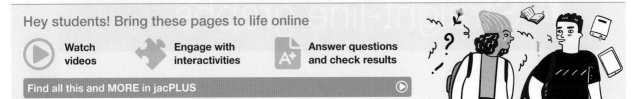
7.1.1 Introduction

A useful application of linear equations is to make predictions about what will happen in the future. For example, if a linear profit equation is modelled, then this equation could be used to predict future profits. One way of modelling a linear equation is to plot the relevant data and then draw a linear line of best fit that is modelled. The graph that the data is plotted on is known as a scatter plot. This enables people to determine the relationship between the two variables.

It is interesting to note that a lot of world records follow a linear trend over time. One event that challenges this is the men's long jump world record. At the 1968 Summer Olympics Bob Beamon (pictured) smashed the record by an amazing 55 cm with a jump of 8.90 m. This jump certainly went against the linear trend. This record stood until 1991 when Mike Powell jumped 8.95 m at the World Championships. If you plot the previous world records and draw a line of best fit, it clearly shows this went against the linear trend over the previous 60 or so years.

7.1.2 Syllabus links

Lesson	Lesson title	Syllabus links
7.2	Constructing straight-line graphs	○ Understand and use the slope-intercept form of a linear function, $y = mx + c$ where m is slope (gradient) and c is y-intercept.
		○ Construct a straight line graph using a linear function of the form, $y = mx + c$ where m is slope (gradient) and c is y-intercept.
		○ Determine the slope (gradient), x-intercept and y-intercept of a straight line from its equation.
7.3	Determine and interpret the slope and intercepts	○ Determine the slope (gradient), x-intercept and y-intercept of a straight line from its graph.
7.4	Modelling with straight-line graphs	○ Interpret, in context, the slope (gradient) and intercept of a linear function used to model and analyse a practical situation.
		○ Construct and analyse a straight-line graph to model a given linear relationship, e.g. modelling the cost of filling a fuel tank of a car against the number of litres of petrol required.

Source: General Mathematics Senior Syllabus 2024 © State of Queensland (QCAA) 2024; licensed under CC BY 4.0.

LESSON
7.2 Constructing straight-line graphs

SYLLABUS LINKS

- Understand and use the slope-intercept form of a linear function, $y = mx + c$ where m is slope (gradient) and c is y-intercept.
- Construct a straight line graph using a linear function of the form, $y = mx + c$ where m is slope (gradient) and c is y-intercept.
- Determine the slope (gradient), x-intercept and y-intercept of a straight line from its equation.

Source: General Mathematics Senior Syllabus 2024 © State of Queensland (QCAA) 2024; licensed under CC BY 4.0.

7.2.1 Linear functions

A function is a relationship between a set of inputs and outputs, such that each input is related to exactly one output. Each input and output of a function can be expressed as an ordered pair, with the first element of the pair being the input and the second element of the pair being the output.

A function of x is denoted as $f(x)$. For example, if we have the function $f(x) = x + 3$, then each output will be exactly 3 greater than each input.

This is shown in the input/output table below.

Input x	Output $f(x)$
0	3
1	4
2	5
3	6

A linear function is a set of ordered pairs that form a straight line when graphed.

The slope (gradient) of a linear function

The **slope** of a straight-line function, also known as the **gradient**, determines the change in the y-value for each change in x-value. The slope can be found by analysing the equation, by examining the graph or by calculating the change in values if two points are given. The slope is typically represented by the pronumeral m.

A positive slope means that the y-value is increasing as the x-value increases, and a negative slope means that the y-value is decreasing as the x-value increases.

A slope of $\dfrac{a}{b}$ means that for every increase of b in the x-value, there is an increase of a in the y-value.

For example, a slope of $\dfrac{2}{3}$ means that for every increase

of 3 in the x-value, the y-value increases by 2.

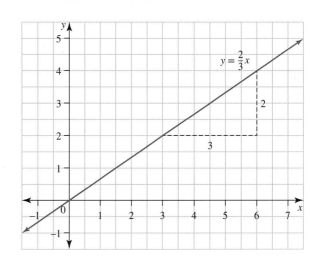

x- and *y*-intercepts

The ***x*-intercept** of a linear function is the point where the graph of the equation crosses the *x*-axis. This occurs when $y = 0$.

The ***y*-intercept** of a linear function is the point where the graph of the equation crosses the *y*-axis. This occurs when $x = 0$.

In the graph of $y = x + 3$, we can see that the *x*-intercept is at $(-3, 0)$ and the *y*-intercept is at $(0, 3)$. These points can also be determined algebraically by putting $y = 0$ and $x = 0$ into the equation.

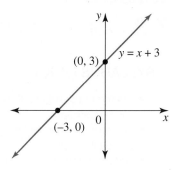

Slope–intercept form

All linear equations relating the variables *x* and *y* can be rearranged into the form $y = mx + c$, where *m* is the slope (gradient). This is known as the slope–intercept form of the equation.

If a linear equation is in **slope–intercept form**, the number and sign in front of the *x*-value gives the value of the slope of the equation. For example, in $y = 4x + 5$, the slope is 4.

The value of *c* in linear equations written in slope–intercept form **is the *y*-intercept** of the equation. This is because the *y*-intercept occurs when $x = 0$, and when $x = 0$ the equation simplifies to $y = c$. The value of *c* in $y = 4x + 5$ is 5.

Slope–intercept form

The slope–intercept form of any linear function is:

$$y = mx + c$$

where *m* is the slope (gradient)

** *c* is the *y*-intercept.**

The slope–intercept form can also be written as $y = a + bx$, where *a* is the *y*-intercept and *b* is the slope. This form is commonly used when solving linear problems using technology.

WORKED EXAMPLE 1 Determining the slope and *y*-intercepts from a linear equation

Identify the slopes (gradients) and *y*-intercepts of the following linear equations.

a. $y = 5x + 2$ b. $y = \dfrac{x}{2} - 3$ c. $y = -2x + 4$

d. $2y = 4x + 3$ e. $3y - 4x = 12$

THINK	WRITE
a. 1. Write the equation. It is in the form $y = mx + c$.	a. $y = 5x + 2$
2. Identify the coefficient of *x*.	The coefficient of *x* is 5.
3. Identify the value of *c*.	The value of *c* is 2.
4. Write the answer.	The slope is 5 and the *y*-intercept is 2.

b. 1. Write the equation. It is in the form $y = mx + c$.

b. $y = \dfrac{x}{2} - 3$

2. Identify the coefficient of x. $\dfrac{x}{2}$ is equivalent to $\dfrac{1x}{2}$ or $\dfrac{1}{2}x$, so the coefficient of x is $\dfrac{1}{2}$.

x has been multiplied by $\dfrac{1}{2}$, so the coefficient is $\dfrac{1}{2}$.

3. Identify the value of c.

The value of c is -3.

4. Write the answer.

The slope is $\dfrac{1}{2}$ and the y-intercept is -3.

c. 1. Write the equation. It is in the form $y = mx + c$.

c. $y = -2x + 4$

2. Identify the coefficient of x.

The coefficient of x is -2 (the coefficient includes the sign).

3. Identify the value of c.

The value of c is 4.

4. Write the answer.

The slope is -2 and the y-intercept is 4.

d. 1. Write the equation. Rearrange the equation so that it is in the form $y = mx + c$.

d. $2y = 4x + 3$

$$\dfrac{2y}{2} = \dfrac{4x}{2} + \dfrac{3}{2}$$

$$y = 2x + \dfrac{3}{2}$$

2. Identify the coefficient of x.

The coefficient of x is 2.

3. Identify the value of c.

The value of c is $\dfrac{3}{2}$.

4. Write the answer.

The slope is 2 and the y-intercept is $\dfrac{3}{2}$.

e. 1. Write the equation. Rearrange the equation so that it is in the form $y = mx + c$.

e.
$$3y - 4x = 12$$
$$3y - 4x + 4x = 12 + 4x$$
$$3y = 4x + 12$$

$$\dfrac{3y}{3} = \dfrac{4}{3}x + \dfrac{12}{3}$$

$$y = \dfrac{4}{3}x + 4$$

2. Identify the coefficient of x.

The coefficient of x is $\dfrac{4}{3}$.

3. Identify the value of c.

The value of c is 4.

4. Write the answer.

The slope is $\dfrac{4}{3}$ and the y-intercept is 4.

7.2.2 Constructing straight-line graphs

Linear graphs can be constructed by plotting the points and then ruling a line between them, as shown in the diagram.

If the points or a table of values are not given, then the points can be found by substituting x-values into the rule and determining the corresponding y-values. If a table of values is provided, then the graph can be constructed by plotting the points given and joining them.

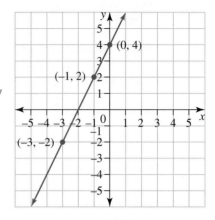

WORKED EXAMPLE 2 Constructing a linear graph from a set of points

Construct a linear graph that passes through the points (−1, 2), (0, 4), (1, 6) and (3, 10):
a. without technology
b. using technology.

THINK	DRAW/DISPLAY
a. 1. Using grid paper, rule up the Cartesian plane (set of axes) and plot the points.	a. 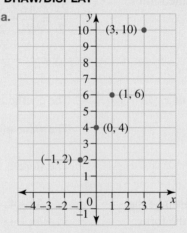
2. Using a ruler, rule a line through the points.	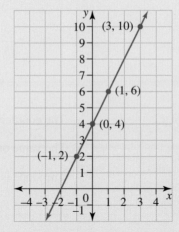

b. 1. Enter the points into your calculator or spreadsheet (the first number corresponds to the *x*-values and the second to the *y*-values).

b.

	A	B
1	**x-values**	**y-values**
2	-1	2
3	0	4
4	1	6
5	3	10

2. Highlight the cells.

	A	B
1	**x-values**	**y-values**
2	-1	2
3	0	4
4	1	6
5	3	10
6		

3. Use the scatterplot function of your calculator or spreadsheet to display the plot and the trend line.

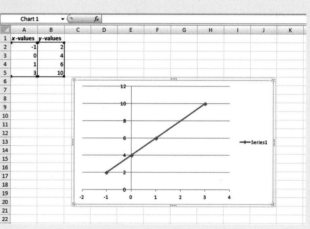

Constructing straight-line graphs using the slope and *y*-intercept method

A linear graph can be constructed by using the slope (gradient) and *y*-intercept. The *y*-intercept is marked on the *y*-axis, and then another point is found by using the slope.

For example, a slope of 2 means that for an increase of 1 in the *x*-value, the *y*-value increases by 2. If the *y*-intercept is $(0, 3)$, then add 1 to the *x*-value $(0 + 1)$ and 2 to the *y*-value $(3 + 2)$ to identify another point that the line passes through, $(1, 5)$.

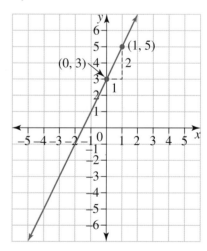

Using the slope (gradient) and the *y*-intercept, sketch the graph of each of the following.

a. **A linear graph with a slope of 3 and a *y*-intercept of 1**

b. $y = -2x + 4$

c. $y = \dfrac{3}{4}x - 2$

THINK

a. 1. Interpret the slope.

   2. Write the coordinates of the *y*-intercept.

   3. Determine the *x*- and *y*-values of another point using the slope.

   4. Construct a set of axes and plot the two points. Using a ruler, rule a line through the points.

WRITE

a. A slope of 3 can be written as $\dfrac{3}{1}$, which means that for an increase of 1 in the *x*-value, there is an increase of 3 in the *y*-value.

y-intercept: $(0, 1)$

New *x*-value $= 0 + 1 = 1$
New *y*-value $= 1 + 3 = 4$
Another point on the graph is $(1, 4)$.

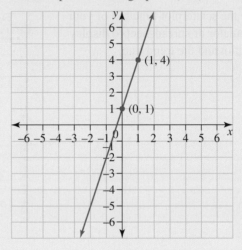

b. 1. Identify the value of the slope and *y*-intercept.

   2. Interpret the slope.

   3. Write the coordinates of the *y*-intercept.

   4. Determine the *x*- and *y*-values of another point using the slope.

b. $y = -2x + 4$ has a slope of -2 and a *y*-intercept of 4.

A slope of -2 means that for an increase of 1 in the *x*-value, there is a decrease of 2 in the *y*-value

y-intercept: $(0, 4)$

New *x*-value $= 0 + 1 = 1$
New *y*-value $= 4 - 2 = 2$
Another point on the graph is $(1, 2)$.

5. Construct a set of axes and plot the two points. Using a ruler, rule a line through the points.

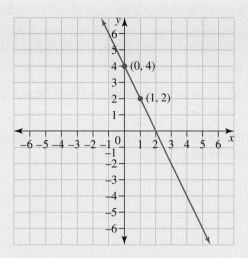

c. 1. Identify the value of the slope and y-intercept.

c. $y = \dfrac{3}{4}x - 2$ has a slope of $\dfrac{3}{4}$ and a y-intercept of -2.

2. Interpret the slope.

A slope of $\dfrac{3}{4}$ means that for an increase of 4 in the x-value, there is an increase of 3 in the y-value

3. Write the coordinates of the y-intercept.

y-intercept: $(0, -2)$

4. Determine the x- and y-values of another point using the slope.

New x-value $= 0 + 4 = 4$
New y-value $= -2 + 3 = 1$
Another point on the graph is $(4, 1)$.

5. Construct a set of axes and plot the two points. Using a ruler, rule a line through the points.

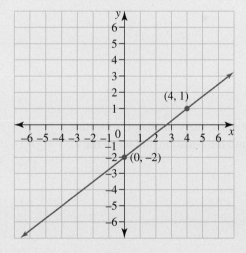

Sketching graphs using the x- and y-intercepts

If the points of a linear graph where the line crosses the x- and y-axes (the x- and y-intercepts) are known, then the graph can be constructed by marking these points and ruling a line through them.

To identify the x-intercept, substitute $y = 0$ into the equation and then solve the equation for x.

To identify the y-intercept, substitute $x = 0$ into the equation and then solve the equation for y.

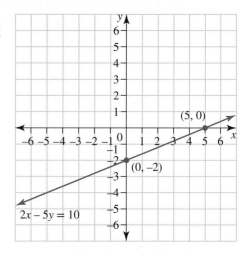

WORKED EXAMPLE 4 Constructing a linear graph using the x- and y-intercepts

Calculate the values of the x- and y-intercepts for the following linear equations, and hence sketch their graphs.

a. $3x + 4y = 12$ **b.** $y = 5x$ **c.** $3y = 2x + 1$

THINK

a. 1. To determine the x-intercept, substitute $y = 0$ and solve for x.

2. To determine the y-intercept, substitute $x = 0$ into the equation and solve for y.

WRITE

a. x-intercept: $y = 0$
$$3x + 4y = 12$$
$$3x + 4 \times 0 = 12$$
$$3x = 12$$
$$\frac{3x}{3} = \frac{12}{3}$$
$$x = 4$$
x-intercept: $(4, 0)$

y-intercept: $x = 0$
$$3x + 4y = 12$$
$$3 \times 0 + 4y = 12$$
$$\frac{4y}{4} = \frac{12}{4}$$
$$y = 3$$
y-intercept: $(0, 3)$

3. Draw a set of axes and plot the *x*- and *y*-intercepts. Draw a line through the two points.

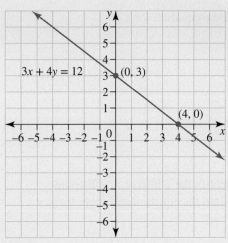

b. 1. To identify the *x*-intercept, substitute $y = 0$ into the equation and solve for *x*.

b. *x*-intercept: $y = 0$
$$y = 5x$$
$$0 = 5x$$
$$x = 0$$
x-intercept: $(0, 0)$

2. To identify the *y*-intercept, substitute $x = 0$ into the equation and solve for *y*.

y-intercept: $x = 0$
$$y = 5x$$
$$= 5 \times 0$$
$$= 0$$
y-intercept: $(0, 0)$

3. As the *x*- and *y*-intercepts are the same, we need to determine another point on the graph. Substitute $x = 1$ into the equation.

$$y = 5x$$
$$= 5 \times 1$$
$$= 5$$
Another point on the graph is $(1, 5)$.

4. Draw a set of axes. Plot the intercept and the second point. Draw a line through the intercepts.

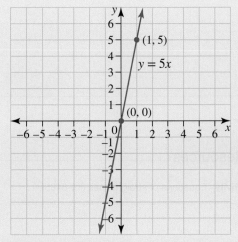

c. 1. To identify the *x*-intercept, substitute $y = 0$ into the equation.

c. *x*-intercept: $y = 0$
$$3y = 2x + 1$$
$$3 \times 0 = 2x + 1$$
$$0 = 2x + 1$$

2. Solve the equation for x.

$$0 - 1 = 2x + 1 - 1$$
$$-1 = 2x$$

$$\frac{-1}{2} = \frac{2x}{2}$$

$$x = \frac{-1}{2}$$

x-intercept: $\left(-\dfrac{1}{2}, 0\right)$

3. To identify the y-intercept, substitute $x = 0$ into the equation and solve for y.

y-intercept: $x = 0$
$$3y = 2 \times 0 + 1$$
$$3y = 1$$

$$\frac{3y}{3} = \frac{1}{3}$$

$$y = \frac{1}{3}$$

y-intercept: $\left(0, \dfrac{1}{3}\right)$

4. Draw a set of axes and mark the x- and y-intercepts. Draw a line through the intercepts.

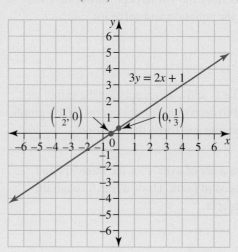

Resources

Interactivity   Equations of straight lines (int-6485)

Exercise 7.2 Constructing straight-line graphs

7.2 Exercise	7.2 Exam questions

Simple familiar	Complex familiar	Complex unfamiliar
1, 2, 3, 4, 5, 6, 7, 8	9, 10, 11	12

Simple familiar

1. **WE1** Identify the slopes (gradients) and y-intercepts of the following linear equations.

   a. $y = 2x + 1$

   b. $y = -x + 3$

   c. $y = \frac{1}{2}x + 4$

   d. $4y = 4x + 1$

2. Determine the slopes and y-intercepts of the following linear equations.

   a. $y = \frac{3x - 1}{5}$

   b. $y = 5(2x - 1)$

   c. $y = \frac{3 - x}{2}$

   d. $2y + 3x = 6$

3. Determine the slope and y-intercept in each of the following linear equations.

   a. $x - y = 4$

   b. $3x - 2y = 6$

   c. $x = 4y - 1$

   d. $5x - 2y = 4$

4. **WE2** Construct a straight-line graph that passes through the points $(2, 5), (4, 9)$ and $(0, 1)$:

   a. without technology

   b. using a spreadsheet or otherwise.

5. **WE3** Using the slope and the y-intercept, sketch the following linear graphs.

   a. Slope $= 2, y$-intercept $= 5$

   b. Slope $= -3, y$-intercept $= 0$

   c. Slope $= \frac{1}{2}, y$-intercept $= 3$

   d. Slope $= 2, y$-intercept $= 4$

6. **WE4** Calculate the values of the x- and y-intercepts for the following linear equations, and hence sketch their graphs.

   a. $2x + 5y = 20$

   b. $4y = 3x + 5$

7. Calculate the values of the x- and y-intercepts for the following linear equations, and hence sketch their graphs.

   a. $2x + y = 6$

   b. $y = 3x + 9$

   c. $2y = 3x + 4$

   d. $3y - 4 = 5x$

8. Using the slope, identify another point in addition to the y-intercept that lies on each the following straight lines. Hence, sketch the graph of each straight line.

   a. Slope $= 4, y$-intercept $= 3$

   b. Slope $= -3, y$-intercept $= 1$

   c. Slope $= \frac{1}{4}, y$-intercept $= 4$

   d. Slope $= -\frac{2}{5}, y$-intercept $= -2$

Complex familiar

9. A line has a slope of -2 and passes through the points $(1, 4)$ and $(a, 8)$. Determine the value of a.

10. A straight line passes through the following points: $(3, 7), (0, a), (2, 5)$ and $(-1, -1)$. Construct a graph and hence determine the value of the unknown, a.

11. A straight line passes through the points $(2, 5), (0, 9), (-1, 11)$ and $(4, a)$. Construct a graph of the straight line and hence determine the value of the unknown, a.

Complex unfamiliar

12. Determine the x- and y-intercepts for the linear function $ax + by - 3x + y = 2$ in terms of a and b.

Fully worked solutions for this chapter are available online.

LESSON
7.3 Determine and interpret the slope and intercepts

SYLLABUS LINKS

- Determine the slope (gradient), x-intercept and y-intercept of a straight line from its graph.

Source: General Mathematics Senior Syllabus 2024 © State of Queensland (QCAA) 2024; licensed under CC BY 4.0.

7.3.1 The equation of a straight line

Given the slope and y-intercept

When we are given the slope and y-intercept of a straight line, we can enter these values into the equation $y = mx + c$ to determine the equation of the straight line. Remember that m is equal to the value of the slope and c is equal to the value of the y-intercept.

For example, if we are given a slope of 3 and a y-intercept of 6, then the equation of the straight line would be $y = 3x + 6$.

Given the slope and one point

When we are given the slope and one point of a straight line, we need to establish the value of the y-intercept to determine the equation of the straight line. This can be done by substituting the coordinates of the given point into the equation $y = mx + c$ and then solving for c.

Remember that m is equal to the value of the slope so this can also be substituted into the equation.

Given two points

When we are given two points of a straight line, we can determine the value of the slope of a straight line between these points as discussed in Lesson 7.2 (by using $m = \dfrac{y_2 - y_1}{x_2 - x_1}$). Once the slope has been found, we can determine the y-intercept by substituting one of the points into the equation $y = mx + c$ and then solving for c.

WORKED EXAMPLE 5 Determining the equation of a straight line

Determine the equations of the following straight lines.
a. **A straight line with a slope of 2 passing through the point (3, 7)**
b. **A straight line passing through the points (1, 6) and (3, 0)**
c. **A straight line passing through the points (2, 5) and (5, 5)**

THINK	WRITE
a. 1. Write the slope–intercept form of a straight line.	**a.** $y = mx + c$
2. Substitute the value of the slope into the equation (in place of m).	Slope $= m = 2$ $y = 2x + c$
3. Substitute the values of the given point into the equation and solve for c.	$(3, 7)$ $7 = 2(3) + c$ $7 = 6 + c$ $c = 1$
4. Substitute the value of c back into the equation and write the answer.	The equation of the straight line is $y = 2x + 1$.
b. 1. Write the formula to determine the slope given two points.	**b.** $m = \dfrac{y_2 - y_1}{x_2 - x_1}$
2. Let one of the given points be (x_1, y_1) and let the other point be (x_2, y_2).	Let $(1, 6) = (x_1, y_1)$. Let $(3, 0) = (x_2, y_2)$.
3. Substitute the values into the equation to determine the value of m.	$m = \dfrac{0 - 6}{3 - 1}$ $= \dfrac{-6}{2}$ $= -3$
4. Substitute the value of m into the equation $y = mx + c$.	$y = mx + c$ $y = -3x + c$
5. Substitute the values of one of the points into the equation and solve for c. *Note:* The point $(1, 6)$ could also be used.	$(3, 0)$ $0 = -3(3) + c$ $0 = -9 + c$ $c = 9$
6. Substitute the value of c back into the equation and write the answer.	The equation of the straight line is $y = -3x + 9$.
c. 1. Write the equation to determine the slope given two points.	**c.** $m = \dfrac{y_2 - y_1}{x_2 - x_1}$
2. Let one of the given points be (x_1, y_1) and let the other point be (x_2, y_2).	Let $(2, 5) = (x_1, y_1)$. Let $(5, 5) = (x_2, y_2)$.
3. Substitute the values into the equation to determine the value of m.	$m = \dfrac{5 - 5}{5 - 2}$ $= \dfrac{0}{3}$ $= 0$
4. A slope of 0 indicates that the straight line is horizontal and the equation of the line is of the form $y = c$.	$y = c$

<table>
<tr>
<td>5. Substitute the values of one of the points into the equation and solve for c.</td>
<td>$(2, 5)$
$5 = c$
$c = 5$</td>
</tr>
<tr>
<td>6. Substitute the value of c back into the equation and write the answer.</td>
<td>The equation of the straight line is $y = 5$.</td>
</tr>
</table>

For part **b** of Worked example 5, try substituting the other point into the equation at step **5**. You will find that the calculated value of c is the same, giving you the same equation as an end result.

7.3.2 Determining the slope (gradient) from a graph

The value of the slope can be found from a graph of a linear function. The slope can be found by selecting two points on the line, then calculating the change in the y-values and dividing by the change in the x-values.

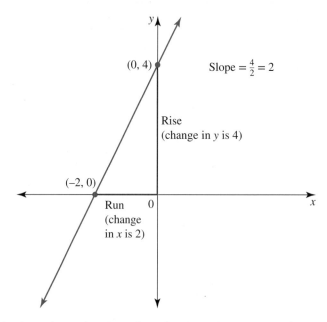

In other words, the general rule to determine the value of a slope that passes through the points (x_1, y_1) and (x_2, y_2) is as follows.

Slope (gradient) formula

$$m = \frac{\text{rise}}{\text{run}} = \frac{y_2 - y_1}{x_2 - x_1}$$

For all horizontal lines, the y-values will be equal to each other, so the numerator of $\frac{y_2 - y_1}{x_2 - x_1}$ will be 0. Therefore, the slope of horizontal lines is 0.

For all vertical lines, the x-values will be equal to each other, so the denominator of $\frac{y_2 - y_1}{x_2 - x_1}$ will be 0. Dividing a value by 0 is undefined; therefore, the slope of vertical lines is undefined.

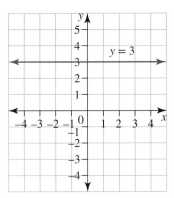

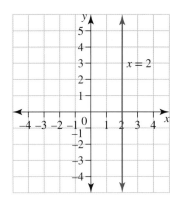

WORKED EXAMPLE 6 Determining the slope from a graph

Determine the values of the slopes of the following graphs.

a.

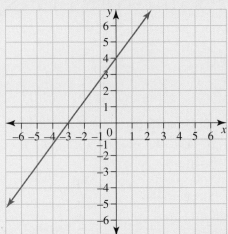

b.

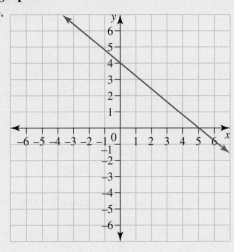

THINK

a. 1. Identify two points on the graph. (Select the *x*- and *y*-intercepts.)

WRITE

a.

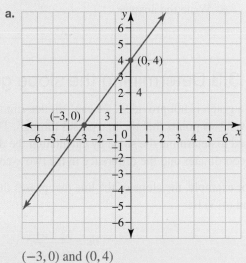

$(-3, 0)$ and $(0, 4)$

2. Determine the rise in the graph (change in *y*-values).

$4 - 0 = 4$

3. Determine the run in the graph (change in *x*-values).

$0 - (-3) = 3$

4. Substitute the values into the formula for the slope.

$$\text{Slope} = \frac{\text{rise}}{\text{run}}$$

$$= \frac{4}{3}$$

5. Write the answer.

The slope is $\frac{4}{3}$.

b. 1. Identify two points on the graph.(Select the x- and y-intercepts.)

b.

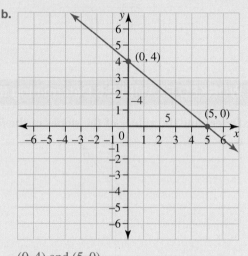

$(0, 4)$ and $(5, 0)$

2. Determine the rise in the graph (change in y-values).

$0 - 4 = -4$

3. Determine the change in the x-values.

$5 - 0 = 5$

4. Substitute the values into the formula for the slope.

$$\text{Slope} = \frac{\text{rise}}{\text{run}}$$

$$= -\frac{4}{5}$$

5. Write the answer.

The slope is $-\frac{4}{5}$.

7.3.3 Determining the slope given two points

If a graph is not provided, we can still determine the slope if we are given two points that the line passes through. The same formula is used to determine the slope by calculating the difference in the two y-coordinates and the difference in the two x-coordinates:

For example, the slope of the line that passes through the points $(1, 1)$ and $(4, 3)$ is:

$$m = \frac{y_2 - y_1}{x_2 - x_1}$$

$$= \frac{3 - 1}{4 - 1}$$

$$= \frac{2}{3}$$

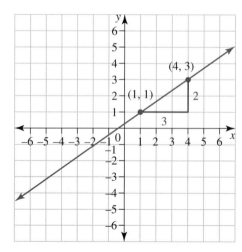

WORKED EXAMPLE 7 Determining the slope from two points

Determine the value of the slopes of the linear graphs that pass through the following points.

a. $(4, 6)$ and $(5, 9)$

b. $(2, -1)$ and $(0, 5)$

c. $(0.5, 1.5)$ and $(-0.2, 1.8)$

THINK	WRITE
a. 1. Number the points.	a. Let $(4, 6) = (x_1, y_1)$ and $(5, 9) = (x_2, y_2)$.
2. Write the formula for the slope and substitute the values.	$m = \dfrac{y_2 - y_1}{x_2 - x_1}$ $= \dfrac{9 - 6}{5 - 4}$ $= \dfrac{3}{1}$
3. Simplify the fraction and write the answer.	The slope is 3.
b. 1. Number the points.	b. Let $(2, -1) = (x_1, y_1)$ and $(0, 5) = (x_2, y_2)$.
2. Write the formula for the slope and substitute the values.	$m = \dfrac{y_2 - y_1}{x_2 - x_1}$ $= \dfrac{5 - (-1)}{0 - 2}$ $= \dfrac{6}{-2}$
3. Simplify the fraction and write the answer.	The slope is -3.
c. 1. Number the points.	c. Let $(0.5, 1.5) = (x_1, y_1)$ and $(-0.2, 1.8) = (x_2, y_2)$.
2. Write the formula for the slope and substitute the values.	$m = \dfrac{y_2 - y_1}{x_2 - x_1}$ $= \dfrac{1.8 - 1.5}{(-0.2) - 0.5}$ $= \dfrac{0.3}{-0.7}$
3. Simplify the fraction and write the answer.	The slope is $-\dfrac{3}{7}$.

Exercise 7.3 Determine and interpret the slope and intercepts learn on

| **7.3 Exercise** | **7.3 Exam questions** on |

Simple familiar	Complex familiar	Complex unfamiliar
1, 2, 3, 4, 5, 6, 7, 8, 9	10, 11, 12, 13, 14	15

These questions are even better in jacPLUS!
- Receive immediate feedback
- Access sample responses
- Track results and progress

Find all this and MORE in jacPLUS ▶

Simple familiar

1. **WE5** Determine the equations of the following straight lines.

   a. A straight line with a slope of 5 passing through the point $(-2, -5)$
   b. A straight line passing through the points $(-3, 4)$ and $(1, 6)$
   c. A straight line passing through the points $(-3, 7)$ and $(0, 7)$

2. **MC** Identify the equation that represents the line that passes through the points $(3, 8)$ and $(12, 35)$.

   A. $y = 3x + 1$ **B.** $y = -3x + 1$ **C.** $y = 3x - 1$ **D.** $y = \dfrac{1}{3}x + 1$

3. **WE6** Determine the value of the slope of each of the following graphs.

   a.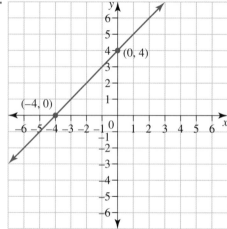
   b.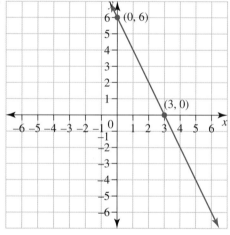

4. Determine the value of the slope and y-intercept of each of the following graphs.

   a.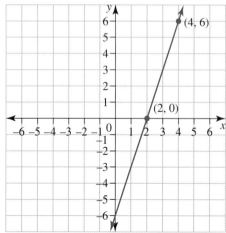
   b.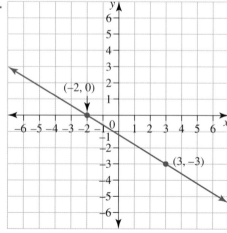

5. **MC** Identify the graph that has a slope of $-\dfrac{1}{4}$.

A.

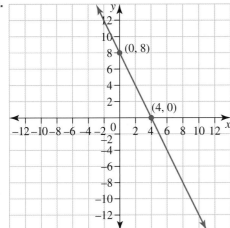

B.

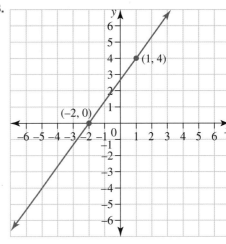

C.

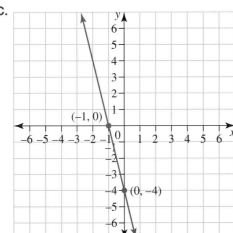

D.

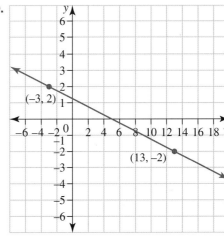

6. **WE7** Determine the value of the slopes of the straight-line graphs that pass through the following points.

   a. $(2, 3)$ and $(5, 12)$ **b.** $(-1, 3)$ and $(2, 7)$

   c. $(-0.2, 0.7)$ and $(0.5, 0.9)$ **d.** $(-2, 0)$ and $(3, 0)$

7. Calculate the values of the slopes of the straight-line graphs that pass through the following points.

   a. $(3, 6)$ and $(2, 9)$ **b.** $(-4, 5)$ and $(1, 8)$

   c. $(-0.9, 0.5)$ and $(0.2, -0.7)$ **d.** $(1.4, 7.8)$ and $(3.2, 9.5)$

   e. $\left(\dfrac{4}{5}, \dfrac{2}{5}\right)$ and $\left(\dfrac{1}{5}, -\dfrac{6}{5}\right)$ **f.** $\left(\dfrac{2}{3}, \dfrac{1}{4}\right)$ and $\left(\dfrac{3}{4}, -\dfrac{2}{3}\right)$

8. Using an appropriate method, determine the slopes of the lines that pass through the following points.

   a. $(0, 5)$ and $(1, 8)$ **b.** $(0, 2)$ and $(1, -2)$

   c. $(0, -3)$ and $(1, -5)$ **d.** $(0, -1)$ and $(2, -3)$

9. A straight line passes through the points $(-2, 2)$ and $(-2, 6)$.

   a. Determine the slope of the line.

   b. Determine the equation of this line.

10. A line has a slope of 5. If it passes through the points $(-2, b)$ and $(-1, 7)$, determine the value of b.

11. Kari is calculating the equation of a straight line passing through the points $(-2, 5)$ and $(3, 1)$. Her working is shown below.

$$\frac{y_2 - y_1}{x_2 - x_1} = \frac{1-5}{-2-3} \qquad y = \frac{4}{5}x + c$$

$$= \frac{-4}{-5} \qquad 1 = \frac{4}{5} \times 3 + c$$

$$= \frac{4}{5} \qquad c = 1 - \frac{12}{5}$$

$$= \frac{-7}{5}$$

$$y = \frac{4}{5}x - \frac{7}{5}$$

   a. Identify the error in Kari's working.
   b. Calculate the correct equation of the straight line passing through these two points.

12. Otis was asked to determine the slope of the line that passes through the points $(3, 5)$ and $(4, 2)$. His response was $\frac{5-2}{4-3} = 3$.

   a. Explain the error in Otis's working out. Hence, calculate the correct slope.
   b. Describe the advice you would give Otis so that he can accurately calculate the slopes of straight lines given two points.

13. A straight-line graph passes through the points $(2, 0), (0, a)$ and $(1, 3)$.

   a. Explain using mathematical reasoning why the value of a must be greater than 3.
   b. i. Identify the two points that can be used to determine the slope of the line. Justify your answer by calculating the value of the slope.
      ii. Using your answer from part i, determine the value of a.

14. Georgio is comparing the cost and distance of various long-distance flights, and after drawing a scatterplot he creates an equation for a line of best fit to represent his data.
Georgio's line of best fit is $y = 0.08x + 55$, where y is the cost of the flight and x is the distance of the flight in kilometres.

   a. Estimate the cost of a flight between Melbourne and Sydney (713 km) using Georgio's equation.
   b. Estimate the cost of a flight between Melbourne and Broome (3121 km) using Georgio's equation.
   c. All of Georgio's data came from flights of distances between 400 km and 2000 km.
   Comment on the suitability of using Georgio's equation for shorter and longer flights than those he analysed. Describe other factors that might affect the cost of these flights.

15. Mariana is a scientist and is collecting data measuring lung capacity (in litres) and time taken to swim 25 metres (in seconds). Unfortunately a spillage in her lab causes all of her data to be erased apart from the records of a person with a lung capacity of 3.5 litres completing the 25 metres in 55.8 seconds and a person with a lung capacity of 4.8 litres completing the 25 metres in 33.3 seconds.

Mariana constructs a linear equation using the remaining data. Mariana uses the equation to predict the time taken to swim 25 metres with a lung capacity of 6 litres as this is the capacity of an elite swimmer.

Mariana is sure her prediction is incorrect as it is 7.45 seconds faster than her best swimmer.

Determine how long it took her best swimmer to swim the 25 metres.

Justify your response using mathematical reasoning.

Fully worked solutions for this chapter are available online.

LESSON
7.4 Modelling with straight-line graphs

SYLLABUS LINKS

- Interpret, in context, the slope (gradient) and intercept of a linear function used to model and analyse a practical situation.
- Construct and analyse a straight-line graph to model a given linear relationship, e.g. modelling the cost of filling a fuel tank of a car against the number of litres of petrol required.

Source: General Mathematics Senior Syllabus 2024 © State of Queensland (QCAA) 2024; licensed under CC BY 4.0.

7.4.1 Recognising a linear model

Practical problems in which there is a constant change over time can be modelled by linear equations. The constant change, such as the rate at which water is leaking or the hourly rate charged by a tradesperson, can be represented by the slope of the equation. Usually the y-value is the changing quantity and the x-value is time.

The starting point or initial point of the problem is represented by the y-intercept, when the x-value is 0. This represents the initial or starting value. In situations where there is a negative slope, the x-intercept represents when there is nothing left, such as the time taken for a leaking water tank to empty.

Identifying the constant change and the starting point can help to construct a linear equation to represent a practical problem. Once this equation has been established, we can use it to calculate specific values or to make predictions as required.

Elle is an occupational therapist who charges an hourly rate of \$35 on top of an initial charge of \$50. Construct a linear equation to represent Elle's charge, C, for a period of t hours.

THINK	WRITE
1. Determine the constant change and the starting point.	Constant change $= 35$ Starting point $= 50$
2. Construct the equation in terms of C by writing the value of the constant change as the coefficient of the pronumeral (t) that affects the change, and writing the starting point as the y-intercept.	$C = 35t + 50$
3. Write the answer.	An equation to represent Elle's charge is $C = 35t + 50$, where C is in dollars and t is in hours.

7.4.2 Solving practical problems

Once an equation is found to represent the practical problem, solutions to the problem can be found by sketching the graph and reading off important information such as the value of the x- and y-intercepts and the slope. Knowing the equation can also help to determine other values related to the problem.

Interpreting the parameters of linear models

When we have determined important values in practical problems, such as the value of the intercepts and slope, it is important to be able to relate these back to the problem and to interpret their meaning.

For example, if we are given the equation $d = -60t + 300$ to represent the distance a car is in kilometres from a major city after t hours, the value of the slope (-60) would represent the speed of the car in km/h (60 km/h), the y-intercept (300) would represent the distance the car is from the city at the start of the problem (300 km), and the x-intercept ($t = 5$) would represent the time it takes for the car to reach the city (5 hours).

Note: In the above example, the value of the slope is negative because the car is heading towards the city, as opposed to away from the city, and we are measuring the distance the car is from the city.

eles-3071

WORKED EXAMPLE 9 Solving a practical problem using a linear model

A bike tyre has 500 cm³ of air in it before being punctured by a nail. After
the puncture, the air in the tyre is leaking at a rate of 5 cm³/minute.
a. Construct an equation to represent the amount of air, A, in the tyre
   t minutes after the puncture occurred.
b. Interpret what the value of the slope in the equation means.
c. Determine the amount of air in the tyre after 12 minutes.
d. By solving your equation from part a, determine how long, in minutes,
   it will take before the tyre is completely flat (i.e. there is no air left).

THINK	WRITE
a. 1. Determine the constant change and the starting point.	a. Constant change $= -5$ Starting point $= 500$
2. Construct the equation in terms of A by writing the value of the constant change as the coefficient of the pronumeral that affects the change, and writing the starting point as the y-intercept.	$A = -5t + 500$
b. 1. Identify the value of the slope in the equation.	b. $A = -5t + 500$ The value of the slope is -5.
2. Identify what this value means in terms of the problem.	The value of the slope represents the rate at which the air is leaking from the tyre. In this case it means that for every minute, the tyre loses 5 cm³ of air.
c. 1. Using the equation found in part a, substitute $t = 12$ and evaluate.	c. $A = -5t + 500$ $\quad = -5 \times 12 + 500$ $\quad = 440$
2. Write the answer.	There is 440 cm³ of air in the tyre after 12 minutes.
d. 1. When the tyre is completely flat, $A = 0$.	d. $0 = -5t + 500$
2. Solve the equation for t.	$0 - 500 = -5t + 500 - 500$ $-500 = -5t$ $\dfrac{-500}{-5} = \dfrac{-5t}{-5}$ $100 = t$
3. Write the answer.	After 100 minutes the tyre will be flat.

Exercise 7.4 Modelling with straight-line graphs

7.4 Exercise	7.4 Exam questions

Simple familiar	Complex familiar	Complex unfamiliar
1, 2, 3, 4, 5, 6, 7, 8, 9, 10, 11, 12, 13	14, 15, 16, 17	18, 19, 20

These questions are even better in jacPLUS!
- Receive immediate feedback
- Access sample responses
- Track results and progress

Find all this and MORE in jacPLUS ▶

Simple familiar

1. **WE8** An electrician charges a call-out fee of $90 plus an hourly rate of $65 per hour.
Construct an equation that determines the electrician's charge, C, for a period of t hours.

2. An oil tanker is leaking oil at a rate of 250 litres per hour. Initially there was 125 000 litres of oil in the tanker. Construct an equation that represents the amount of oil, A, in litres in the oil tanker t hours after the oil started leaking.

3. **WE9** A children's swimming pool is being filled with water. The amount of water in the pool at any time can be found using the equation $A = 20t + 5$, where A is the amount of water in litres and t is the time in minutes.

   a. Explain using mathematical reasoning why this equation can be represented by a straight line.
   b. State the value of the y-intercept and what it represents.
   c. Construct the graph of $A = 20t + 5$ on a set of axes.
   d. The pool holds 500 litres. By solving an equation, determine how long it will take to fill the swimming pool. Write your answer correct to the nearest minute.

4. A yoga ball is being pumped full of air at a rate of $40\,\text{cm}^3/\text{second}$. Initially there is $100\,\text{cm}^3$ of air in the ball.

   a. Construct an equation that represents the amount of air, A, in the ball after t seconds.
   b. Interpret what the value of the y-intercept in the equation means.
   c. Determine how much air, in cm^3, is in the ball after 2 minutes.
   d. When fully inflated the ball holds $100\,000\,\text{cm}^3$ of air. Determine how long, in minutes, it takes to fully inflate the ball. Write your answer to the nearest minute.

5. Kirsten is a long-distance runner who can run at a rate of $12\,\text{km/h}$. The distance, d, in km she travels from the starting point of a race can be represented by the equation $d = at - 0.5$.

   a. Write the value of a.
   b. Write the y-intercept. In the context of this problem, explain what this value means.
   c. Determine how far Kirsten is from the starting point after 30 minutes.
   d. The finish point of this race is 21 km from the starting point. Determine how long, in hours and minutes, it takes Kirsten to run the 21 km. Give your answer correct to the nearest minute.

6. Petrol is being pumped into an empty tank at a rate of 15 litres per minute.

   a. Construct an equation to represent the amount of petrol in litres, P, in the tank after t minutes.
   b. Interpret what the value of the slope in the equation represents.
   c. If the tank holds 75 litres of petrol, determine the time taken, in minutes, to fill the tank.
   d. The tank had 15 litres of petrol in it before being filled. Write another equation to represent the amount of petrol, P, in the tank after t minutes.
   e. State the limitations of P and t using mathematical reasoning.

7. Gert rides to and from work on his bike. The distance and time taken for him to ride home can be modelled using the equation $d = 37 - 22t$, where d is the distance from home in km and t is the time in hours.

   a. Determine the distance, in km, between Gert's work and home.
   b. Explain why the slope of the line in the graph of the equation is negative.
   c. By solving an equation determine the time, in hours and minutes, taken for Gert to ride home. Write your answer correct to the nearest minute.
   d. Explain using mathematical reasoning the limitations of the values of t and d.
   e. Sketch the graph of the equation.

8. Fred deposits $40 in his bank account each week. At the start of the year he had $120 in his account. The amount in dollars, A, that Fred has in his account after t weeks can be found using the equation $A = at + b$.

   a. State the values of a and b.
   b. In the context of this problem, interpret what the y-intercept represents.
   c. Determine how many weeks it will take Fred to save $3000.

9. Michaela is a real estate agent. She receives a commission of 1.5% on house sales plus a payment of $800 each month. Michaela's monthly wage can be modelled by the equation $W = ax + b$, where W represents Michaela's total monthly wage and x represents her house sales in dollars.

   a. State the values of a and b.
   b. Explain using mathematical reasoning whether there is an upper limit to the x-values of the model.
   c. In March Michaela's total house sales were $452\,000. Determine her monthly wage for March.
   d. In September Michaela earned $10\,582.10. Determine the amount of house sales she made in September.

10. An electrician charges a call-out fee of $175 on top of an hourly rate of $60.
   a. Construct an equation to represent the electrician's fee in terms of his total charge, C, and hourly rate, h.
   b. Claire is a customer and is charged $385 to install a hot water system. Determine how many hours she was charged for.

   The electrician changes his fee structure. The new fee structure is summarised in the following table.

Time	Call-out fee	Quarter-hourly rate
Up to two hours	$100	$20
2–4 hours	$110	$25
Over 4 hours	$115	$50

   c. Using the new fee structure, determine how much, in dollars, Claire would be charged for the same job.
   d. Construct an equation that models the new fee structure for between 2 and 4 hours.

11. Express the following situations as linear models.
   a. Julie works at a department store and is paid $19.20 per hour. She has to work for a minimum of 10 hours per week, but due to her study commitments she can work for no more than 20 hours per week.
   b. The results in a driving test are marked out of 100, with 4 marks taken off for every error made on the course. The lowest possible result is 40 marks.

12. The Dunn family departs from home for a caravan trip. They travel at a rate of 80 km/h. The distance they travel from home, in km, can be modelled by a linear equation.

   a. Write the value of the slope of the graph of the linear equation.
   b. Write the value of the y-intercept. Explain what this value means in the context of this problem.
   c. Using your values from parts **a** and **b**, write an equation to represent the distance the Dunn family are from home at any given point in time.
   d. Determine for how long, in hours, the family have travelled when they are 175 km from home. Give your answer to the nearest minute.
   e. The Dunns travel for 2.5 hours before stopping. Determine the distance they are from home.

13. The table shows the amount of money in Kim's savings account at different dates. Kim withdraws the same amount of money every five days.

Date	26/11	1/12	6/12
Amount	$1250	$1150	$1050

   a. The amount of money at any time, t days, in Kim's account can be modelled by a linear equation. Explain why.
   b. Using a calculator, spreadsheet or otherwise, construct a straight line graph to represent the amount of money Kim has in her account from 26 November.
   c. Determine the slope of the line of the graph, and explain the meaning of the slope in the context of this problem.
   d. Determine the linear equation that models this situation.
   e. In the context of this problem, explain the meaning of the x-intercept.
   f. Kim will need at least $800 to go on a beach holiday over the Christmas break (starting on 21 December). Show that Kim will not have enough money for her holiday.

14. Monique is setting up a new business selling T-shirts through an online auction site. Her supplier in China agrees to a deal whereby they will supply each T-shirt for $3.50 providing she buys a minimum of 100 T-shirts. The deal is valid for up to 1000 T-shirts.

    a. Set up a linear model to represent this situation.
    b. Explain the limitations of the x- and y-values for this model.

15. A large fish tank is being filled with water. After 1 minute the height of the water is 2 cm and after 4 minutes the height of the water is 6 cm. The height of the water in cm, h, after t minutes can be modelled by a linear equation.

    Determine whether the fish tank was empty of water before being filled. Justify your answer using calculations.

16. There are two advertising packages for Get2Msg.com. Package A charges per cm^2 and package B charges per letter. The costs for both packages increase at a constant rate. The table shows the costs for package A for areas from 4 to 10 cm^2.

Area, in cm^2	Cost (excluding administration charge of $25)
4	30
6	45
8	60
10	75

    Package B costs 58 cents per standard letter plus an administration cost of $55.
    Betty and Boris of B'n'B Bedding want to place the advertisement shown on Get2Msg.com.
    Explain which package would be the better option for them. Justify your answer by calculating the costs they would pay for both packages.

17. Kazem has saved $4500 for a holiday and estimates his spending on the holiday will be 15% of his initial savings per week. Determine how many weeks Kazem has before he runs out of money on his holiday. Use mathematical reasoning to justify your answer.

18. Jeff sells luxury cars. He receives a wage of $400 per week. Each month he makes no commission on the first $15 000 worth of sales but then makes 15% commission on the rest.
    Determine how much Jeff needs to sell to earn a total of $20 000 in a month.

19. Curly lives in a country town and needs to travel regularly for work. She is always trying to find cheaper fuel for her car. Curly uses an app to find the best price in the area for fuel. The app says that ServoXP is selling fuel at $2.10/L and UltraXX is selling the cheapest fuel at $1.99/L.
    She estimates that it will cost $2.00 to drive to ServoXP, and as UltraXX is out of her way, it will cost $6.00 to drive there.

Determine how much fuel Curly needs to buy to make it worth her while to drive to UltraXX. Give your answer to the nearest litre.

20. Frankie was told that the circumference of a circle divided by the diameter was always the same ratio. She decided to test this by measuring the circumference and the diameter of 5 circular objects.
    Her results are summarised below.

Diameter (cm)	Circumference (cm)
52	165
33	112
13	39
7	21

Use Frankie's measurements to calculate a mean ratio and use this to estimate the diameter of a circle with a circumference of 40 cm. (Give your answer correct to 2 decimal places.)

Fully worked solutions for this chapter are available online.

LESSON
7.5 Review

7.5.1 Summary

doc-
41486

Hey students! Now that it's time to revise this chapter, go online to:

 Access the chapter summary

 Review your results

 A+ **Practise exam questions**

Find all this and MORE in jacPLUS ▶

7.5 Exercise

learn on

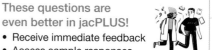

7.5 Exercise	7.5 Exam questions on

Simple familiar	Complex familiar	Complex unfamiliar
1, 2, 3, 4, 5, 6, 7, 8, 9, 10, 11, 12	13, 14, 15, 16	17, 18, 19, 20

These questions are even better in jacPLUS!
- Receive immediate feedback
- Access sample responses
- Track results and progress

Find all this and MORE in jacPLUS ▶

Simple familiar

1. **MC** The slope of the line passing through the points $(4, 6)$ and $(-2, -6)$ is:

   A. -2 **B.** -0.5 **C.** 0 **D.** 2

2. **MC** The x- and y-intercepts of the linear graph with equation $3x - y = 6$ are:

   A. $(2, 6)$ **B.** $(0, 2)$ and $(-6, 0)$ **C.** $(0, 2)$ and $(6, 0)$ **D.** $(2, 0)$ and $(0, -6)$

3. **MC** The slope of the graph shown in the following diagram is:

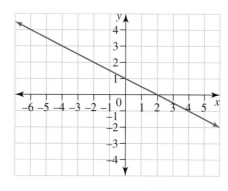

   A. -2 **B.** -1 **C.** $-\dfrac{1}{2}$ **D.** 1

4. **MC** Identify which of the following is a sketch of the graph with equation $y = 4x + 2$.

A.
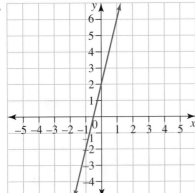

B.
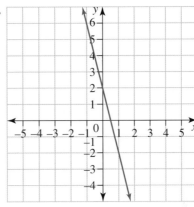

C.
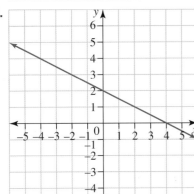

D.
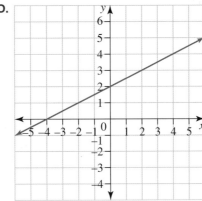

5. **MC** Bertha knits teddy bears and sells them at the local farmers' market. Bertha spends $120 in wool, and it costs her an additional $4.50 to make each teddy bear. She sells the bears for $14.50 each. Determine how many teddy bears Bertha needs to sell to cover her costs.

    A. 8   **B.** 9   **C.** 10   **D.** 12

6. **MC** The line that passes through the point $(2, -1)$ is:

    A. $y = -2x + 5$ **B.** $y = 2x - 1$ **C.** $y = -2x + 1$ **D.** $x + y = 1$

The following information relates to questions 7 and 8.

An inflated party balloon has a small hole and is slowly deflating. The initial volume of the balloon is $1000\,\text{cm}^3$ and the balloon loses $5\,\text{cm}^3$ of air every minute.

7. **MC** If V represents the volume of the balloon in cm^3 and t represents the time in minutes, an equation to represent the volume of the balloon after t minutes is:

    A. $V = \dfrac{1000}{5t}$ **B.** $V = \dfrac{1000 - 5t}{60}$

    C. $V = 1000 + 5t$ **D.** $V = 1000 - 5t$

8. **MC** The time taken for the balloon to have lost $650\,\text{cm}^3$ of air is:

    A. 70 minutes   **B.** 130 minutes

    C. 270 minutes   **D.** 330 minutes

9. **MC** The equation of a straight line passing through the points $(-2, 3)$ and $(5, 1)$ is:

    A. $y = -\dfrac{2}{7}x + 2\dfrac{3}{7}$ **B.** $y = 2\dfrac{3}{7}x - \dfrac{2}{7}$ **C.** $y = \dfrac{2}{7}x + 2\dfrac{3}{7}$ **D.** $y = \dfrac{2}{7}x - 2\dfrac{3}{7}$

10. **MC** The linear equation $d = 5.6h - 4.56$ gives an estimate of the relationship between the height of an athlete and the distance that they can long jump in metres.

Using this equation, the estimated distance that a 1.8-metre-tall athlete could long jump would be:

A. 5 m
B. 6.02 m
C. 5.46 m
D. 5.52 m

11. Sketch the following graphs by calculating the x- and y-intercepts. Hence, state the slope of each graph.

   a. $2x + y = 5$

   b. $y - 4x = 8$

   c. $4(x + 3y) = 16$

   d. $3x + 4y - 10 = 0$

12. Determine the slope of the lines passing through the following pairs of points.

   a. $(3, -2)$ and $(0, 4)$

   b. $(5, 11)$ and $(-2, 18)$

   c. $(0.3, 4.1)$ and $(1.2, 5.3)$

   d. $\left(\dfrac{2}{5}, \dfrac{1}{4} \right)$ and $\left(-\dfrac{1}{4}, \dfrac{3}{5} \right)$

Complex familiar

13. A line has a slope of $-\dfrac{3}{4}$. If the line passes through the points $(-a, 3)$ and $(-2, 6)$, determine the value of a.

14. Complete the following table.

	Equation	Slope	y-intercept	x-intercept
a	$y = 5x - 3$	5		
b	$y = 3x + 1$			
c	$3y = 6x - 9$			
d	$2y + 4x = 8$			
e			5	−5
f		2		2

15. Miriam has a sweet tooth, and her favourite sweets are strawberry twists and chocolate ripples. The local sweet shop sells both as part of their pick and mix selection, so Miriam fills a bag with them.

Each strawberry twist weighs 5 g and each chocolate ripple weighs 9 g. In Miriam's bag there are 28 sweets, weighing a total of 188 g.

Determine the number of each type of sweet that Miriam bought by forming and solving a linear equation.

16. Tommy is saving for a remote-controlled car that is priced at $49. He has $20 in his piggy bank. Tommy saves $3 of his pocket money every week and puts it in his piggy bank.

Determine how many weeks it will take for Tommy to have saved enough money to purchase the remote-controlled car.

17. The recommended maximum heart rate for an adult during exercise can be found by calculating 85% of the difference between 220 and the age of the adult.

    Sketch a graph that shows the recommended maximum heart rate for a person aged 20 to 70 years.

    Using your graph or otherwise, determine the recommended maximum heart rate for a 25-year-old person.

18. A currency converter is showing that one Australian dollar would buy 50.48 Indian rupees. Ben is a cricket fan and plans to fly to India for a test match.

    Ben has a budget of $A2000 while he is in India. He has estimated his costs as 5500 rupees per night for accommodation, 2200 rupees a day for food, 550 rupees a day for transport and 2000 rupees a day for extras.

    If he is in Delhi for 6 nights and 7 days, determine whether his budget will cover his expenses. Use mathematical reasoning to justify your answer.

19. Trudy is unaware that there is a small hole in the petrol tank of her car. Petrol is leaking out of the tank at a constant rate of 5 mL/ min.

    Trudy has parked her car in a long-term carpark at the airport and gone on a holiday. Initially there is 45 litres of petrol in the tank.

    Determine how many hours it will take for the petrol tank to become empty.

20. The median price for a block of land ($'000) from fifteen suburbs and the distance in kilometres from the centre of a city were collected. When graphed, the data showed a linear pattern.

    To determine an equation that could be used to determine the median house price, p, at a distance, d km from the city centre, the following points were used: (3, 808) and (20, 468).

    Using this information, estimate the distance in kilometres from the city centre of a block of land costing $650 800. (Give your answer correct to 1 decimal place.)

Fully worked solutions for this chapter are available online.

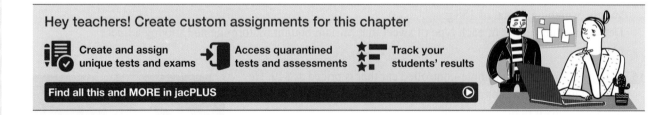

Hey teachers! Create custom assignments for this chapter

- Create and assign unique tests and exams
- Access quarantined tests and assessments
- Track your students' results

Find all this and MORE in jacPLUS ▶

Answers

Chapter 7 Straight-line graphs

7.2 Constructing straight-line graphs

7.2 Exercise

1. a. Slope $= 2$, y-intercept $= 1$

   b. Slope $= -1$, y-intercept $= 3$

   c. Slope $= \dfrac{1}{2}$, y-intercept $= 4$

   d. Slope $= 1$, y-intercept $= \dfrac{1}{4}$

2. a. Slope $= \dfrac{3}{5}$, y-intercept $= -\dfrac{1}{5}$

   b. Slope $= 10$, y-intercept $= -5$

   c. Slope $= -\dfrac{1}{2}$, y-intercept $= \dfrac{3}{2}$

   d. Slope $= -\dfrac{3}{2}$, y-intercept $= 3$

3. a. $m = 1$, $c = -4$ b. $m = \dfrac{3}{2}$, $c = -3$

   c. $m = \dfrac{1}{4}$, $c = \dfrac{1}{4}$ d. $m = \dfrac{5}{2}$, $c = -2$

4. a, b
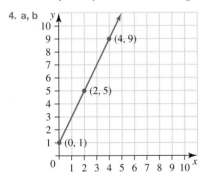

5. a. $(1, 7)$
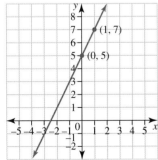

   b. $(1, -3)$
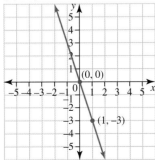

c. $(2, 4)$
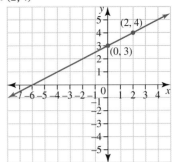

d. $(0, 4)$ and $(1, 6)$
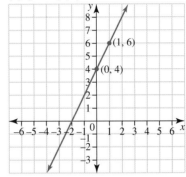

6. a. $(10, 0)$ and $(0, 4)$
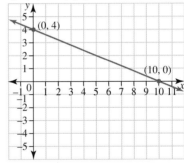

   b. $\left(-\dfrac{5}{3}, 0\right)$ and $\left(0, \dfrac{5}{4}\right)$
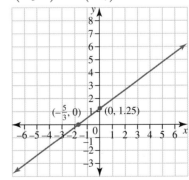

7. a. $(3, 0)$ and $(0, 6)$

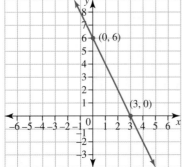

b. $(-3, 0)$ and $(0, 9)$

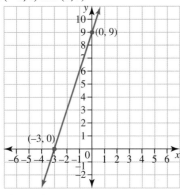

c. $\left(-\dfrac{4}{3}, 0\right)$ and $(0, 2)$

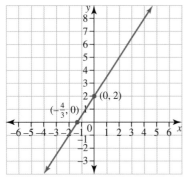

d. $\left(-\dfrac{4}{5}, 0\right)$ and $\left(0, \dfrac{4}{3}\right)$

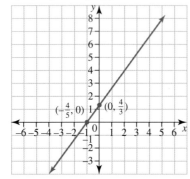

8. a. $(1, 7)$

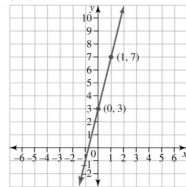

b. $(1, -2)$

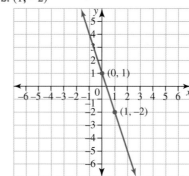

c. $(4, 5)$

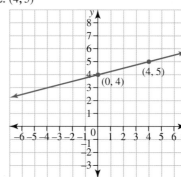

d. $(5, -4)$

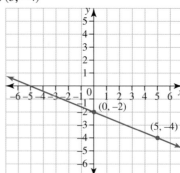

9. -1

10. $a = 1$

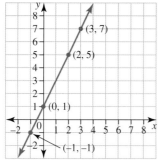

11.

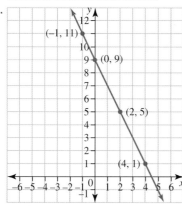

$a = 1$

12. x-intercept: $\left(\dfrac{2}{a-3}, 0\right)$

y-intercept: $\left(\dfrac{2}{b+1}, 0\right)$

7.3 Determine and interpret the slope and intercepts

7.3 Exercise

1. **a.** $y = 5x + 5$ **b.** $y = 0.5x + 5.5$
   c. $y = 7$

2. C

3. **a.** 1   **b.** -2

4. **a.** Slope $= 3$, y-intercept $= -6$
   b. Slope $= -\dfrac{3}{5}$, y-intercept $= -\dfrac{6}{5}$

5. D

6. **a.** 3   **b.** $\dfrac{4}{3}$ **c.** $\dfrac{2}{7}$ **d.** 0

7. **a.** -3 **b.** $\dfrac{3}{5}$ **c.** $-\dfrac{12}{11}$
   d. $\dfrac{17}{18}$ **e.** $\dfrac{8}{3}$ **f.** -11

8. **a.** 3   **b.** -4 **c.** -2 **d.** -1

9. **a.** The slope is undefined.
   b. $x = -2$

10. $b = 2$

11. **a.** Kari did not assign the x- and y-values for each point before calculating the slope, and she mixed up the values.
    b. $y = -\dfrac{4}{5}x + \dfrac{17}{5}$

12. **a.** Otis swapped the x- and y-values, calculating $\dfrac{y_2 - y_1}{x_1 - x_2}$.
    The correct slope is -3.
    b. Label each x and y pair before substituting them into the formula.

13. **a.** The points $(1, 3)$ and $(2, 0)$ tell us that the graph has a negative slope so the y-intercept must have a greater value than 3.
    b. **i.** $(2, 0)$ and $(1, 3)$; slope $= -3$
    ii. $a = 6$

14. **a.** \$112
    b. \$305
    c. All estimates outside the parameters of Georgio's original data set (400 km to 2000 km) will be unreliable, with estimates further away from the data set being more unreliable than those closer to the data set.
    Other factors that might affect the cost of flights include air taxes, fluctuating exchange rates and the choice of airlines for various flight paths.

15. 20 seconds

7.4 Modelling with straight-line graphs

7.4 Exercise

1. $C = 65t + 90$

2. $A = -250t + 125\,000$

3. **a.** Both variables in the equation have a power of 1.
   b. y-intercept $= 5$. This represents the amount of water initially in the pool.
   c.

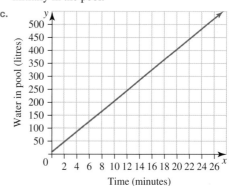

   d. 25 minutes

4. **a.** $A = 40t + 100$
   b. How much air was initially in the ball
   c. 4900 cm^3
   d. 42 seconds

5. **a.** 12
   b. y-intercept $= -0.5$. This means that Kirsten starts 0.5 km before the starting point of the race.
   c. 5.5 km
   d. 1 hour, 48 minutes

6. **a.** $P = 15t$
   b. The additional amount of petrol in the tank each minute
   c. 5 minutes
   d. $P = 15t + 15$
   e. $0 \le t \le 4$

7. a. 37 km

b. The distance to Gert's home is reducing as time passes.

c. 1 hour, 41 minutes

d. $0 \le t \le 1.682$ h

e.

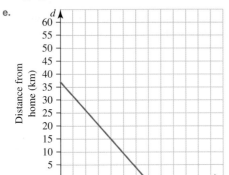

8. a. $a = 40, b = 120$

b. The amount of money in Fred's account at the start of the year

c. 72 weeks

9. a. $a = 0.015, b = 800$

b. No, there is no limit to how much Michaela can earn in a month.

c. $7580

d. $652 140

10. a. $C = 60h + 175$

b. 3.5 hours

c. $460

d. $C = 110 + 100h, 2 \le h \le 4$

11. a. $P = 19.2t$

b. $R = 100 - 4e$

12. a. 80

b. y-intercept $= 0$. This means that they start from home.

c. $d = 80t$

d. 2 hours, 11 minutes

e. 200 km

13. a. Kim withdraws the same amount each 5 days, so there is a constant decrease.

b.

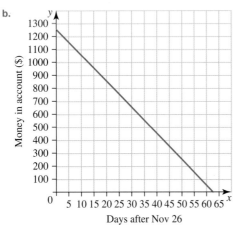

c. Slope $= -20$. This means that Kim withdraws an average of $20 each day.

d. $M = 1250 - 20t$

e. The x-intercept represents when there will be no money left in Kim's account.

f. After 25 days (on 21 December) Kim will have $750 in her account, so she will not have enough for her holiday.

14. a. $C = 3.5t, 100 \le t \le 1000$

b. The x-values represent the number of T-shirts Monique can buy.
There is an upper limit as the deal is valid only up to 1000 T-shirts.

15. No, the y-intercept is $\dfrac{2}{3}$, so there was water in the tank to start with.

16. Package A $= \$126.25$; Package B $= \$113.00$. Package B is the better option.

17. Week 7

18. $136 777.80

19. 37 L

20. 12.73 cm

7.5 Review

7.5 Exercise

1. D

2. D

3. C

4. A

5. D

6. D

7. D

8. B

9. A

10. D

11. a. x-intercept: $(2.5, 0)$, y-intercept: $(0, 5)$

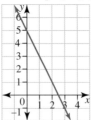

b. x-intercept: $(-2, 0)$, y-intercept: $(0, 8)$

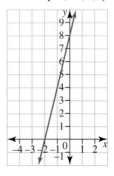

c. *x*-intercept: $(4, 0)$, *y*-intercept: $\left(0, \dfrac{4}{3}\right)$

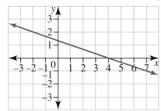

d. *x*-intercept: $\left(\dfrac{10}{3}, 0\right)$, *y*-intercept: $(0, 2.5)$

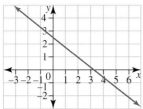

12. a. -2 **b.** -1

    c. $\dfrac{4}{3}$ **d.** $-\dfrac{7}{13}$

13. -2

14.

	Equation	Slope	*y*-intercept	*x*-intercept
a	$y = 5x - 3$	5	-3	0.6
b	$y = 3x + 1$	3	1	$-\dfrac{1}{3}$
c	$3y = 6x - 9$	2	-3	1.5
d	$2y + 4x = 8$	-2	4	2
e	$y = x + 5$	1	5	-5
f	$y = 2x - 4$	2	-4	2

15. 16 strawberry twists and 12 chocolate ripples.

16. 10 weeks

17. 166

18. Ben's total expenses will be \$A1312.40, so he will be \$687.60 under budget.

19. 150 hours

20. 10.9 km

UNIT

2

Applications of linear equations and trigonometry, matrices and univariate data analysis

Source: General Mathematics Senior Syllabus 2024 © State of Queensland (QCAA) 2024; licensed under CC BY 4.0.

8 Simultaneous equations and their applications

LESSON SEQUENCE

Fully worked solutions for this chapter are available online.

EXAM PREPARATION

Access exam-style questions in every lesson, available online.

on Resources

Solutions	Solutions — Chapter 8 (sol-1249)
Exam questions	Exam question booklet — Chapter 8 (eqb-0265)
Digital documents	Learning matrix — Chapter 8 (doc-41488)
	Chapter summary — Chapter 8 (doc-41489)

LESSON
8.1 Overview

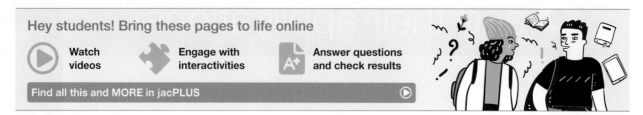
8.1.1 Introduction

Ways of determining the solutions to simultaneous linear equations have important roles in engineering, physics, chemistry, computer science and economics, but also have many simple applications. For example, imagine you decide to make and sell handmade bracelets at a local market. You will want to know how many bracelets you need to sell to make a profit, which will depend on how much it will cost you to hire and set up your stall and the cost of the materials. The cost of the number of bracelets and the profit you make from selling them can be set up as linear equations and solved using a graph or algebraically.

Piecewise linear graphs, which are linear graphs defined on a sequence of intervals, and in particular step graphs, which take constant y values over intervals on the x-axis, are useful tools in real-life situations. For instance, an airport car park might have a fee for the first 15 minutes, another fee for the next 15 minutes, another fee for a stay longer than 30 minutes but shorter than an hour, and so on. A step graph will help you determine how much a person would be charged depending on how long they parked there.

In the modern world, these mathematical tools have even broader applications. In the tech industry, algorithms for solving linear equations are fundamental to machine learning and data science, helping companies make predictions and optimise operations. In finance, they are used to model and predict market behaviours, manage risks and optimise investment portfolios. The rise of smart cities also relies on these methods to efficiently manage resources such as electricity and water, plan traffic flow and improve public services.

8.1.2 Syllabus links

Lesson	Lesson title	Syllabus links
8.2	**Solving simultaneous linear equations graphically**	○ Solve a pair of simultaneous linear equations graphically.
8.3	**Solving simultaneous equations algebraically**	○ Solve a pair of simultaneous linear equations, algebraically using substitution and elimination.
8.4	**Solving practical problems using simultaneous equations**	○ Solve practical problems involving simultaneous linear equations.
8.5	**Piecewise linear graphs and step graphs**	○ Sketch piece-wise linear graphs and step graphs.
		○ Interpret piece-wise linear graphs and step graphs used to model practical situations.

LESSON
8.2 Solving simultaneous linear equations graphically

SYLLABUS LINKS

• Solve a pair of simultaneous linear equations graphically.

Source: Adapted from General Mathematics Senior Syllabus 2024 © State of Queensland (QCAA) 2024; licensed under CC BY 4.0.

8.2.1 Solutions to simultaneous equations

Simultaneous equations are sets of equations that can be solved together. They often represent practical problems that have two or more unknowns. For example, you can use simultaneous equations to calculate the cost of individual apples and oranges when different amounts of each are bought. These equations can be written in the form $y = mx + c$.

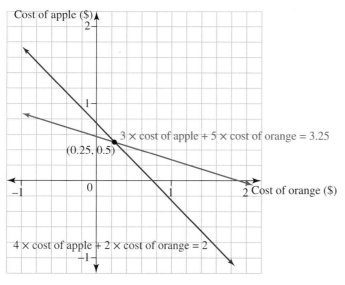

By graphing a pair of simultaneous equations, a point of intersection can be found. The coordinates of the point of intersection are the x- and y-values that satisfy both of the equations.

WORKED EXAMPLE 1 Identifying the solutions to a pair of simultaneous equations

a. **Identify the coordinates of the point that simultaneously solves the two equations shown in the graph**
b. **Check, by substitution, that the solution to part a is correct.**

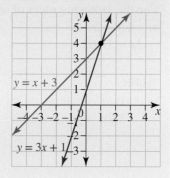

THINK	WRITE
a. Identify the coordinates of the point of intersection (the point where the two lines meet).	a. $(1, 4)$

---

b. The point of intersection $(1, 4)$ represents an x-value of 1 and a y-value of 4. Check that the point of intersection satisfies both equations by substituting $x = 1$ and $y = 4$ into each equation to see if the LHS of the equation is equal to the RHS of the equation.

b. $y = x + 3$
Let $x = 1$ and $y = 4$.
$$\text{LHS} = y$$
$$= 4$$
$$\text{RHS} = x + 3$$
$$= 1 + 3$$
$$= 4$$
$$\text{LHS} = \text{RHS}$$
The solution is correct.

$y = 3x + 1$
Let $x = 1$ and $y = 4$.
$$\text{LHS} = y$$
$$= 4$$
$$\text{RHS} = 3x + 1$$
$$= 3(1) + 1$$
$$= 4$$
$$\text{LHS} = \text{RHS}$$
The solution is correct.

Two lines are **coincident** if they lie one on top of the other. For example, the line and line segment shown in the graph are coincident.

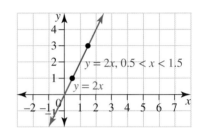

For coincident lines, every point where the lines coincide satisfies both equations and hence is a solution to the simultaneous equations. Therefore, there are an infinite number of solutions to simultaneous equations of coincident lines.

Coincident lines have essentially the same equation, although the equations may have been altered or multiplied by a constant so they appear different. For example, $y = 2x + 3$ and $2y - 4x = 6$ are coincident lines; $y = 2x + 3$ and $2y - 4x = 6$ have the same gradient and y-intercept.

8.2.2 Equations with no solutions

If two linear lines do not intersect, there is no solution to the equations.

For linear equations, which are represented by straight lines, the only situation in which the lines do not intersect is if the lines are parallel. Parallel lines have the same gradient but different y-intercepts. For example, $y = 2x - 1$ and $y = 2x + 1$ are equations of parallel lines.

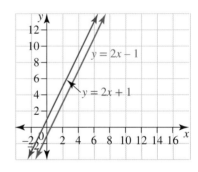

WORKED EXAMPLE 2 Determining if a solution exists

For the following pairs of equations, state, with reasons, whether the two lines will intersect.
a. $y + x = 2$
$3x - 5 = 2y$

b. $y = 2x + 5$
$3y - 6x = 15$

c. $5y = 4x + 6$
$10y - 8x = 15$

THINK

a. 1. Both equations are for straight lines, since the highest power of x is 1.

WRITE

a. $y + x = 2$ $3x - 5 = 2y$

2. Transpose both formulas into the form $y = mx + c$, where m is the gradient and c is the y-intercept. (We studied transposing formula in Lesson 5.4.)

$$y + x = 2$$
$$y + x - x = 2 - x$$
$$y = 2 - x$$
$$y = 2 + -x$$
$$y = -x + 2$$

$$2y = 3x - 5$$
$$\frac{2y}{2} = \frac{3x - 5}{2}$$
$$y = \frac{3x}{2} - \frac{5}{2}$$

3. The gradients are not the same, so the lines are not parallel.

$$m = -1, \ c = 2 \qquad m = \frac{3}{2}, \ c = -\frac{5}{2}$$

4. Write the answer.

The two lines will intersect as their gradients are not the same.

b. 1. Both equations are for straight lines. Transpose both formulas into the form $y = mx + c$.

b. $y = 2x + 5$

$$3y - 6x = 15$$
$$3y - 6x + 6x = 15 + 6x$$
$$3y = 15 + 6x$$
$$\frac{3y}{3} = \frac{15 + 6x}{3}$$
$$y = \frac{15}{3} + \frac{6x}{3}$$
$$y = 5 + 2x$$

2. The gradients are the same, so the lines are parallel. In fact, the lines also have the same y-intercept ($c = 5$), so the lines are coincident.

$$m = 2, \ c = 5 \qquad m = 2, \ c = 5$$

3. Write the answer.

The two lines are coincident (they lie one on top of the other), since they have the same value for their gradient ($m = 2$) and the same value for their y-intercept ($c = 5$).
The lines intersect along their whole length.

c. 1. Both equations are equations for straight lines. Transpose both formulas into the form $y = mx + c$.

c. $5y = 4x + 6$
$$\frac{5y}{5} = \frac{4x + 6}{5}$$
$$y = \frac{4x}{5} + \frac{6}{5}$$

$$10y - 8x = 15$$
$$10y - 8x + 8x = 15 + 8x$$
$$10y = 15 + 8x$$
$$\frac{10y}{10} = \frac{15 + 8x}{10}$$
$$y = \frac{15}{10} + \frac{8x}{10}$$
$$y = \frac{3}{2} + \frac{4x}{5}$$

2. The gradients are the same, so the lines are parallel. The y-intercepts are different, so the lines do not lie one on top of the other. Parallel lines do not intersect.

$$m = \frac{4}{5}, \ c = \frac{6}{5} \qquad m = \frac{4}{5}, \ c = \frac{3}{2}$$

3. Write the answer.

The two lines will not intersect. Their gradients are the same but their y-intercepts are different and hence they are parallel lines.

8.2.3 Graphical solutions

The solution to a pair of simultaneous equations can be found by graphing the two equations and identifying the coordinates of the point of intersection.

To sketch a straight line, you need to identify two points. There are a variety of ways to sketch a straight line:

- Calculate the x- and y-intercepts.
- Use the gradient and y-intercept method.
- Calculate the value of the two points by substituting values into the equations.
- Use technology.

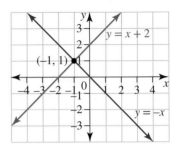

The coordinates of the point of intersection $(-1, 1)$ are shown, and this is determined by graphing the straight lines $y = x + 2$ and $y = -x$.

eles-6387

WORKED EXAMPLE 3 Solving simultaneous equations graphically

Solve the simultaneous equations $y = 5x + 2$ and $y = 2x - 3$ graphically.

THINK	WRITE

1. To solve the simultaneous equations, sketch each of the equations on the same axes.

Both equations are in the form $y = mx + c$, so identify the gradient and y-intercept in each equation.

$$y = 5x + 2 \qquad\qquad y = 2x - 3$$

$$m = 5 \qquad\qquad\qquad m = 2$$
$$c = 2 \qquad\qquad\qquad c = -3$$

2. For the gradient, identify the rise and the run.

$$m = \frac{\text{rise}}{\text{run}} = \frac{5}{1} \qquad\qquad m = \frac{\text{rise}}{\text{run}} = \frac{2}{1}$$

$$\text{rise} = 5, \text{run} = 1 \qquad \text{rise} = 2, \text{run} = 1$$

3. Draw a set of axes and sketch both lines.
For the line $y = 5x + 2$:
- place a point on the y-axis at the y-intercept ($c = 2$), as shown in blue
- from the y-intercept, run 1 and rise 5 and place a point, as shown in blue
- draw a line between the two points and label the line, as shown in blue.

For the line $y = 2x - 3$:
- place a point on the y-axis at the y-intercept ($c = -3$), as shown in pink
- from the y-intercept, run 1 and rise 2 and place a point, as shown in pink
- draw a line between the two points and label the line, as shown in pink.

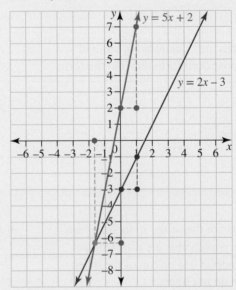

4. Carefully read the coordinates of the point of intersection. The point of intersection appears to be $\left(-1\frac{2}{3}, -6\frac{1}{3}\right)$, as shown in green.

5. Write the solution.

The solution is $x = -1\dfrac{2}{3}$ and $y = -6\dfrac{1}{3}$.

6. Check that the solution satisfies both equations.

$y = 5x + 2$
LHS $= y$
$\quad = -6\dfrac{1}{3}$

RHS $= 5x + 2$
$\quad = 5\left(-1\dfrac{2}{3}\right) + 2$
$\quad = \dfrac{5}{1} \times \dfrac{-5}{3} + 2$
$\quad = \dfrac{-25}{3} + 2$
$\quad = \dfrac{-25}{3} + \dfrac{6}{3}$
$\quad = \dfrac{-19}{3}$
$\quad = -6\dfrac{1}{3}$
$\quad = $ LHS
LHS $=$ RHS
The solution is correct.

$y = 2x - 3$
LHS $= y$
$\quad = -6\dfrac{1}{3}$

RHS $= 2x - 3$
$\quad = 2\left(-1\dfrac{2}{3}\right) - 3$
$\quad = \dfrac{2}{1} \times \dfrac{-5}{3} - 3$
$\quad = \dfrac{-10}{3} - 3$
$\quad = \dfrac{-10}{3} - \dfrac{9}{3}$
$\quad = \dfrac{-19}{3}$
$\quad = -6\dfrac{1}{3}$
$\quad = $ LHS
LHS $=$ RHS
The solution is correct.

eles-3167

WORKED EXAMPLE 4 Constructing and solving simultaneous equations graphically

Your friend is three years older than twice the age of her brother. The sum of their ages is 18. Calculate the age of both your friend and her brother using two simultaneous equations.

THINK	WRITE
1. To create two equations, allocate a variable for your friend's age and another variable for her brother's age.	Let $y =$ your friend's age. Let $x =$ her brother's age.
2. Use the information in the question to form two equations.	
• Your friend (y) is three years older than twice the age of her brother (x).	$y = 2x + 3$
• The sum of their ages is 18.	$y + x = 18$
• Transpose the equation to make y the subject	$y = -x + 18$

3. Graph the two equations.

4. Identify the coordinates of the point of intersection (the point where the two lines meet).

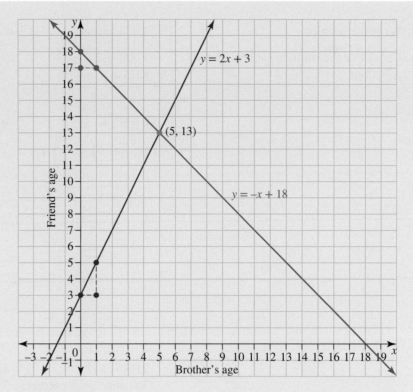

5. Check that the point of intersection satisfies both equations by substituting $x = 5$ and $y = 13$ into each equation to see if the LHS of the equation is equal to the RHS of the equation.

Point of intersection: $(5, 13)$

$y = 2x + 3$
Let $x = 5$ and $y = 13$.
LHS $= 13$
RHS $= 2(5) + 3$
$\quad\;\; = 13$
$\quad\;\; =$ LHS
LHS $=$ RHS
The solution is correct

$y + x = 18$
Let $x = 5$ and $y = 13$.
LHS $= 13 + 5$
$\quad\;\; = 18$
RHS $= 18$
$\quad\;\; =$ LHS
LHS $=$ RHS
The solution is correct.

6. Write the answer.

The friend is 13 years old and her brother is 5 years old.

---

on Resources

Interactivity Solving simultaneous equations graphically (int-6452)

Exercise 8.2 Solving simultaneous linear equations graphically

learn on

8.2 Exercise | **8.2 Exam questions** on

Simple familiar	Complex familiar	Complex unfamiliar
1, 2, 3, 4, 5, 6, 7, 8, 9, 10, 11	12, 13, 14	N/A

Simple familiar

1. **WE1** Identify the coordinates of the point that simultaneously solves the two equations in each of the graphs below.

   a.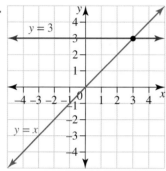

   b.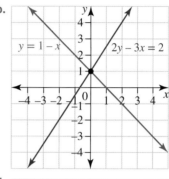

   c.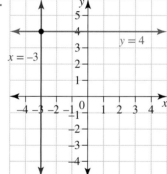

   d.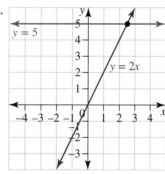

2. State, with reasons, whether the following values for x and y are solutions for the given pair of simultaneous equations.

   a. $x + y = 5 \quad x = 4, \ y = 1$
   $2x + 4y = 12$

   b. $x + 3y = -1 \quad x = 3, \ y = -1$
   $3x - 2y = 14$

3. a. For the graph shown, identify the time at which the two transport companies, T and O, charge the same amount.

   b. Calculate the equations for the lines that represent each company.

   c. Use the coordinates of the point of intersection to check that your equations for each company are correct.

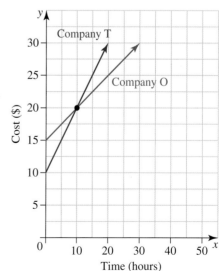

4. **WE2** For the following pairs of equations, state, with reasons, whether the two lines will intersect.

   a. $3y = x + 12$
   $y - 5x = 7.5$

   b. $y = 4x - 7$
   $8x - 2y - 14 = 0$

   c. $5y = 4x + 6$
   $1.2y - x = 15.5$

d. Copy and complete the following statements.

   i. A system of linear equations will have a unique solution if the equations have _____ gradient(s) and _____ y-intercept(s).

   ii. A system of linear equations will have an infinite number of solutions if the equations have _____ gradient(s) and _____ y-intercept(s).

5. The graph below shows the distance of two cars from a particular town.

  a. Determine how far each car travels in one hour.

  b. Calculate the gradient of each line.

  c. Determine whether the cars are ever at the same spot at the same time. Explain.

  d. Describe what the y-intercepts indicate about the two cars.

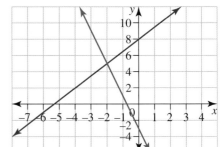

6. **WE3** Solve the following pairs of simultaneous equations graphically.

  a. $y = -2$
     $y = 2x - 4$

  b. $x = 2$
     $y = 4$

  c. $y = x - 3$
     $y = 1 - x$

  d. $y = 3x + 6$
     $y = 1 - 2x$

  e. $x - y = 0$
     $2y + x = 0$

  f. $x - y = 3$
     $3x + y = 7$

7. Determine the coordinates of the point of intersection of the lines $9x + 3y = 27$ and $y = 3x - 3$.

8. A pair of simultaneous equations is solved graphically as shown in the diagram. From the diagram, determine the solution for this pair of simultaneous equations.

9. The following equations represent a pair of simultaneous equations.

$$y = 5x + 1 \text{ and } y = 2x - 5$$

Sketch both graphs on the same set of axes and solve the equations graphically.

10. Sketch and solve the following three simultaneous equations.

$$y = 3x + 7, y = 2x + 8 \text{ and } y = -2x + 12$$

11. Solve the following groups of simultaneous equations graphically.

  a. $y = 4x + 1$ and $y = 3x - 1$

  b. $y = x - 5$ and $y = -3x + 3$

  c. $y = 3(x - 1)$ and $y = 2(2x + 1)$

  d. $y = \dfrac{x}{2} - 1$ and $y = \dfrac{x}{2} + 4$

Complex familiar

12. **WE4** Your friend is nine years younger than three times the age of her brother. The sum of their ages is 11. Calculate the age of both your friend and her brother using two simultaneous equations.

13. In a fruit shop, one shopper has 3 apples and 5 oranges in their basket and another shopper has 4 apples and 2 oranges. The total cost of those pieces of fruit is shown.

Total cost = $2.00   Total cost = $3.25

  a. Determine how much the apples cost.

  b. Determine the difference in cost between an apple and an orange.

14. Consider the following groups of graphs.
   i. $y_1 = 5x - 4$ and $y_2 = 6x + 8$
   ii. $y_1 = -3x - 5$ and $y_2 = 3x + 1$
   iii. $y_1 = 2x + 6$ and $y_2 = 2x - 4$
   iv. $y_1 = -x + 3$, $y_2 = x + 5$ and $y_3 = 2x + 6$

   a. Determine whether there are solutions for all of these groups of graphs. If not, identify the group(s) of graphs for which there are no solutions and explain why this is the case.
   b. Where possible, determine the point of intersection for each group of graphs using any method.

Fully worked solutions for this chapter are available online.

LESSON
8.3 Solving simultaneous equations algebraically

SYLLABUS LINKS

- Solve a pair of simultaneous linear equations, algebraically using substitution and elimination.

Source: Adapted from General Mathematics Senior Syllabus 2024 © State of Queensland (QCAA) 2024; licensed under CC BY 4.0.

8.3.1 Solving simultaneous equations using substitution

Simultaneous equations can also be solved algebraically. One algebraic method is known as substitution. The substitution method is used when one (or both) of a pair of simultaneous equations is presented in a form where one of the two variables is the subject of the equation. In the equation $y = 2x - 1$, the variable y is the subject.

The substitution method involves replacing a variable in one equation with the other equation. This produces a new third equation expressed in terms of a single variable.

For the pair of simultaneous equations $c = 12b - 15$ and $2c + 3b = -3$, the first equation is presented with c as the subject, so substitution should be used.

eles-3100

WORKED EXAMPLE 5 Solving simultaneous equations using substitution

Solve the following pairs of simultaneous equations using substitution.
a. $c = 12b - 15$ and $2c + 3b = -3$
b. $y = 4x + 6$ and $y = 6x + 2$
c. $3x + 2y = -1$ and $y = x - 8$

THINK	WRITE
a. 1. Identify a variable that has been written as the subject of the equation, that is in terms of the other variable.	a. $c = 12b - 15$
2. Substitute the variable $c = 12b - 15$ into the equation.	$2c + 3b = -3$ $2(12b - 15) + 3b = -3$

3. Expand and simplify the left-hand side, and solve the equation for the unknown variable.

$$24b - 30 + 3b = -3$$
$$27b - 30 = -3$$
$$27b = -3 + 30$$
$$27b = 27$$
$$b = 1$$

4. Substitute the value for the unknown back into one of the equations.

$$c = 12b - 15$$
$$= 12(1) - 15$$
$$= -3$$

5. Write the answer.

The solution is $b = 1$ and $c = -3$.

b. 1. Both equations are presented with y as the subject, so let them equal each other.

b. $4x + 6 = 6x + 2$

2. Move all of the variables to one side.

$$4x - 4x + 6 = 6x - 4x + 2$$
$$6 = 6x - 4x + 2$$
$$6 = 2x + 2$$

3. Solve for the unknown.

$$6 - 2 = 2x + 2 - 2$$
$$4 = 2x$$

$$\frac{4}{2} = \frac{2x}{2}$$
$$2 = x$$

4. Substitute the value found, $x = 2$, into either of the original equations.

$$y = 4x + 6$$
$$= 4 \times 2 + 6$$
$$= 8 + 6$$
$$= 14$$

5. Write the answer.

The solution is $x = 2$ and $y = 14$.

c. 1. The second equation is written with y as the subject, so substitute $y = x - 8$ into the other equation.

c. $\qquad 3x + 2y = -1$
$$3x + 2(x - 8) = -1$$

2. Expand and simplify the equation.

$$3x + 2x - 16 = -1$$
$$5x - 16 = -1$$

3. Solve for the unknown.

$$5x - 16 + 16 = -1 + 16$$
$$5x = 15$$
$$x = 3$$

4. Substitute the value found, $x = 3$, into either of the original equations.

$$y = x - 8$$
$$= 3 - 8$$
$$= -5$$

5. Write the answer.

The solution is $x = 3$ and $y = -5$.

 Resources

Interactivity Solving simultaneous equations using substitution (int-6453)

8.3.2 Solving simultaneous equations using elimination

Simultaneous equations that have both variables on the same side of the equals sign in both equations can be solved using **elimination**.

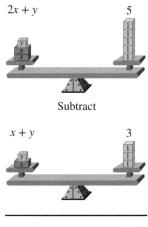

If the coefficients of the variable have the same sign, we subtract one equation from the other to eliminate the variable.

The process of elimination is carried out by adding (or subtracting) the left-hand sides and the right-hand sides of each equation together.

Consider the equations $2x + y = 5$ and $x + y = 3$. The process of subtracting each side of the equation from each other is visualised on the scales shown.

To represent this process algebraically, the setting out would look like:

$$\begin{array}{r} 2x + y = 5 \\ -(x + y = 3) \\ \hline x = 2 \end{array}$$

Once the value of x has been determined, it can be substituted into either original equation to calculate y.

$$2(2) + y = 5 \Rightarrow y = 1$$

If the coefficients of the variable have the opposite sign, we add the equations to eliminate the variable.

$$\begin{array}{ll} 3x + 4y = 14 & \qquad 6x - 2y = 12 \\ 5x - 4y = 2 & \qquad 6x + 3y = 27 \\ \text{(add equations to eliminate } y) & \text{(subtract equations to eliminate } x) \end{array}$$

---

eles-6388

WORKED EXAMPLE 6 Solving simultaneous equations using elimination

Solve the following pairs of simultaneous equations using elimination.
a. $3x + y = 5$ and $4x - y = 2$
b. $2a + b = 7$ and $a + b = 5$
c. $3c + 4d = 5$ and $2c + 3d = 4$

THINK	WRITE
a. 1. Write the simultaneous equations with one on top of the other.	a. $3x + y = 5$ [1] $4x - y = 2$ [2]
2. Select one pronumeral to be eliminated.	Select y.
3. Check the coefficients of the pronumeral being eliminated.	The coefficients of y are 1 and -1.
4. If the coefficients are the same number but with different signs, add the equations together.	$[1] + [2]$: $3x + 4x + y - y = 5 + 2$ $\qquad\qquad\qquad 7x = 7$
5. Solve the equation for the unknown pronumeral.	$\dfrac{7x}{7} = \dfrac{7}{7}$ $\qquad x = 1$
6. Substitute the pronumeral back into one of the equations.	$3x + y = 5$ $3(1) + y = 5$

7. Solve the equation to determine the value of the other pronumeral.

$$3 + y = 5$$
$$3 - 3 + y = 5 - 3$$
$$y = 2$$

8. Write the answer.

The solution is $x = 1$ and $y = 2$.

b. 1. Write the simultaneous equations with one on top of the other.

b. $2a + b = 7 \qquad [1]$
$a + b = 5 \qquad [2]$

2. Select one pronumeral to be eliminated.

Select b.

3. Check the coefficients of the pronumeral being eliminated.

The coefficients of b are both 1.

4 If the coefficients are the same number with the same sign, subtract one equation from the other.

$[1] - [2]$:
$2a - a + b - b = 7 - 5$
$a = 2$

5. Solve the equation for the unknown pronumeral.

$a = 2$

6. Substitute the pronumeral back into one of the equations.

$a + b = 5$
$2 + b = 5$

7. Solve the equation to determine the value of the other pronumeral.

$b = 5 - 2$
$b = 3$

8. Write the answer.

The solution is $a = 2$ and $b = 3$.

c. 1. Write the simultaneous equations with one on top of the other.

c. $3c + 4d = 5 \qquad [1]$
$2c + 3d = 4 \qquad [2]$

2. Select one pronumeral to be eliminated.

Select c.

3. Check the coefficients of the pronumeral being eliminated.

The coefficients of c are 3 and 2.

4. If the coefficients are different numbers, then multiply them both by another number, so they both have the same coefficient value.

$3 \times 2 = 6$
$2 \times 3 = 6$

5. Multiply the equations (all terms in each equation) by the numbers selected in step **4**.

$[1] \times 2$:
$6c + 8d = 10$
$[2] \times 3$:
$6c + 9d = 12$

6. Check the sign of each coefficient for the selected pronumeral.

$6c + 8d = 10 \qquad [3]$
$6c + 9d = 12 \qquad [4]$
Both coefficients of c are positive 6.

7. If the signs are the same, subtract one equation from the other and simplify.

$[3] - [4]$:
$6c - 6c + 8d - 9d = 10 - 12$
$-d = -2$

8. Solve the equation for the unknown.

$d = 2$

9. Substitute the pronumeral back into one of the equations.

$2c + 3d = 4$
$2c + 3(2) = 4$

10. Solve the equation to determine the value of the other pronumeral.

$$2c + 6 = 4$$
$$2c + 6 - 6 = 4 - 6$$
$$2c = -2$$
$$\frac{2c}{2} = \frac{-2}{2}$$
$$c = -1$$

11. Write the answer.

The solution is $c = -1$ and $d = 2$.

Exercise 8.3 Solving simultaneous equations algebraically   learn on

8.3 Exercise	8.3 Exam questions on

Simple familiar	Complex familiar	Complex unfamiliar
1, 2, 3, 4, 5, 6, 7, 8, 9, 10	11, 12	N/A

These questions are even better in jacPLUS!
- Receive immediate feedback
- Access sample responses
- Track results and progress

Find all this and MORE in jacPLUS ▶

Simple familiar

1. **WE5** Solve the following pairs of simultaneous equations using substitution.
   a. $y = 2x + 1$ and $2y - x = -1$
   b. $m = 2n + 5$ and $m = 4n - 1$

2. Calculate the solutions to the following pairs of simultaneous equations using substitution.
   a. $2(x + 1) + y = 5$ and $y = x - 6$
   b. $\dfrac{x + 5}{2} + 2y = 11$ and $y = 6x - 2$

3. **MC** Identify which one of the following pairs of simultaneous equations would best be solved using the substitution method.
   A. $4y - 5x = 7$ and $3x + 2y = 1$
   B. $3c + 8d = 19$ and $2c - d = 6$
   C. $12x + 6y = 15$ and $9x - y = 13$
   D. $n = 9m + 12$ and $3m + 2n = 7$

4. Using the substitution method, solve the following pairs of simultaneous equations
   a. $y = 2x + 5$ and $y = 3x - 2$
   b. $y = 5x - 2$ and $y = 7x + 2$
   c. $y = 2(3x + 1)$ and $y = 4(2x - 3)$
   d. $y = 5x - 9$ and $3x - 5y = 1$
   e. $3(2x + 1) + y = -19$ and $y = x - 1$
   f. $\dfrac{3x + 5}{2} + 2y = 2$ and $y = x - 2$

5. Solve the following pair of simultaneous equations using the substitution method.

$$3x + y = 8 \text{ and } 2x - y = 7$$

6. **WE6** Solve the following pairs of simultaneous equations using elimination:

   a. $4x + y = 6$ and $x - y = 4$ **b.** $x + y = 7$ and $x - 2y = -5$

   c. $2x - y = -5$ and $x - 3y = -10$ **d.** $4x + 3y = 29$ and $2x + y = 13$

   e. $5x - 7y = -33$ and $4x + 3y = 8$ **f.** $\dfrac{x}{2} + y = 7$ and $3x + \dfrac{y}{2} = 20$

7. Consider the following pair of simultaneous equations:

$$ax - 3y = -16 \text{ and } 3x + y = -2.$$

If $y = 4$, calculate the values of a and x.

8. **MC** The first step when solving the following pair of simultaneous equations using the elimination method is:

$$2x + y = 3 \quad [1]$$
$$3x - y = 2 \quad [2]$$

   A. equations [1] and [2] should be added together.
   B. both equations should be multiplied by 2.
   C. equation [1] should be subtracted from equation [2].
   D. equation [1] should be multiplied by 2 and equation [2] should be multiplied by 3.

9. Brendon and Marcia were each asked to solve the following pair of simultaneous equations.

$$3x + 4y = 17 \quad [1]$$
$$4x - 2y = 19 \quad [2]$$

Marcia decided to use the elimination method. Her solution steps were:

Step 1: $[1] \times 4$:
      $12x + 16y = 68 \quad [3]$
      $[2] \times 3$:
      $12x - 6y = 57 \quad [4]$
Step 2: $[3] + [4]$:
      $10y = 125$
Step 3: $y = 12.5$
Step 4: Substitute $y = 12.5$ into [1]:
      $3x + 4(12.5) = 17$
Step 5: Solve for x:
      $3x = 17 - 50$
      $3x = -33$
      $x = -11$
Step 6: The solution is $x = -11$ and $y = 12.5$.

   a. Marcia has made an error in step 2. Explain where she has made her error, and hence correct her mistake.
   b. Using the correction you made in part **a**, determine the correct solution to this pair of simultaneous equations.

Brendon decided to eliminate y instead of x.

   c. Using Brendon's method of eliminating y first, show all the appropriate steps involved to reach a solution.

10. Solve the following groups of simultaneous equations. Write your answers correct to 2 decimal places.

   a. $4(x + 6) = y - 6$ and $2(y + 3) = x - 9$
   b. $6x + 5y = 8.95$, $y = 3x - 1.36$ and $2x + 3y = 4.17$

11. In a ball game, a player can kick the ball into the net to score a goal or place the ball over the line to score a behind. The scores in a game between the Rockets and the Comets were:

    Rockets: 6 goals 12 behinds, total score 54
    Comets: 7 goals 5 behinds, total score 45
    The two simultaneous equations that can represent this information are shown.
    Rockets: $6x + 12y = 54$
    Comets: $7x + 5y = 45$

    a. By solving the two simultaneous equations, determine the number of points that are awarded for a goal and a behind.
    b. Using the results from part **a**, determine the scores for the game between the Jetts, who scored 4 goals and 10 behinds, and the Meteorites, who scored 6 goals and 9 behinds.

12. Mick and Minnie both work part time at an ice-cream shop. The simultaneous equations shown represent the number of hours Mick (x) and Minnie (y) work each week.

    Equation 1: Total number of hours worked by Minnie and Mick: $x + y = 15$
    Equation 2: Number of hours worked by Minnie in terms of Mick's hours: $y = 2x$

    a. Explain why substitution would be the best method to use to solve these equations.
    b. Using substitution, determine the number of hours worked by Mick and Minnie each week.
    c. To ensure that he has time to do his Mathematics homework, Mick changes the number of hours he works each week. He now works $\frac{1}{3}$ of the number of hours worked by Minnie.

    An equation that can be used to represent this information is $x = \frac{y}{3}$.

    Calculate the number of hours worked by Mick, given that the total number of hours that Mick and Minnie work does not change.

Fully worked solutions for this chapter are available online.

LESSON
8.4 Solving practical problems using simultaneous equations

SYLLABUS LINKS

- Solve practical problems involving simultaneous linear equations.

Source: General Mathematics Senior Syllabus 2024 © State of Queensland (QCAA) 2024; licensed under CC BY 4.0.

8.4.1 Setting up simultaneous equations

The solutions to a set of simultaneous equations satisfy all equations that were used. Simultaneous equations can be used to solve problems involving two or more variables or unknowns, such as the cost of 1 kg of apples and bananas, or the number of adults and children attending a show.

WORKED EXAMPLE 7 Setting up a pair of simultaneous equations

At a fruit shop, 2 kg of apples and 3 kg of bananas cost $13.16, and 3 kg of apples and 2 kg of bananas cost $13.74.
Represent this information in the form of a pair of simultaneous equations.

THINK	WRITE
1. Identify the two variables.	The cost of 1 kg of apples and the cost of 1 kg of bananas
2. Select two pronumerals to represent these variables. Define the variables.	a = cost of 1 kg of apples b = cost of 1 kg of bananas
3. Identify the key information and rewrite it using the pronumerals selected.	2 kg of apples can be written as $2a$. 3 kg of bananas can be written as $3b$.
4. Construct two equations using the information.	$2a + 3b = 13.16$ $3a + 2b = 13.74$

A **break-even point** is a point where the costs equal the selling price. It is also the point where there is zero profit.

For example, if the equation $C = 45 + 3t$ represents the production cost to sell t shirts, and the equation $R = 14t$ represents the revenue from selling the shirts for $14 each, then the break-even point is the number of shirts that need to be sold to cover all costs.

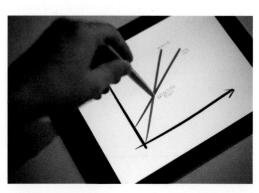

eles-3033

WORKED EXAMPLE 8 Solving a practical problem using simultaneous equations

Santo sells shirts for $25. The revenue, R, for selling n shirts
is represented by the equation $R = 25n$. The cost to make n
shirts is represented by the equation $C = 2200 + 3n$.

a. Solve the equations simultaneously to determine the
   break-even point.
b. Determine the profit or loss, in dollars, for the following
   shirt orders.
   i. 75 shirts
   ii. 220 shirts

THINK	WRITE
a. **1.** Write the two equations.	**a.** $C = 2200 + 3n$ $R = 25n$
2. Equate the equations ($R = C$).	$2200 + 3n = 25n$
3. Solve for the unknown.	$2200 + 3n = 25n$ $2200 + 3n - 3n = 25n - 3n$ $2200 = 22n$ $\dfrac{2200}{22} = n$ $n = 100$
4. Substitute back into either equation to determine the values of C and R.	$R = 25n$ $= 25 \times 100$ $= 2500$
5. Write the answer in the context of the problem.	The break-even point is $(100, 2500)$. Therefore, 100 shirts need to be sold to cover the production cost, which is $2500.
b. i. **1.** Write the two equations.	**b. i.** $C = 2200 + 3n$ $R = 25n$
2. Substitute the given value into both equations.	$n = 75$ $C = 2200 + 3 \times 75$ $= 2425$ $R = 25 \times 75$ $= 1875$
3. Determine the profit/loss by subtracting the cost, C, from the revenue, R.	Profit/loss $= R - C$ $= 1875 - 2425$ $= -550$
4. Write the answer.	Since the answer is negative, it means that Santo lost $550 (i.e. selling 75 shirts did not cover the cost to produce the shirts).
ii. **1.** Write the two equations.	**ii.** $C = 2200 + 3n$ $R = 25n$

2. Substitute the given value into both equations.	$n = 220$ $C = 2200 + 3 \times 220$ $\quad = 2860$ $R = 25 \times 220$ $\quad = 5500$
3. Determine the profit/loss by subtracting the cost, C, from the revenue, R.	Profit/loss $= R - C$ $\quad = 5500 - 2860$ $\quad = 2640$
4. Write the answer.	Since the answer is positive, it means that Santo made $2640 profit from selling 220 shirts.

 Resources

 Interactivity Break-even points (int-6454)

Exercise 8.4 Solving practical problems using simultaneous equations

learn on

8.4 Exercise	**8.4 Exam questions** on

Simple familiar	**Complex familiar**	**Complex unfamiliar**
1, 2, 3, 4, 5, 6, 7, 8	9, 10, 11, 12, 13, 14	15, 16

These questions are even better in jacPLUS!

• Receive immediate feedback
• Access sample responses
• Track results and progress

Find all this and MORE in jacPLUS ▶

Simple familiar

1. **WE7** Mary bought 4 donuts and 3 cupcakes for $10.55, and Sharon bought 2 donuts and 4 cupcakes for $9.90.
   Letting d represent the cost of a donut and c represent the cost of a cupcake, set up a pair of simultaneous equations to represent this information.

2. A pair of simultaneous equations representing the number of adults (a) and children (c) attending the zoo is shown below.
   Equation 1: $a + c = 350$
   Equation 2: $25a + 15c = 6650$

   a. By solving the pair of simultaneous equations, determine the total number of adults and children attending the zoo.
   b. In the context of this problem, explain what equation 2 represents.

3. **WE8** Yolanda sells handmade bracelets at a market for $12.50.
The revenue, R, for selling n bracelets is represented by the equation $R = 12.50n$.
The cost to make n bracelets is represented by the equation $C = 80 + 4.50n$.

   a. i. By solving the equations simultaneously, determine the break-even point.
      ii. In the context of this problem, explain what the break-even point means.
   b. Determine the profit or loss, in dollars, if Yolanda sells:

      i. 8 bracelets
      ii. 13 bracelets.

4. The entry fee for a charity fun run event is $18. It costs event organisers $2550 for the hire of the tent and $3 per entry for administration. Any profit will be donated to local charities. An equation to represent the revenue for the entry fee is $R = an$, where R is the total amount collected in entry fees, in dollars, and n is the number of entries.

   a. Write an equation for the value of a.
   b. The equation that represents the cost for the event is
      $C = 2550 + bn$.
      Write an equation for the value of b.
   c. By solving the equations simultaneously, determine the number of entries needed to break even.
   d. A total of 310 entries are received for this charity event. Show that the organisers will be able to donate $2100 to local charities.
   e. Determine the number of entries needed to donate $5010 to local charities.

5. A school group travelled to the city by bus and returned by train. The two equations show the adult, a, and student, s, ticket prices to travel on the bus and train.
   Bus: $3.5a + 1.5s = 42.50$
   Train: $4.75a + 2.25s = 61.75$

   a. Write the cost of a student bus ticket, s, and an adult bus ticket, a.
   b. Solve the simultaneous equations and hence determine the number of adults and the number of students in the school group.

6. The following pair of simultaneous equations represents the number of adult and concession tickets sold and the respective ticket prices for the premiere screening of the blockbuster *Aliens attack*.
   Equation 1: $a + c = 544$
   Equation 2: $19.50a + 14.50c = 9013$

   a. Determine the costs, in dollars, of an adult ticket, a, and a concession ticket, c.
   b. In the context of this problem, explain what equation 1 represents.
   c. By solving the simultaneous equations, determine how many adult and concession tickets were sold for the premiere.

7. Charlotte has a babysitting service and charges $12.50 per hour. After Charlotte calculated her set-up and travel costs, she constructed the cost equation $C = 45 + 2.50h$, where C represents the cost in dollars per job and h represents the hours Charlotte babysits for.

   a. Write an equation that represents the revenue, R, earned by Charlotte in terms of number of hours, h.
   b. By solving the equations simultaneously, determine the number of hours Charlotte needs to babysit to cover her costs (that is, the break-even point).
   c. In one week, Charlotte had four babysitting jobs as shown in the table.

Babysitting job	1	2	3	4
Number of hours (h)	5	3.5	4	7

      i. Determine whether Charlotte made a profit or loss for each individual babysitting job.
     ii. Determine whether Charlotte made a profit this week. Justify your answer using calculations.

   d. Charlotte made a $50 profit on one job. Determine the total number of hours she babysat for.

8. Sally and Nem decide to sell cups of lemonade from their front yard to the neighbourhood children. The cost to make the lemonade using their own lemons can be represented using the equation $C = 0.25n + 2$, where C is the cost in dollars and n is the number of cups of lemonade sold.

   a. If they sell cups of lemonade for 50 cents, write an equation to represent the selling price, S, for n number of cups of lemonade.
   b. By solving two simultaneous equations, determine the number of cups of lemonade Sally and Nem need to sell in order to break even (i.e. cover their costs).
   c. Sally and Nem increase their selling price. If they make a $7 profit for selling 20 cups of lemonade, determine the new selling price.

FRESH SQUEEZED
LEMONADE
100% NATURAL

Complex familiar

9. Trudi and Mia work part time at the local supermarket after school. The following table shows the number of hours worked for both Trudi and Mia and the total wages, in dollars, paid over two weeks.

Week	Trudi's hours worked	Mia's hours worked	Total wages
Week 1	15	12	$400.50
Week 2	9	13	$328.75

   By solving a pair of simultaneous equations, determine the hourly rate of pay for Trudi and the hourly rate of pay for Mia.

10. Brendan uses carrots and apples to make his special homemade fruit juice. One week he buys 5 kg of carrots and 4 kg of apples for $31.55. The next week he buys 4 kg of carrots and 3 kg of apples for $24.65.

   a. By solving a pair of simultaneous equations, determine how much Brendan spends on 1 kg each of carrots and apples.
   b. Determine the amount Brendan spends the following week when he buys 2 kg of carrots and 1.5 kg of apples. Give your answer correct to the nearest 5 cents.

11. The table shows the number of 100-g serves of strawberries and grapes and the total kilojoule intake.

Fruit	100 g serves	
Strawberry, s	3	4
Grapes, g	2	3
Total kilojoules	**1000**	**1430**

By solving a pair of simultaneous equations, determine the number of kilojoules (kJ) for a 100-g serve of strawberries.

12. Two budget car hire companies offer the following deals for hiring a medium-size family car.

Car company	Deal
FreeWheels	$75 plus $1.10 per km travelled
GetThere	$90 plus $0.90 per km travelled

a. By solving two equations simultaneously, determine the value at which the cost of hiring a car will be the same.

b. Rex and Jan hire a car for the weekend. They expect to travel a distance of 250 km over the weekend. Explain which car hire company they should use and why. Justify your answer using calculations.

13. The CotX T-Shirt Company produces T-shirts at a cost of $7.50 each after an initial set-up cost of $810.

a. Determine the cost to produce 100 T-shirts.

b. Complete the following table that shows the cost of producing T-shirts.

n	0	20	30	40	50	60	80	100	120	140
C										

c. Write an equation that represents the cost, C, to produce n T-shirts.

d. CotX sells each T-shirt for $25.50. Write an equation that represents the amount of sales, S, in dollars for selling n T-shirts.

e. By solving two simultaneous equations, determine the number of T-shirts that must be sold for CotX to break even.

f. If CotX needs to make a profit of at least $5000, determine the minimum number of T-shirts they will need to sell to achieve this outcome.

14. There are three types of fruit for sale at the market: starfruit, s, mango, m, and papaya, p. The following table shows the amount of fruit bought and the total cost in dollars.

Starfruit, s	Mango, m	Papaya, p	Total cost, $
5	3	4	19.40
4	2	5	17.50
3	5	6	24.60

Using simultaneous equations, determine the cost of 2 starfruit, 4 mangoes and 4 papayas.

15. The following table shows the number of boxes of three types of cereal bought each week for a school camp, as well as the total cost for each week.

Cereal	Week 1	Week 2	Week 3
Corn Pops, c	2	1	3
Rice Crunch, r	3	2	4
Muesli, m	1	2	1
Total cost, $	**27.45**	**24.25**	**36.35**

Wen is the cook at the camp. She decides to work out the cost of each box of cereal using simultaneous equations. She incorrectly sets up the following equations:

$$2c + c + 3c = 27.45$$
$$3r + 2r + 4r = 24.25$$
$$m + 2m + m = 36.35$$

a. Explain why these simultaneous equations will not determine the cost of each box of cereal.

b. Determine the total cost for cereal for week 4's order of 3 boxes of Corn Pops, 2 boxes of Rice Crunch and 2 boxes of muesli.

16. The Comet Cinema offers four types of tickets to the movies: adult, concession, senior and member. The table shows the number and types of tickets bought to see four different movies and the total amount of tickets sales in dollars.

Movie	Adult, a	Concession, c	Seniors, s	Members, m	Total sales, $
Wizard boy	24	52	12	15	1071.00
Champions	35	8	45	27	1105.50
Pixies on ice	20	55	9	6	961.50
Horror nite	35	15	7	13	777.00

The blockbuster movie *Love hurts* took the following tickets sales: 77 adults, 30 concessions, 15 seniors and 45 members. Determine the total ticket sales for *Love hurts* in dollars and cents.

Fully worked solutions for this chapter are available online.

LESSON
8.5 Piecewise linear graphs and step graphs

SYLLABUS LINKS

- Sketch piece-wise linear graphs and step graphs.
- Interpret piece-wise linear graphs and step graphs used to model practical situations.

Source: General Mathematics Senior Syllabus 2024 © State of Queensland (QCAA) 2024; licensed under CC BY 4.0.

8.5.1 Open and closed end points

Piecewise graphs are formed by two or more linear graphs that are joined at points of intersection. A piecewise graph is continuous, which means there are no breaks or gaps in the graph, as shown in the diagram.

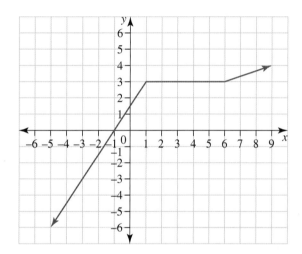

Step graphs are formed by two or more linear graphs that have zero gradients. Step graphs have breaks, as shown in the second diagram.

For this step graph, the following equations could be used to describe it.

$$y = -4, -5 < x \leq -1$$
$$y = -1, -1 < x \leq 5$$
$$y = 2, 5 < x \leq 9$$

The end points of each line depend on whether the point is included in the interval.

For example, the interval $-1 < x \leq 5$ will have an open end point at $x = -1$, because x does not equal -1 in this case. The same interval will have a closed end point at $x = 5$, because x is less than or equal to 5.

A closed end point means that the x-value is also 'equal to' the value. An open end point means that the x-value is not equal to the value; that is, it is less than or greater than only.

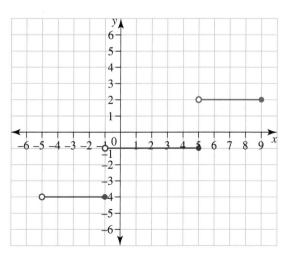

WORKED EXAMPLE 9 Constructing and solving a piecewise linear graph

A piecewise linear graph is constructed from the following linear graphs.

$$y = 2x + 1, x \le a$$
$$y = 4x - 1, x > a$$

a. By solving the equations simultaneously, determine the point of intersection and hence state the value of a.

b. Sketch the piecewise linear graph.

THINK

a. 1. Determine the intersection point of the two graphs by solving the equations simultaneously.

2. The x-value of the point of intersection determines the x-intervals for where the linear graphs meet.

b. 1. By hand, or using technology such as a spreadsheet or mathematical software, sketch the two graphs without taking into account the intervals.

WRITE

a. $y = 2x + 1$
$y = 4x - 1$

Solve by substitution:
$$2x + 1 = 4x - 1$$
$$2x - 2x + 1 = 4x - 2x - 1$$
$$1 = 2x - 1$$
$$1 + 1 = 2x - 1 + 1$$
$$2 = 2x$$
$$x = 1$$
Substitute $x = 1$ to calculate y:
$$y = 2(1) + 1$$
$$= 3$$
The point of intersection is $(1, 3)$.
$x = 1$ and $y = 3$

$x = 1$; therefore, $a = 1$.

b.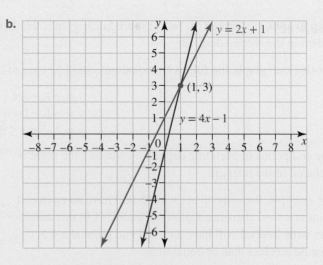

2. Identify which graph exists within the stated x-intervals to sketch the piecewise linear graph.

$y = 2x + 1$ exists for $x \leq 1$.
$y = 4x - 1$ exists for $x > 1$.
Remove the sections of each graph that do not exist for these values of x.

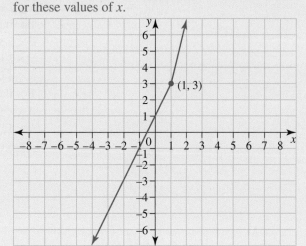

 Resources

 Interactivity Piecewise linear graphs (int-6486)

eles-6389

WORKED EXAMPLE 10 Constructing a step graph

Construct a step graph from the following equations, making sure to take note of the relevant end points.

$$y = 1, \ -3 < x \leq 2$$
$$y = 4, \ 2 < x \leq 4$$
$$y = 6, \ 4 < x \leq 6$$

THINK

1. Construct a set of axes and draw each line within the stated x-intervals.

WRITE

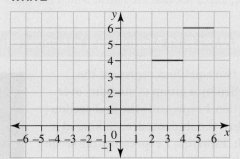

2. Draw in the end points.

For the line $y = 1$:
$-3 < x \leq 2$
$x > -3$ is an open circle.
$x \leq 2$ is a closed circle.
For the line $y = 4$:
$2 < x \leq 4$
$x > 2$ is an open circle.
$x \leq 4$ is a closed circle.
For the line $y = 6$:
$4 < x \leq 6$
$x > 4$ is an open circle.
$x \leq 6$ is a closed circle.

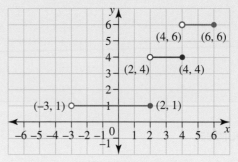

 Resources

 Interactivity Step functions (int-6281)

8.5.2 Modelling with piecewise linear and step graphs

Consider the real-life situation of a leaking water tank. For the first 3 hours it leaks at a constant rate of 12 litres per minute; after 3 hours the rate of leakage slows down (decreases) to 9 litres per minute. The water leaks at a constant rate in both situations and can therefore be represented as a linear graph. However, after 3 hours the slope of the line changes because the rate at which the water is leaking changes.

eles-6390

WORKED EXAMPLE 11 Modelling a practical situation using a piecewise linear graph

The following two equations represent the distance travelled by a group of students over 5 hours. Equation 1 represents the first section of the hike, when the students are walking at a pace of 4 km/h. Equation 2 represents the second section of the hike, when the students change their walking pace.

Equation 1: $d = 4t, 0 \leq t \leq 2$
Equation 2: $d = 2t + 4, 2 \leq t \leq 5$

The variable d is the distance in km from the campsite, and t is the time in hours.

a. **Determine the time, in hours, for which the group travelled in the first section of the hike.**
b. i. **Determine their walking pace in the second section of their hike.**
   ii. **Determine for how long, in hours, they walked at this pace.**
c. **Sketch a piecewise linear graph to represent the distance travelled by the group of students over the five hour hike.**

THINK		WRITE	
a.	1. Determine which equation the question applies to.	a.	This question applies to equation 1.
	2. Look at the time interval for this equation.		$0 \le t \le 2$
	3. Interpret the information and write the answer.		The group travelled for 2 hours.
b. i.	1. Determine which equation the question applies to.	b. i.	This question applies to equation 2.
	2. Interpret the equation. The walking pace is found by the coefficient of t, as this represents the gradient.		$d = 2t + 4, 2 \le t \le 5$ The coefficient of t is 2
	3. Write the answer.		The walking pace is 2 km/h.
ii.	1. Look at the time interval shown.	ii.	$2 \le t \le 5$
	2. Interpret the information and write the answer.		They walked at this pace for 3 hours.
c.	1. Determine the distance travelled before the change of pace.	c.	Change after $t = 2$ hours: $d = 4t$ $d = 4 \times 2$ $d = 8$ km
	2. Using a calculator, spreadsheet or otherwise, sketch the graph of $d = 4t$ between $t = 0$ and $t = 2$.		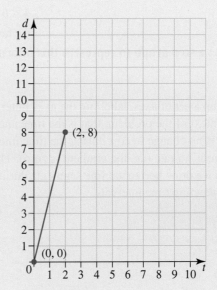
	3. Solve the simultaneous equations by substitution to determine the point of intersection: $d = 4t$ and $d = 2t + 4$; $\therefore 4t = 2t + 4$.		$4t = 2t + 4$ $4t - 2t = 2t - 2t + 4$ $2t = 4$ $t = 2$ Substitute $t = 2$ into $d = 4t$: $d = 4 \times 2 = 8$

4. Using a spreadsheet or otherwise, sketch the graph of $d = 2t + 4$ between $t = 2$ and $t = 5$.

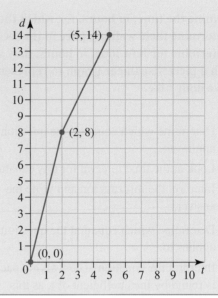

The following sign shows the car parking fees in a shopping carpark.

CARPARK FEES

$0 < 2$ hours	$1.00
$2 < 4$ hours	$2.50
$4 < 6$ hours	$5.00
$6+$ hours	$6.00

Construct a step graph to represent this information.

THINK

1. Draw up a set of axes, labelling the axes in terms of the context of the problem, that is the time and cost. There is no change in cost during the time intervals, so there is no rate (i.e. the gradient is zero). This means we draw horizontal line segments during the corresponding time intervals.

WRITE

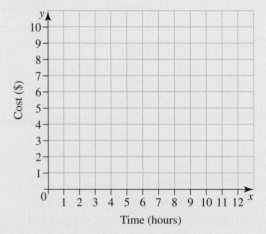

2. Draw segments to represent the different time intervals. As the cost changes at the start of each time interval, this is represented by a closed circle. Hence, the end of a time period must be represented by an open circle.

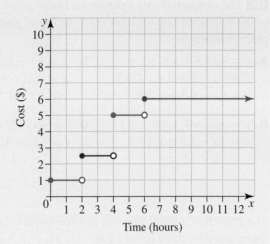

Exercise 8.5 Piecewise linear graphs and step graphs

learn on

8.5 Exercise	8.5 Exam questions on

These questions are even better in jacPLUS!
- Receive immediate feedback
- Access sample responses
- Track results and progress

Find all this and MORE in jacPLUS ▶

Simple familiar	Complex familiar	Complex unfamiliar
1, 2, 3, 4, 5, 6, 7, 8, 9, 10, 11	12, 13, 14, 15, 16, 17	18, 19, 20

Simple familiar

1. **WE9** A piecewise linear graph is constructed from the following linear graphs.

$$y = -3x - 3, \ x \leq a$$
$$y = x + 1, \ x \geq a$$

a. By solving the equations simultaneously, determine the point of intersection and hence state the value of a.
b. Sketch the piecewise linear graph.

2. Consider the following linear graphs that make up a piecewise linear graph.

$$y = 2x - 3, x \leq a$$
$$y = 3x - 4, a \leq x \leq b$$
$$y = 5x - 12, x \geq b$$

a. Using a spreadsheet or otherwise, sketch the three linear graphs.
b. Determine the two points of intersection.
c. Using the points of intersection, determine the values of a and b.
d. Sketch the piecewise linear graph.

3. **MC** The diagram shows a piecewise linear graph. Identify which one of the following options represents the linear graphs that make up the piecewise graph.

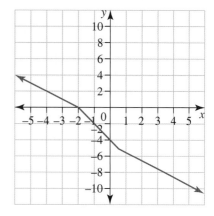

A. $y = -2x - 4, x \leq -2$
$y = -x - 2, -2 \leq x \leq 0.5$
$y = -x - 4.5, x \geq 0.5$

B. $y = -x - 2, x \leq -2$
$y = -2x - 4, -2 \leq x \leq 0.5$
$y = -x - 4.5, x \geq 0.5$

C. $y = -2x - 4, x \leq 0$
$y = -x - 2, 0 \leq x \leq -5$
$y = -x - 4.5, x \geq -5$

D. $y = -x - 2, x \leq 0$
$y = -2x - 4, 0 \leq x \leq -5$
$y = -x - 4.5, x \geq -5$

4. The growth of a small tree was recorded over 6 months. It was found that the tree's growth could be represented by three linear equations, where h is the height in centimetres and t is the time in months.

Equation 1: $h = 2t + 20, 0 \leq t \leq a$
Equation 2: $h = t + 22, a \leq t \leq b$
Equation 3: $h = 3t + 12, b \leq t \leq c$

a. i. By solving equations 1 and 2 simultaneously, determine the value of a.
   ii. By solving equations 2 and 3 simultaneously, determine the value of b.
b. Explain why $c = 6$.
c. Identify the time interval during which the tree grew the most.
d. Sketch the piecewise linear graph that shows the height of the tree over the 6-month period.

5. **WE10** Construct a step graph from the following equations, making sure to take note of the relevant end points.

$$y = 3, 1 < x \leq 4$$
$$y = 1.5, 4 < x \leq 6$$
$$y = -2, 6 < x \leq 8$$

6. The following table shows the costs to hire a plumber.

Time (minutes)	Cost ($)
$0 < 15$	45
$15 < 30$	60
$30 < 45$	80
$45 < 60$	110

a. Represent this information on a step graph.
b. Anton hired the plumber for a job that took 23 minutes. Determine how much Anton will be charged for this job.

7. Airline passengers are charged an excess for any luggage that weighs 20 kg or over. The following graph shows these charges for luggage weighing over 20 kg.

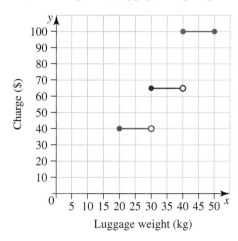

a. Determine how much excess a passenger would be charged for luggage that weighs 31 kg.
b. Nerada checks in her luggage and is charged $40. Determine the maximum excess luggage she could have without having to pay any more.
c. Hilda and Hanz have two pieces of luggage between them. Neither of them are full. One piece weighs 32 kg and the other piece weighs 25 kg. Explain how they could minimise their excess luggage charges.

8. **WE11** The following two equations represent water being added to a water tank over 15 hours, where w is the water in litres and t is the time in hours.

Equation 1: $w = 25t, 0 \leq t \leq 5$
Equation 2: $w = 30t - 25, 5 \leq t \leq 15$

a. Determine how many litres of water are in the tank after 5 hours.
b. i. Determine the rate at which the water is being added to the tank after 5 hours.
   ii. Determine for how long the water added to the tank at this rate.
c. Sketch a piecewise graph to represent the water in the tank at any time, t, over the 15-hour period.

9. **WE12** The costs to hire a paddle boat are listed in the following table. Construct a step graph to represent the cost of hiring a paddle boat for up to 40 minutes.

Time (minutes)	Hire cost ($)
0 < 20	15
20 < 30	20
30 < 40	25

10. The postage costs to send parcels from the Northern Territory to Sydney are shown in the following table.

Weight of parcel (kg)	Cost ($)
$0 < 0.5$	6.60
$0.5 < 1$	16.15
$1 < 2$	21.35
$2 < 3$	26.55
$3 < 4$	31.75
$4 < 5$	36.95

a. Represent this information in a step graph.
b. Pammie has two parcels to post to Sydney from the Northern Territory. One parcel weighs 450 g and the other weighs 525 g. Explain whether it is cheaper to send the parcels individually or together. Justify your answer using calculations.

11. The following linear equations represent the distance sailed by a yacht from the yacht club during a race, where d is the distance in kilometres from the yacht club and t is the time in hours from the start of the race.

Equation 1: $d = 20t, 0 \leq t \leq 0.75$
Equation 2: $d = 15t + 3.75, 0.75 \leq t \leq 1.25$
Equation 3: $d = -12t + 37.5, 1.25 \leq t \leq b$

a. Using a spreadsheet or otherwise, determine the points of intersection.
b. In the context of this problem, explain why equation 3 has a negative gradient.
c. Calculate how far the yacht is from the starting point before it turns and heads back to the yacht club.
d. Determine the duration, to the nearest minute, of the yacht's sailing time for this race. Hence, determine the value for b. Write your answer correct to 2 decimal places.

Complex familiar

12. A step graph is shown below. Write the equations that make up the graph.

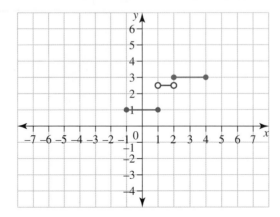

13. A car hire company charges a flat rate of $50 plus 75 cents per kilometre up to and including 150 kilometres. An equation to represent this cost, C, in dollars is given as $C = 50 + ak$, $0 \leq k \leq b$, where k is the distance travelled in kilometres.

 a. Write the values of a and b.
 b. Using a spreadsheet or otherwise, sketch this equation on a set of axes, using appropriate values.
 c. The cost charged for distances over 150 kilometres is given by the equation $C = 87.50 + 0.5k$. Determine the charge in cents per kilometre for distances over 150 kilometres.
 d. By solving the two equations simultaneously, determine the point of intersection and hence show that the graph will be continuous.
 e. Sketch the equation $C = 87.50 + 0.5k$ for $150 \leq k \leq 300$ on the same set of axes as part b.

14. The temperature of a wood-fired oven, $T\,°C$, steadily increases until it reaches $200\,°C$. Initially the oven has a temperature of $18\,°C$ and it reaches the temperature of $200\,°C$ in 10 minutes.

 a. Construct an equation that determines the temperature of the oven during the first 10 minutes. Include the time interval, t, in your answer.

 Once the oven has heated up for 10 minutes, a loaf of bread is placed in the oven to cook for 20 minutes. An equation that represents the temperature of the oven during the cooking of the bread is $T = 200$, $a \leq t \leq b$.

 b. i. Write the values of a and b.
   ii. In the context of this problem, explain what a and b represent.

 After the 20 minutes of cooking, the oven's temperature is lowered. The temperature decreases steadily, and after 30 minutes the oven's temperature reaches $60\,°C$. An equation that determines the temperature of the oven during the last 30 minutes is $T = mt + 340$, $d \leq t \leq e$.

 c. Determine the values of m, d and e.

 d. Explain what m represents in this equation.

 e. Using your values from the previous parts, sketch the graph that shows the changing temperature of the wood fired oven during the 60-minute interval.

15. The amount of money in a savings account over 12 months is shown in the following piecewise graph, where A is the amount of money in dollars and t is the time in months.

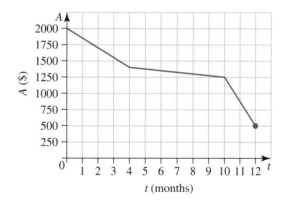

t (months)

 One of the linear graphs that make up the piecewise linear graph is $A = 2000 - 150t$, $0 \leq t \leq a$.

 a. Determine the value of a.
 b. The equation that intersects with $A = 2000 - 150t$ is given by $A = b - 50t$. If the two equations intersect at the point $(4, 1400)$, show that $b = 1600$.
 c. The third equation is given by the rule $A = 4100 - 300t$. By solving a pair of simultaneous equations, determine the time interval for this equation.
 d. Using an appropriate equation, determine the amount of money in the account at the end of the 12 months.

16. Stamp duty is a government charge on the purchase of items such as cars and houses. The table shows the range of stamp duty charges for purchasing a car in South Australia.

Car price (P)	Stamp duty (S)
$0 - 1000$	1%
$1000 - 2000$	$10 + 2\%(P - 1000)$
$2000 - 3000$	$30 + 3\%(P - 2000)$
$3000+$	$60 + 4\%(P - 3000)$

a. Explain why the stamp duty costs for cars can be modelled by a piecewise linear graph.

The stamp duty charge for a car purchased for $1000 or less can be expressed by the equation $S = 0.01P$, where S is the stamp duty charge and P is the purchase price of the car for $0 \leq P \leq 1000$. Similar equations can be used to express the charges for cars with higher prices.

Equation 1: $S = 0.01P, 0 \leq P \leq 1000$
Equation 2: $S = 0.02P - 10, a < P \leq b$
Equation 3: $S = 0.03P - c, 2000 < P \leq d$
Equation 4: $S = fP - e, P > 3000$

b. For equations 2, 3 and 4, determine the values of a, b, c, d, e and f.
c. Using a spreadsheet or otherwise, determine the points of intersection for the equations in part b.
d. Suki and Boris purchase a car and pay $45 in stamp duty. Determine the price they paid for their car.

17. A small inflatable swimming pool that holds 1500 litres of water is being filled using a hose. The amount of water, A, in litres in the pool after t minutes is shown in the following graph.

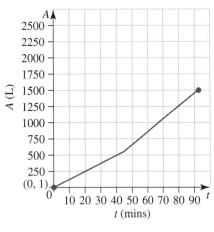

a. Estimate the amount of water, in litres, in the pool after 45 minutes.
b. Determine the amount of water being added to the pool each minute during the first 45 minutes.
c. After 45 minutes the children become impatient and turn the hose up.
   The equation $A = 20t - 359$ determines the amount of water, A, in the pool t minutes after 45 minutes. Using this equation, determine the time taken, in minutes, to fill the pool. Give your answer to the nearest whole minute.

18. The distance of a dancer's right foot from the floor during a dance recital can be found using the following linear graphs, where h is the height in centimetres from the floor and t is the duration of the recital in seconds.

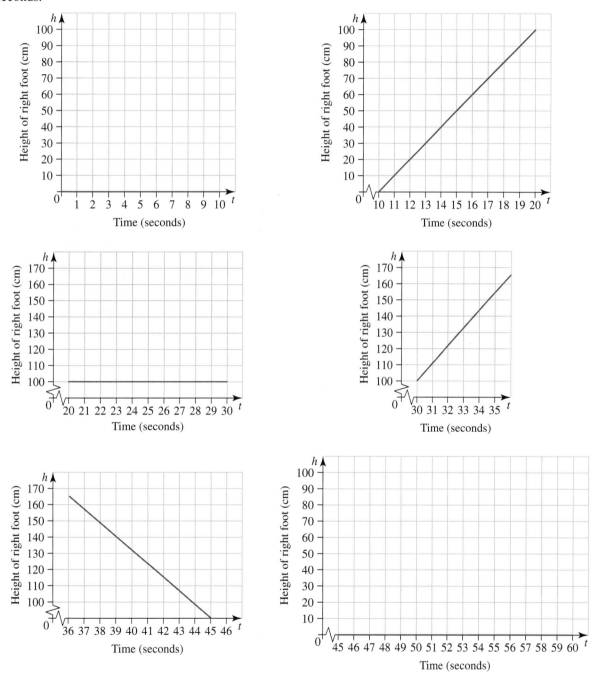

Construct a graph that shows the distance of the dancer's right foot from the floor at any time during the recital. Clearly state all key features.

19. Determine the values of a, b and c, and sketch the piecewise linear graph formed by these four linear graphs. Describe any issues you encounter.

   i. $y = x + 4$, $x \leq a$

   ii. $y = 2x + 3$, $a \leq x \leq b$

   iii. $y = x + 6$, $b \leq x \leq c$

   iv. $y = 3x + 1$, $x \geq c$

20. The Slippery Slide ride is a new addition to a famous theme park. The slide has a horizontal distance of 20 metres and is comprised of four sections.

The first section is described by the equation $h = -3x + 12$, $0 \leq x \leq a$, where h is the height in metres from the ground and x is the horizontal distance in metres from the start. In the first section, the slide drops 3 metres over a horizontal distance of 1 metre before meeting the second section.

The remaining sections of the slides are modelled by the following equations.

Section 2: $h = -\dfrac{2x}{3} + \dfrac{29}{3}$, $a \leq x \leq b$

Section 3: $h = -2x + 13$, $b \leq x \leq c$

Section 4: $h = -\dfrac{5x}{16} + \dfrac{25}{4}$, $c \leq x \leq d$

Sketch the graph that shows the height at any horizontal distance from the start of the slide and explain why $d = 20$.

Fully worked solutions for this chapter are available online.

LESSON
8.6 Review

8.6.1 Summary

8.6 Exercise

8.6 Exercise	8.6 Exam questions on

Simple familiar	Complex familiar	Complex unfamiliar
1, 2, 3, 4, 5, 6, 7, 8, 9, 10	11, 12, 13, 14, 15, 16	17, 18, 19, 20

Simple familiar

1. For each of the following graphs, determine the coordinates of the point of intersection. Justify your answers.

a.
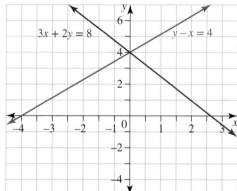

b.
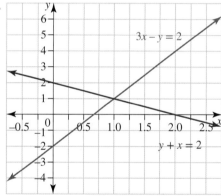

c.
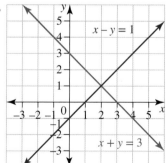

2. The solution to the following pair of simultaneous equations is:

$$x + 5y = 7$$
$$y = 5 - 2x$$

A. $x = -8, y = 3$ **B.** $x = 2.25, y = 0.5$ **C.** $x = 2, y = 1$ **D.** $x = 1, y = 2$

3. For each of the following graphs, determine the coordinates of the point of intersection. Justify your answers.

a.
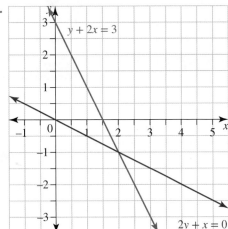

b.
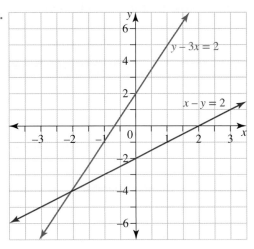

c.
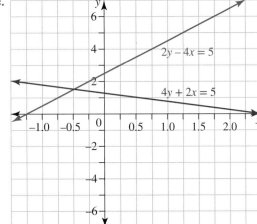

4. Solve the following pairs of simultaneous equations by graphing the equations and identifying the point of intersection.

a. $y = 2x + 3$
$y = 8 - 3x$

b. $y + 2x = -8$
$y = 2x + 4$

c. $2y = 12x - 16$
$3y = -6x - 24$

d. $2x - y = -1$
$3x + y = 11$

5. For each of the pairs of equations below, calculate the gradients and indicate whether the lines are parallel or perpendicular or neither.

a. $y = 2x - 4$
$6y - 12x = 20$

b. $2y = 5 - 6x$
$3y = -9x + 18$

c. $3y + 2x = 9$
$6y + 4x = 18$

d. $y = -2x + 3$
$y = \dfrac{1}{2}x - 4$

e. $y = 4x + 5$
$2y - 15x = 10$

f. $4y = 6 - x$
$y = 4x + 6$

6. Solve the system of equations $y = 5x$ and $y = -3x + 8$ by substitution.

7. Solve the system of equations $x + 3y = 1$ and $y = 2x + 5$ by elimination by first rearranging the equations.

8. Solve the following pairs of simultaneous equations by substitution and identifying the point of intersection.

   a. $31 = y - 2x$
      $2x + 2y = 14$

   b. $x + 2y = 4$
      $y = 2x - 3$

   c. $x + y = 6$
      $2x + 4y = 20$

   d. $y = \dfrac{2x}{3} + 2$
      $y = 2x - 2$

9. Solve the following simultaneous equations by using elimination.

   a. $y = 3x - 5$ and $3x + 2y = 17$

   b. $2x + y = 7$ and $3x - y = 3$

   c. $3x - 2y = -16$ and $y = 2(x + 9)$

   d. $4x + 3y = 17$ and $3x + 2y = 13$

10. Determine the break-even points for the following cost and revenue equations. Where appropriate, give your answers correct to 2 decimal places.

    a. $C = 150 + 2x$ and $R = 7.5x$

    b. $C = 13.5x + 25$ and $R = 19.7x$

Complex familiar

11. Solve each of the following pairs of simultaneous equations.

    a. $5x + 2y = 17$
       $y = \dfrac{3x - 7}{2}$

    b. $2x + 7y = 17$
       $x = \dfrac{1 - 3y}{4}$

    c. $2x + 3y = 13$
       $y = \dfrac{4x - 15}{5}$

    d. $-2x - 3y = -14$
       $x = \dfrac{2 + 5y}{3}$

12. Solve the simultaneous equations $y + x = 10$ and $-2y + x = -5$ using technology.

13. **MC** Two adults and four children went to the circus. They paid a total of $55.00 for their tickets. One adult and three children paid $35.00 to enter the same circus. Identify which one of the following sets of simultaneous equations represents this situation, where a is the cost for an adult and c is the cost of a child's entry.

    A. $a + c = 55$
       $a + c = 35$

    B. $2a + c = 55$
       $a + 3c = 35$

    C. $2a + 3c = 55$
       $a + 4c = 35$

    D. $2a + 4c = 55$
       $a + 3c = 35$

14. **MC** Bertha knits teddy bears and sells them at the local farmers' market. Bertha spends $120 in wool, and it costs her an additional $4.50 to make each teddy bear. She sells the bears for $14.50 each. Determine how many teddy bears Bertha needs to sell to break even.

    A. 6

    B. 8

    C. 9

    D. 12

15. A step graph is formed by the following equations:

    $$y = -3, 0 < x \leq 3$$
    $$y = 2, 3 < x < 5$$
    $$y = 5, x \geq 5$$

    Construct the step graph that represents this information.

16. The following two linear equations make a piecewise linear graph.

    $$y = -2x + 1, x \leq a$$
    $$y = -3x + 2, x \geq a$$

    a. Solve the equations simultaneously, and hence determine the value of a.
    b. Sketch the piecewise linear graph.

17. Solve each of the following pairs of simultaneous equations for x and y in terms of m and n.

a. $mx - y = n$
   $y = nx$

b. $mx - ny = n$
   $y = x$

c. $mx - ny = -m$
   $x = y - n$

d. $mx - y = m$
   $x = \dfrac{y + m}{n}$

18. Determine the values of a and b so that the pair of equations $ax + by = 17$ and $2ax - by = -11$ has only one solution of $(-2, 3)$.

19. Suzanne is starting a business selling homemade cupcakes. It will cost her $250 to buy all of the equipment, and each cupcake will cost $2.25 to make. Suzanne sells each cupcake for $6.50.

Calculate the total number of cupcakes Suzanne needs to sell to break even.

Give your answer correct to the nearest whole number and verify your solution graphically.

20. Jerri and Samantha have both entered a 10-km fun run for charity. The distance travelled by Jerri can be modelled by the linear equation

$$d = 6t - 0.1$$

where d is the distance in km from the starting point and t is time in hours.

The distance Samantha is from the starting point at any time, t hours, can be modelled by the piecewise linear graph

$$d = 4t, \ 0 \le t \le \frac{1}{2}$$

$$d = 8t - 2, \ \frac{1}{2} \le t \le b.$$

Show that Samantha crossed the finishing line ahead of Jerri by 11 minutes.

Fully worked solutions for this chapter are available online.

Hey teachers! Create custom assignments for this chapter

Create and assign unique tests and exams

Access quarantined tests and assessments

Track your students' results

Find all this and MORE in jacPLUS ⊙

Answers

Chapter 8 Simultaneous equations and their applications

8.2 Solving simultaneous linear equations graphically

8.2 Exercise

1. a. $(3, 3)$ b. $(0, 1)$

   c. $(-3, 4)$ d. $\left(\dfrac{5}{2}, 5\right)$

2. a. Yes, because the coordinates of the point satisfy both equations.

   b. No, because the point doesn't satisfy both equations.

3. a. 10 hours

   b. Company T: $y = x + 10$

      Company O: $y = \dfrac{1}{2}x + 15$

   c. By substitution, $y = 20$ when $x = 10$.

4. a. Since the gradients are different, the lines will intersect.

   b. The graphs intersect along the whole length, since the lines have the same gradient and y-intercept.

   c. Since the gradients are different, the lines will intersect.

   d. i. A system of linear equations will have a unique solution if the equations have **different** gradients (they are not parallel).

      ii. A system of linear equations will have an infinite number of solutions if the equations have **the same** gradient and **the same** y-intercept.

5. a. Car A travels 60 km, Car B travels 60 km.

   b. Car A: $m = \dfrac{60}{1} = 60$

      Car B: $m = \dfrac{60}{1} = 60$

   c. The cars are never in the same spot at the same time because Car A is always 20 km ahead of Car B.

   d. The y-intercepts indicate that Car A started at 20 km from the town and Car B started from the town.

6. a. $x = 1, y = -2$ b. $x = 2, y = 4$
   c. $x = 2, y = -1$ d. $x = -1, y = 3$
   e. $x = 0, y = 0$ f. $x = \dfrac{5}{2}, y = -\dfrac{1}{2}$

7. $x = 2, y = 3$

8. $x = -2, y = 5$

9.

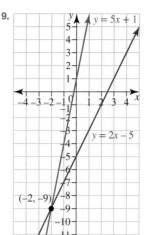

$x = -2, y = -9$

10.

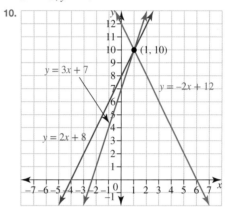

$x = 1, y = 10$

11. a. $x = -2, y = -7$

    b. $x = 2, y = -3$

    c. $x = -5, y = -18$

    d. No solution, lines are parallel

12. Your friend is 6 years old and her brother is 5 years old.

13. a. Apples cost $0.25 each.

    b. Oranges cost $0.50 each, so the difference is $0.25.

14. a. No, the graphs in part iii are parallel but not coincidental (they have the same gradient and different intercepts).

    b. i. $(-12, -64)$ ii. $(-1, -2)$
       iii. No solution   iv. $(-1, 4)$

8.3 Solving simultaneous equations algebraically

8.3 Exercise

1. a. $x = -1, y = -1$ b. $m = 11, n = 3$

2. a. $x = 3, y = -3$ b. $x = 1, y = 4$

3. D

4. a. $x = 7, y = 19$ b. $x = -2, y = -12$
   c. $x = 7, y = 44$ d. $x = 2, y = 1$
   e. $x = -3, y = -4$ f. $x = 1, y = -1$

5. Add the two equations and solve for x, then substitute x into one of the equations to solve for y.
$x = 3, y = -1$

6. a. $x = 2, y = -2$ b. $x = 3, y = 4$
   c. $x = -1, y = 3$ d. $x = 5, y = 3$
   e. $x = -1, y = 4$ f. $x = 6, y = 4$

7. $a = 2$ and $x = -2$

8. A

9. a. Marcia added the equations instead of subtracting. The correct result for step 2 is $22y = 11$

   b. $x = 5, y = \dfrac{1}{2}$

   c. $x = 5 \; y = \dfrac{1}{2}$

10. a. $x = -10.71, y = -12.86$
   b. $x = 0.75, y = 0.89$

11. a. Goal = 5 points
     Behind = 2 points

   b. Jetts = 40 points
     Meteorites = 48 points

12. a. Equation 2 has unknowns on each side of the equal sign and can be substituted into equation 1.

   b. Mick works 5 hours and Minnie works 10 hours.

   c. 3 hours 45 minutes (3.75 hours)

8.4 Solving practical problems using simultaneous equations

8.4 Exercise

1. $4d + 3c = 10.55$ and $2d + 4c = 9.90$

2. a. 140 adults and 210 children

   b. The cost of an adult's ticket is $25, the cost of a children's ticket is $15, and the total ticket sales is $6650.

3. a. i. 10
    ii. Yolanda needs to sell 10 bracelets to cover her costs.

   b. i. $16 loss
    ii. $24 profit

4. a. $a = 18$
   b. $b = 3$
   c. 170 entries
   d. $R = \$5580, C = \$3480, P = \$2100$
   e. 504 entries

5. a. The price per adult ticket is $3.50 and the price per student ticket is $1.50.

   b. 4 adults and 19 students

6. a. $a = \$19.50, c = \$14.50$

   b. The total number of tickets sold (both adult and concession)

   c. 225 adult tickets and 319 concession tickets

7. a. $R = 12.50h$
   b. 4.5 hours
   c. i. Charlotte made a profit for jobs 1 and 4, and a loss for jobs 2 and 3.
    ii. Yes, she made $15 profit.
     $(25 + 5 - (10 + 5)) = 30 - 15 = \$15$
   d. 9.5 hours

8. a. $S = 0.5n$ b. 8 cups of lemonade
   c. 70 cents

9. $t = \$14.50, m = \$15.25$

10. a. Cost of 1 kg of carrots = $3.95,
     cost of 1 kg of apples = $2.95
   b. $12.35

11. 140 kJ

12. a. $k = 75$ km
   b. $C_{FreeWheels} = \$350, C_{GetThere} = \$315$. They should use GetThere.

13. a. $1560
   b. See the table at the bottom of the page.*
   c. $C = 810 + 7.5n$
   d. $S = 25.50n$
   e. 45 T-shirts
   f. 323 T-shirts

14. $17.90

15. a. The cost is for the three different types of cereal, but the equations only include one type of cereal.
   b. $34.15

16. $1789.50

8.5 Piecewise linear graphs and step graphs

8.5 Exercise

1. a. Point of intersection $= (-1, 0), a = -1$

   b.

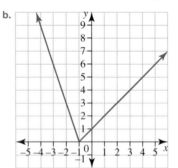

---

*13. b.

n	0	20	30	40	50	60	80	100	120	140
C	810	960	1035	1110	1185	1260	1410	1560	1710	1860

2. a.

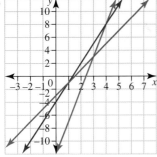

b. $(1, -1)$ and $(4, 8)$

c. $a = 1$ and $b = 4$

d.

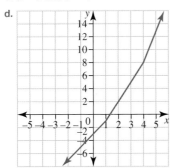

3. B

4. a. i. $a = 2$

    ii. $b = 5$

b. The data is only recorded over 6 months.

c. $5 \leq t \leq 6$ (between 5 and 6 months)

d.

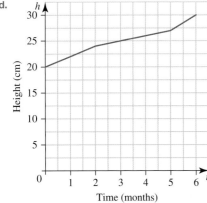

5.

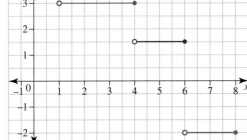

6. a.

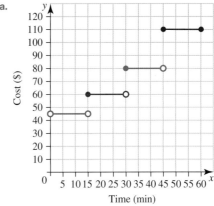

b. $60

7. a. $65

b. 10 kg

c. Place 2–3 kg from the 32-kg bag into the 25-kg bag and pay $80 rather than $105.

8. a. 125 L

b. i. 30 L/h

    ii. 10 h

c.

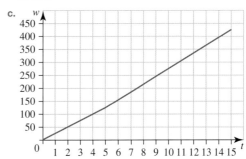

9.

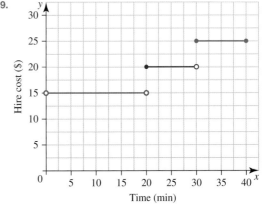

10. a.

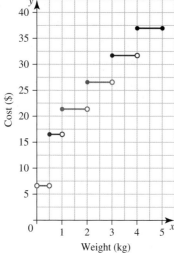

Weight (kg)

b. It is cheaper to post them together ($16.15 together versus $22.75 individually).

11. a. $(0.75, 15)$ and $(1.25, 22.5)$

b. The yacht is returning to the yacht club during this time period.

c. 22.5 km

d. 3 hours, 8 minutes; $b = 3.13$

12. $y = 1, -1 \leq x \leq 1$; $y = 2.5, 1 < x < 2$; $y = 3, 2 \leq x \leq 4$

13. a. $a = 0.75, b = 150$

b.

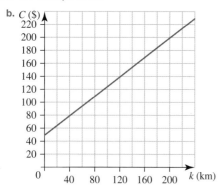

c. 50 cents/km

d. $k = 150, C = 162.50$. This means that the point of intersection $(150, 162.5)$ is the point where the charges change. At this point both equations will have the same value, so the graph will be continuous.

e.

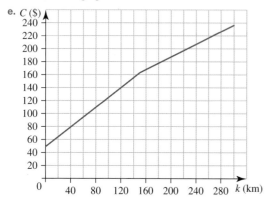

14. a. $T = 18 + 18.2t, 0 \leq t \leq 10$

b. i. $a = 10, b = 30$

ii. a is the time the oven first reaches $200\,°C$ and b is the time at which the bread stops being cooked.

c. $m = \dfrac{-14}{3}, d = 30, e = 60$

d. The change in temperature for each minute in the oven

e.

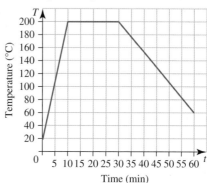

Time (min)

15. a. $a = 4$

b. Sample responses can be found in the worked solutions in the online resources.

c. $10 \leq t \leq 12$

d. 500

16. a. There is a change in the rate for different x-values (i.e. different car prices).

b. $a = 1000, b = 2000, c = 30, d = 3000, e = 60, f = 0.04$

c. $(1000, 10), (2000, 30)$ and $(3000, 60)$

d. 2500

17. a. 540 L **b.** 12 L/min **c.** 93 min

18. Key features of the graph include:
- the intervals when $y = 0$ (the dancer's right foot is on the floor): $0 \leq t \leq 10$ and $45 \leq t \leq 60$
- the maximum height the dancer's right foot is from the floor: 165 cm
- the length of the recital: 60 seconds.

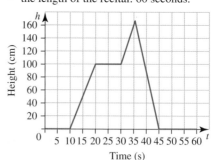

Time (s)

19. $a = 1, b = 3, c = 2.5$

i. $x \leq 1$ **ii.** $1 \leq x \leq 3$

iii. $3 \leq x \leq 2.5$ **iv.** $x \geq 2.5$

$b > c$, which means that graph iii is not valid and the piecewise linear graph cannot be sketched.

20. h (cm)

[graph showing height vs distance, decreasing curve from 12 at x=0 down to 0 at x=20]

The horizontal distance of the slide is 20 m; thus, $d = 20$.

8.6 Review

8.6 Exercise

1. a. $(0, 4)$ **b.** $(1, 1)$ **c.** $(2, 1)$

2. C

3. a. $(2, -1)$ **b.** $(-2, -4)$ **c.** $(-0.5, 1.5)$

4. a. $(1, 5)$ **b.** $(-3, -2)$
    c. $(-2, -4)$ **d.** $(2, 5)$

5. a. $m_1 = 2$ $m_2 = 2$ Parallel

    b. $m_1 = -3$ $m_2 = -3$ Parallel

    c. $m_1 = -\dfrac{2}{3}$ $m_2 = -\dfrac{2}{3}$ Parallel

    d. $m_1 = -2$ $m_2 = \dfrac{1}{2}$ Perpendicular

    e. $m_1 = 4$ $m_2 = 7.5$ Neither

    f. $m_1 = -\dfrac{1}{4}$ $m_2 = 4$ Perpendicular

6. $(1, 5)$

7. $(-2, 1)$

8. a. $(-8, 15)$ **b.** $(2, 1)$
    c. $(2, 4)$ **d.** $(3, 4)$

9. a. $x = 3, y = 4$

    b. $x = 2, y = 3$

    c. $x = -20, y = -22$

    d. $x = 5, y = -1$

10. a. $x = 27.27$ **b.** $x = 4.03$

11. a. $(3, 1)$ **b.** $(-2, 3)$
    c. $(5, 1)$ **d.** $(4, 2)$

12. $(5, 5)$

13. D

14. D

15.

[graph with step segments; a segment at y = 5 from x = 5 onward, a segment at y = 2 from x = 3 to x = 5 with open circles, a segment at y = -3 from x = 1 to x = 3]

16. a. 1

    b.

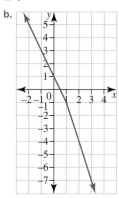

17. a. $\left(\dfrac{n}{m-n}, \dfrac{n^2}{m-n}\right)$ **b.** $\left(\dfrac{n}{m-n}, \dfrac{n}{m-n}\right)$

    c. $\left(\dfrac{n^2-m}{m-n}, \dfrac{mn-m}{m-n}\right)$ **d.** $(0, -m)$

18. $a = -1, b = 5$.

19. Break-even point: 59 cupcakes, \$387.50

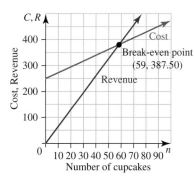

20. Sample responses can be found in the worked solutions in the online resources.

9 Applications of trigonometry

LESSON SEQUENCE

Fully worked solutions for this chapter are available online.

EXAM PREPARATION

Access exam-style questions in every lesson, available online.

on Resources

Solutions	Solutions — Chapter 9 (sol-1250)
Exam questions	Exam question booklet — Chapter 9 (eqb-0266)
Digital documents	Learning matrix — Chapter 9 (doc-41491)
	Chapter summary — Chapter 9 (doc-41492)

LESSON
9.1 Overview

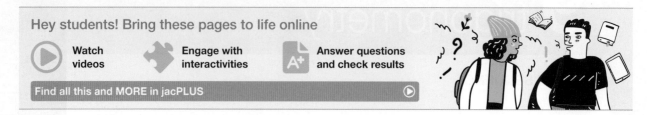
9.1.1 Introduction

Trigonometry is a branch of mathematics that describes the relationship between angles and side lengths of triangles. Trigonometry is widely used in many areas:

- **Architecture and engineering:** Much of architecture and engineering relies on the formation of triangles for support structures. When an architect wants to correctly lay out a curved wall, or work out the slope of a roof height, trigonometry is used. The construction of the Golden Gate Bridge (shown in the photo) involved hundreds of thousands of trigonometric calculations.

- **Music theory:** Music theory involves sound waves, and sound waves travel in a repeating wave pattern. This repeating pattern can be represented graphically by sine and cosine functions. A single note can be modelled on a sine curve and a chord can be modelled with multiple sine curves.

- **Electrical engineers:** The electricity sent to our house requires an understanding of trigonometry. Power companies use what is called alternating current (AC) to send electricity over long distances. This is due to the use of transformers, which require the use of alternating current to function. The alternating current signal has a sinusoidal behaviour.

- **Video games:** When you see a character smoothly glide over obstacles, they don't jump vertically straight up the y-axis; they follow a slightly curved path or a parabolic path instead. Trigonometric calculations help animators ensure their characters jump over obstacles in a realistic manner.

Trigonometry is not limited to the areas listed above. It is used in many others, such as flight engineering, physics, archaeology, criminology and marine biology. This shows how important an understanding of trigonometry is, since it is used in so many different fields.

9.1.2 Syllabus links

Lesson	Lesson title	Syllabus links
9.2	**Trigonometric ratios**	○ Understand and use the trigonometric ratios to find the size of an unknown angle, θ, or the length of an unknown side in a right-angled triangle. • $\cos \theta = \dfrac{\text{adjacent}}{\text{hypotenuse}}$ • $\sin \theta = \dfrac{\text{opposite}}{\text{hypotenuse}}$ • $\tan \theta = \dfrac{\text{opposite}}{\text{adjacent}}$
9.3	**Areas of triangles**	○ Calculate the area of a non-right-angled triangle and solve related practical problems. • $\text{area} = \dfrac{1}{2} bc \sin A$, given two sides, b and c, and an included angle, A. • Heron's rule: $\text{area} = \sqrt{s(s-a)(s-b)(s-c)}$ where $s = \dfrac{a+b+c}{2}$, given three sides, a, b and c.
9.4	**Angles of elevation and depression**	○ Solve two-dimensional practical problems involving the trigonometry of right-angled and non-right-angled triangles, including problems involving angles of elevation and depression and the use of true bearings.
9.5	**The sine rule**	○ Solve two-dimensional problems involving a non-right-angled triangle, ΔABC, with sides, a, b, and c, and corresponding angles A, B and C. • Sine rule: $\dfrac{a}{\sin A} = \dfrac{b}{\sin B} = \dfrac{c}{\sin C}$ (ambiguous case excluded)
9.6	**The cosine rule**	○ Solve two-dimensional problems involving a non-right-angled triangle, ΔABC, with sides, a, b, and c, and corresponding angles A, B and C. • Cosine rule: $c^2 = a^2 + b^2 - 2ab \cos C$
9.7	**True bearings**	○ Solve two-dimensional practical problems involving the trigonometry of right-angled and non-right-angled triangles, including problems involving angles of elevation and depression and the use of true bearings.

Source: General Mathematics Senior Syllabus 2024 © State of Queensland (QCAA) 2024; licensed under CC BY 4.0.

LESSON
9.2 Trigonometric ratios

9.2.1 Sine, cosine and tangent ratios

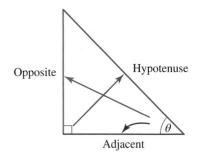

A ratio of the lengths of two sides of a right-angled triangle is called a **trigonometric ratio**. The three most common trigonometric ratios are **sine**, **cosine** and **tangent**. They are abbreviated as sin, cos and tan respectively. Trigonometric ratios are used to calculate the unknown length or acute angle size in right-angled triangles.

It is important to identify and label the features in a right-angled triangle. The labelling convention of a right-angled triangle is as follows:

The longest side of a right-angled triangle is always called the hypotenuse and is opposite the right angle. The other two sides are named in relation to the reference angle, θ. The opposite side is opposite the reference angle, and the adjacent side is next to the reference angle.

9.2.2 The sine ratio

The sine ratio is used when we want to calculate an unknown value given two out of the three following values: opposite, hypotenuse and reference angle.

The sine ratio of θ is written as $\sin(\theta)$ and is defined as follows.

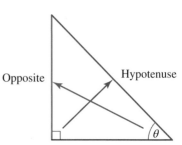

> **Sine ratio**
>
> $$\sin(\theta) = \frac{\text{opposite}}{\text{hypotenuse}} \quad \text{or} \quad \sin(\theta) = \frac{O}{H}$$

The inverse sine function is used to determine the value of the unknown reference angle given the lengths of the hypotenuse and opposite side.

> **Sine ratio for calculating an angle**
>
> $$\theta = \sin^{-1}\left(\frac{O}{H}\right)$$

WORKED EXAMPLE 1 Using the sine ratio to calculate an unknown side length

Calculate the value of the pronumeral x correct to 2 decimal places.

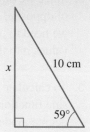

THINK

1. Label all the given information on the triangle.

WRITE

2. Since we have been given the combination of opposite, hypotenuse and the reference angle θ, we need to use the sine ratio. Substitute the given values into the ratio equation.

$\sin(\theta) = \dfrac{O}{H}$

$\sin(59°) = \dfrac{x}{10}$

3. Rearrange the equation to make the unknown the subject and solve.
Make sure your calculator is in degree mode.

$x = 10\sin(59°)$
$= 8.57$

4. Write the answer.

The opposite side length is 8.57 cm.

WORKED EXAMPLE 2 Using the sine ratio to calculate an unknown angle

Calculate the value of the unknown angle, θ, correct to 2 decimal places.

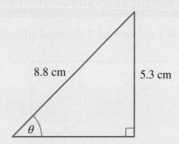

THINK

1. Label all the given information on the triangle.

WRITE

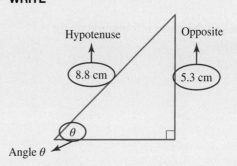

2. Since we have been given the combination of opposite, hypotenuse and the reference angle θ, we need to use the sine ratio. Substitute the given values into the ratio equation.

$$\sin(\theta) = \frac{O}{H}$$
$$= \frac{5.3}{8.8}$$

3. To calculate the angle θ, we need to use the inverse sine function. Make sure your calculator is in degree mode.

$$\theta = \sin^{-1}\left(\frac{5.3}{8.8}\right)$$
$$= 37.03°$$

9.2.3 The cosine ratio

The cosine ratio is used when we want to calculate an unknown value given two out of the three following values: adjacent, hypotenuse and reference angle.

The cosine ratio of θ is written as $\cos(\theta)$ and is defined as follows.

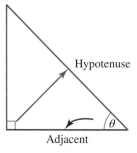

Cosine ratio

$$\cos(\theta) = \frac{\text{adjacent}}{\text{hypotenuse}} \quad \text{or} \quad \cos(\theta) = \frac{A}{H}$$

The inverse cosine function is used to calculate the value of the unknown reference angle when given lengths of the hypotenuse and adjacent side.

Cosine ratio for calculating an angle

$$\theta = \cos^{-1}\left(\frac{A}{H}\right)$$

eles-3063

WORKED EXAMPLE 3 Using the cosine ratio to calculate an unknown side length

Calculate the value of the pronumeral y correct to 2 decimal places.

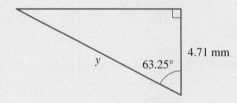

THINK

1. Label all the given information on the triangle.

WRITE

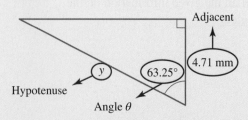

2. Since we have been given the combination of adjacent, hypotenuse and the reference angle θ, we need to use the cosine ratio. Substitute the given values into the ratio equation.

$$\cos(\theta) = \frac{A}{H}$$
$$\cos(63.25°) = \frac{4.71}{y}$$

3. Rearrange the equation to make the unknown the subject and solve. Make sure your calculator is in degree mode.

$$y = \frac{4.71}{\cos(63.25°)}$$
$$= 10.46$$

4. Write the answer.

The length of the hypotenuse is 10.46 mm.

WORKED EXAMPLE 4 Using the cosine ratio to calculate an unknown angle

Calculate the value of the unknown angle, θ, correct to 2 decimal places.

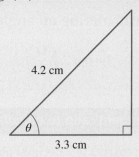

4.2 cm

θ

3.3 cm

THINK

1. Label all the given information on the triangle.

WRITE

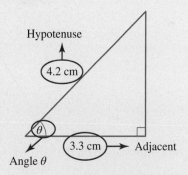

Hypotenuse

4.2 cm

θ

3.3 cm → Adjacent

Angle θ

2. Since we have been given the combination of adjacent, hypotenuse and the reference angle θ, we need to use the cosine ratio. Substitute the given values into the ratio equation.

$$\cos(\theta) = \frac{\text{adjacent}}{\text{hypotenuse}} = \frac{A}{H}$$
$$\cos(\theta) = \frac{3.3}{4.2}$$

3. To calculate the angle θ, we need to use the inverse cosine function. Make sure your calculator is in degree mode.

$$\theta = \cos^{-1}\left(\frac{3.3}{4.2}\right)$$
$$= 38.21°$$

9.2.4 The tangent ratio

The tangent ratio is used when we want to calculate an unknown value given two out of the three following values: opposite, adjacent and reference angle.

The tangent ratio of θ is written as $\tan(\theta)$ and is defined as follows.

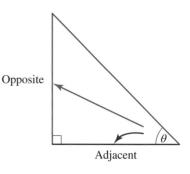

Opposite

Adjacent

θ

> **Tangent ratio**
>
> $$\tan(\theta) = \frac{\text{opposite}}{\text{adjacent}} \quad \text{or} \quad \tan(\theta) = \frac{O}{A}$$

The inverse tangent function is used to calculate the value of the unknown reference angle given the lengths of the adjacent and opposite sides.

> **Tangent ratio for calculating an angle**
>
> $$\theta = \tan^{-1}\left(\frac{O}{A}\right)$$

eles-3064

WORKED EXAMPLE 5 Using the tangent ratio to calculate an unknown side length

Calculate the value of the pronumeral x correct to 2 decimal places.

x

$58°$

9.4 cm

THINK

1. Label all the given information on the triangle.

WRITE

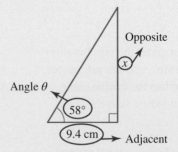

Opposite

x

Angle θ

$58°$

9.4 cm → Adjacent

2. Since we have been given the combination of opposite, adjacent and the reference angle θ, we need to use the tangent ratio. Substitute the given values into the ratio equation.

$$\tan(\theta) = \frac{O}{A}$$

$$\tan(58°) = \frac{x}{9.4}$$

3. Rearrange the equation to make the unknown the subject and solve. Make sure your calculator is in degree mode.

$x = 9.4 \ \tan(58°)$

$x = 15.04$

The opposite side length is 15.04 cm.

Determine the value of the unknown angle, θ, correct to 2 decimal places.

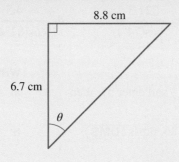

THINK

1. Label all the given information on the triangle.

WRITE

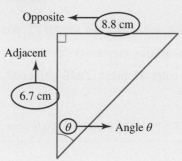

2. Since we have been given the combination of opposite, adjacent and the reference angle θ, we need to use the tangent ratio. Substitute the given values into the ratio equation.

$$\tan(\theta) = \frac{O}{A}$$
$$= \frac{8.8}{6.7}$$

3. To calculate the angle θ, we need to use the inverse tangent function. Make sure your calculator is in degree mode.

$$\theta = \tan^{-1}\left(\frac{8.8}{6.7}\right)$$
$$= 52.72°$$

9.2.5 Calculating trigonometric ratios using a calculator

Calculators can be used to calculate trigonometric ratios. However, calculators require a particular sequence of button presses in order to perform this calculation. Investigate the sequence required for your particular calculator.

In this chapter all angles will be measured in degrees, so make sure that your calculator is in **DEGREE MODE**.

To calculate the **sine ratio** for a given angle on your calculator, use the **sin** function.

To calculate the **cosine ratio** for a given angle on your calculator, use the **cos** function.

To calculate the **tangent ratio** for a given angle on your calculator, use the **tan** function.

The worked examples shown provide correct steps to calculate the trigonometric ratios.

Note: Check the sequence of button presses required by your calculator.

WORKED EXAMPLE 7 Determining a trigonometric ratio correct to 3 decimal places

Using your calculator, calculate the following, correct to 3 decimal places.

a. $\sin(57°)$ b. $9.6\sin(26°)12'$ c. $\dfrac{21.3}{\cos(74°)}$ d. $\dfrac{4.5}{\cos(82°46')}$ e. $\tan(49°32')$

THINK	WRITE
a. With a scientific calculator, press (sin) and enter 57, then press (=).	a. $\sin(57°) = 0.839$
b. Enter 9.6, press (x) and (sin), enter 26, press (DMS), enter 12, press (DMS), then press (=).	b. $9.6\sin(26°)12' = 4.238$
c. Enter 21.3, press (÷) and (cos), enter 74, then press (=).	c. $\dfrac{21.3}{\cos(74°)} = 77.275$
d. Enter 4.5, press (÷) and (cos), enter 82, press (DMS), enter 46, press (DMS), then press (=).	d. $\dfrac{4.5}{\cos(82°)46'} = 35.740$
e. Press (tan), enter 49, press (DMS), enter 32, press (DMS), then press (=).	e. $\tan(49°32') = 1.172$

Note: Some calculators require that the angle size be entered before the trigonometric functions.

9.2.6 Calculating the angle size using a calculator

Similarly, if we are given the sine, cosine or tangent of an angle, we are able to calculate the size of that angle using the calculator. We do this using the inverse functions. On most calculators these are the second function of the sin, cos and tan functions and are denoted $\sin^{-1}$, $\cos^{-1}$ and $\tan^{-1}$.

Problems sometimes supply angles in degrees, minutes and seconds, or require answers to be written in the form of degrees, minutes and seconds. On scientific calculators, you will use the **DMS** (Degrees, Minutes, Seconds) function or the (° ' ") function. *Note:* There are 60 minutes in a degree and 60 seconds in a minute.

WORKED EXAMPLE 8 Determining the size of an angle, correct to the nearest degree or minute

a. **Calculate θ, correct to the nearest degree, given that $\sin(\theta) = 0.738$.**
b. **Given that $\tan(\theta) = 1.647$, calculate θ to the nearest minute.**

THINK	WRITE
a. 1. With a scientific calculator, press (2nd F) (or SHIFT) [sin^{-1}] and enter .738, then press (=).	
2. Round your answer to the nearest degree.	a. $\theta = 48°$
b. 1. With a scientific calculator, press (2nd F) (or SHIFT) [tan^{-1}] and enter 1.647, then press (=).	
2. Convert your answer to degrees and minutes by pressing (DMS).	b. $\theta = 58°44'$

Note: Check the sequence requirements for your calculator.

To calculate $9\sin(45°)$, enter 9, press ⓧ and (**sin**), enter 45, then press (=).

```
         ᴰ      Math ▲
9×sin(45)
           6.363961031
```

To calculate $\dfrac{18}{\sin(44°)}$, enter 18, press (÷) and (**sin**), enter 44, then press (=).

```
         ᴰ      Math ▲
18÷sin(44)
           25.91201771
```

To calculate $\cos(27°)$, with a scientific calculator, press (**cos**) and enter 27, then press (=).

```
         ᴰ      Math ▲
cos(27)
          0.8910065242
```

To calculate $6\cos(55°)$, enter 6, press ⓧ and (**cos**), enter 55, then press (=).

```
         ᴰ      Math ▲
6×cos(55)
           3.441458618
```

To calculate $\tan(60°)$, with a scientific calculator, press (**tan**) and enter 60, then press (=).

```
         ᴰ      Math ▲
tan(60)
           1.732050808
```

To calculate $15\tan(75°)$, enter 15, press ⓧ and (**tan**), enter 75, then press (=).

```
         ᴰ      Math ▲
15×tan(75)
           55.98076211
```

To calculate $\dfrac{18}{\tan(69°)}$, enter 8, press (÷) and (**tan**), enter 69, then press (=).

```
         ᴰ      Math ▲
18÷tan(69)
           6.909552631
```

Note: Some calculators require that the angle size be entered before the trigonometric functions.

9.2.7 The unit circle

If we draw a circle of radius 1 in the Cartesian plane with its centre located at the origin, then we can locate the coordinates of any point on the circumference of the circle by using right-angled triangles.

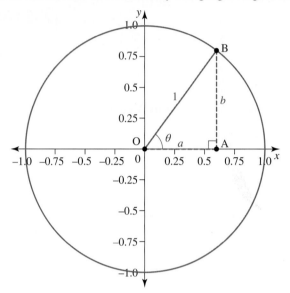

In this diagram, the length of the hypotenuse is 1 and the coordinates of B can be found using the trigonometric ratios.

$$\cos(\theta) = \frac{A}{H} \quad \text{and} \quad \sin(\theta) = \frac{O}{H}$$
$$= \frac{a}{1} \qquad\qquad = \frac{b}{1}$$
$$= a \qquad\qquad = b$$

Therefore, the base length of the triangle, a, is equal to $\cos(\theta)$, and the height of the triangle, b, is equal to $\sin(\theta)$. This gives the coordinates of B as $(\cos(\theta), \sin(\theta))$.

For example, if we have a right-angled triangle with a reference angle of 30° and a hypotenuse of length 1, then the base length of the triangle will be 0.87 and the height of the triangle will be 0.5, as shown in the following triangle.

Similarly, if we calculate the value of $\cos(30°)$ and $\sin(30°)$, we get 0.87 and 0.5 respectively.

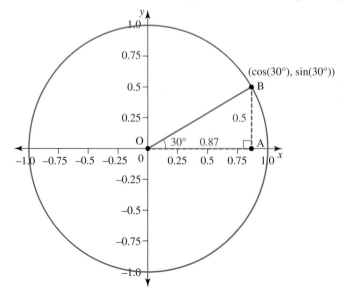

We can actually extend this definition to any point B on the unit circle as having the coordinates $(\cos(\theta), \sin(\theta))$, where θ is the angle measured in an anticlockwise direction.

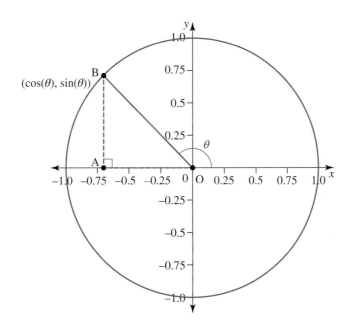

Extending sine and cosine to $180°$

We can place any right-angled triangle with a hypotenuse of 1 in the unit circle so that one side of the triangle lies on the positive x-axis.

The following diagram shows a triangle with base length 0.64, height 0.77 and reference angle 50°.

The coordinates of point B in this triangle are (0.64, 0.77) or $(\cos(50°), \sin(50°))$.

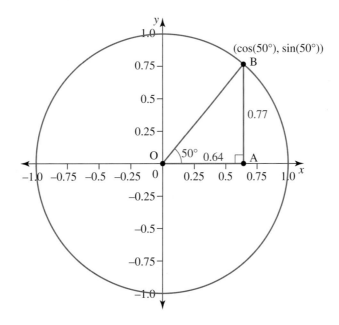

Now reflect the triangle in the *y*-axis as shown in the following diagram.

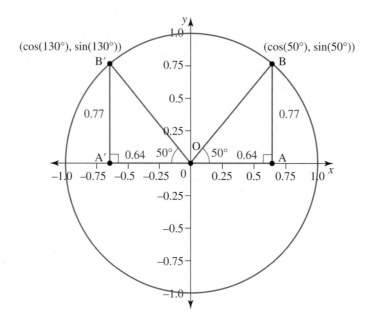

We can see that the coordinates of point B′ are (−0.64, 0.77) or (−cos(50°), sin(50°)).

We have previously determined the coordinates of any point B on the circumference of the unit circle as (cos(θ), sin(θ)), where θ is the angle measured in an anticlockwise direction. In this instance the value of $\theta = 180° − 50 = 130°$.

Therefore, the coordinates of point B′ are (cos(130°), sin(130°)).

This discovery can be extended when we place any right-angled triangle with a hypotenuse of length 1 inside the unit circle.

As previously determined, the point B has coordinates (cos(θ), sin(θ)).

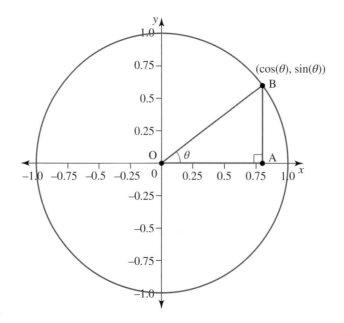

Reflecting this triangle in the y-axis gives:

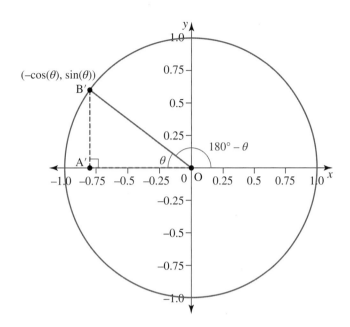

So the coordinates of point B′ are $(-\cos(\theta), \sin(\theta))$. We also know that the coordinates of B′ are $(\cos(180 - \theta), \sin(180 - \theta))$ from the general rule for the coordinates of any point on the unit circle.

Equating the two coordinates for B′ gives us the following equations:

$$-\cos(\theta) = \cos(180 - \theta)$$
$$\sin(\theta) = \sin(180 - \theta)$$

So, to calculate the values of the sine and cosine ratios for angles up to 180°, we can use the following equations.

Sine and cosine ratios for obtuse angles

$$\cos(\theta) = \cos(180 - \theta)$$
$$\sin(\theta) = \sin(180 - \theta)$$

Remember that if two angles sum to 180°, then they are supplements of each other. So if we are calculating the sine or cosine of an angle between 90° and 180°, we start by finding the supplement of the given angle.

eles-3065

WORKED EXAMPLE 9 Determining the sine ratio for an obtuse angle

Calculate the values of:
a. $\sin(140°)$ 　　　　　　　　　　　　　　b. $\cos(160°)$

giving your answers to 2 decimal places.

THINK	WRITE
a. 1. Calculate the supplement of the given angle.	a. $180° - 140° = 40°$
2. Calculate the sine of the supplement angle correct to 2 decimal places.	$\sin(40°) = 0.642\,787\ldots$ $= 0.64$ (to 2 decimal places)

3. The sine of an obtuse angle is equal to the sine of its supplement.

$$\sin(140°) = \sin(40°)$$
$$= 0.64 \text{ (to 2 decimal places)}$$

b. 1. Calculate the supplement of the given angle.

b. $180° - 160° = 20°$

2. Calculate the cosine of the supplement angle correct to 2 decimal places.

$$\cos(20°) = 0.939\,692\ldots$$
$$= 0.94 \text{ (to 2 decimal places)}$$

3. The cosine of an obtuse angle is equal to the negative cosine of its supplement.

$$\cos(160°) = -\cos(20°)$$
$$= -0.94 \text{ (to 2 decimal places)}$$

SOH–CAH–TOA

Trigonometric ratios are relationships between the sides and angles of a right-angled triangle.

In solving trigonometric ratio problems for sine, cosine and tangent, we need to:

1. determine which ratio to use
2. write the relevant equation
3. substitute values from the given information
4. make sure the calculator is in degree mode
5. solve the equation for the unknown lengths or use the inverse trigonometric functions to calculate unknown angles.

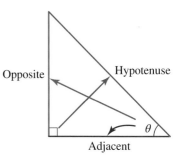

The mnemonic SOH–CAH–TOA

To assist in remembering the trigonometric ratios, the mnemonic SOH–CAH–TOA has been developed.

SOH–CAH–TOA stands for:

- **Sine is Opposite over Hypotenuse**
- **Cosine is Adjacent over Hypotenuse**
- **Tangent is Opposite over Adjacent.**

on Resources

Interactivities Finding the angle when two sides are known (int-6046)
Trigonometric ratios (int-2577)

Exercise 9.2 Trigonometric ratios

9.2 Exercise	9.2 Exam questions on

Simple familiar	Complex familiar	Complex unfamiliar
1, 2, 3, 4, 5, 6, 7, 8, 9, 10, 11, 12, 13, 14, 15, 16	17, 18, 19, 20	21, 22

These questions are even better in jacPLUS!
- Receive immediate feedback
- Access sample responses
- Track results and progress

Find all this and MORE in jacPLUS ▶

Simple familiar

1. **WE1** Calculate the value of *x* correct to 2 decimal places.

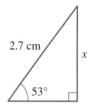

2. Calculate the value of *x*.

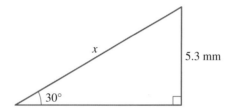

3. **WE2** Calculate the value of the unknown angle, θ, correct to 2 decimal places.

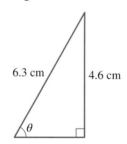

4. Calculate the value of the unknown angle, θ, correct to 2 decimal places.

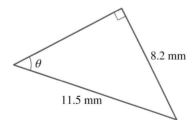

5. **WE3** Calculate the value of *y* correct to 2 decimal places.

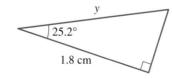

6. Calculate the value of y correct to 2 decimal places.

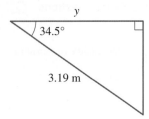

7. **WE4** Calculate the value of the unknown angle, θ, correct to 2 decimal places.

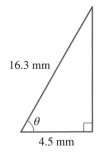

8. Calculate the value of the unknown angle, θ, correct to 2 decimal places.

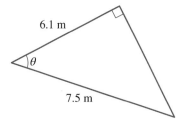

9. **WE5** Calculate the value of x correct to 2 decimal places.

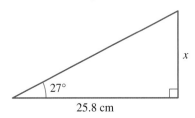

10. Calculate the value of y correct to 2 decimal places.

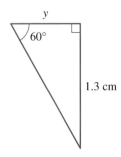

11. **WE6** Calculate the value of the unknown angle, θ, correct to 2 decimal places.

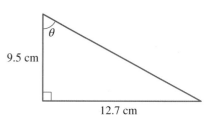

9.5 cm

12.7 cm

12. Calculate the value of the unknown angle, θ, correct to 2 decimal places.

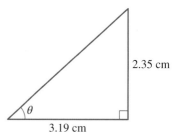

2.35 cm

3.19 cm

13. **WE7** Calculate the value of each of the following, correct to 3 decimal places.

a. $\sin(37°)$

b. $\dfrac{48}{\sin(67°40')}$

c. $\dfrac{6}{\cos(24°)}$

d. $5.9 \cos(2°3')$

e. $9 \tan(63°)$

f. $\tan(33°19')$

14. **WE8** Calculate θ, correct to the units indicated.

a. $\sin(\theta) = 0.167$ (correct to the nearest degree)

b. $\sin(\theta) = 0.277$ (correct to the nearest degree)

c. $\cos(\theta) = 0.058$ (correct to the nearest minute)

d. $\tan(\theta) = 1.517$ (correct to the nearest minute)

15. **WE9** Calculate the value of each of the following, correct to 2 decimal places.

a. $\sin(125°)$

b. $\cos(152°)$

c. $\sin(99.2°)$

d. $\cos(146.7°)$

16. Calculate the value of the unknown angle, θ, correct to 2 decimal places.

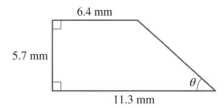

6.4 mm

5.7 mm

11.3 mm

θ

Complex familiar

17. A yacht race follows a triangular course as shown below. Calculate, correct to 1 decimal place:

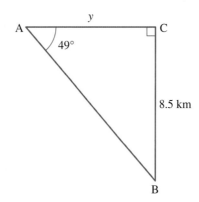

A

y

49°

C

8.5 km

B

a. the distance of the final leg, y

b. the total distance of the course.

18. Consider the following diagram.

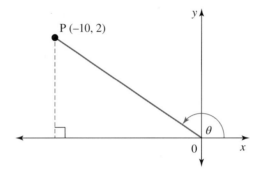

Determine the value of θ. Give your answer correct to the nearest degree.

19. A 2.5-m ladder is placed against a wall. The base of the ladder is 1.7 m from the wall.

   a. Calculate the angle, correct to 2 decimal places, that the ladder makes with the ground.
   b. Calculate how far the ladder reaches up the wall, correct to 2 decimal places.

20. A kitesurfer has a kite of length 2.5 m and strings of length 7 m as shown.

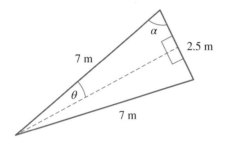

   Calculate the values of the angles θ and α, correct to 2 decimal places.

Complex unfamiliar

21. A truss is used to build a section of a roof. If the vertical height of the truss is 1.5 metres and the span (horizontal distance between the walls) is 8 metres wide, calculate the pitch of the roof (its angle with the horizontal) correct to 2 decimal places.

22. A dog training obstacle course ABCDEA is shown in the diagram with point B vertically above point D. The line BC is horizontal.
   If a dog accidentally shortcuts the course by running directly between B and D, determine how much shorter their course is, giving your answer correct to 2 decimal places.

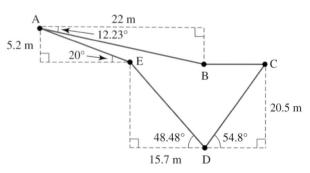

Fully worked solutions for this chapter are available online.

LESSON
9.3 Area of triangles

SYLLABUS LINKS

- Calculate the area of a non-right-angled triangle and solve related practical problems.

  - area $= \frac{1}{2}bc \sin A$, given two sides, b and c, and an included angle, A.

  - Heron's rule: area $= \sqrt{s(s-a)(s-b)(s-c)}$ where $s = \frac{a+b+c}{2}$, given three sides, a, b and c.

Source: General Mathematics Senior Syllabus 2024 © State of Queensland (QCAA) 2024; licensed under CC BY 4.0.

9.3.1 Area of triangles

You should be familiar with calculating the area of a triangle using the rule area $= \frac{1}{2}bh$, where b is the base length and h is the perpendicular height of the triangle. However, for many triangles we are not given the perpendicular height, so this rule cannot be directly used.

Consider the triangle ABC as shown.

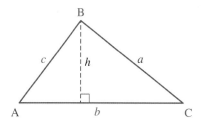

Triangles are conventionally labelled with three pairs of letters: a, b and c, and A, B and C. The lowercase letters, a, b and c, represent the side lengths, and the uppercase letters represent the angles. For each pair of letters, the length letters are opposite the angle letters.

If h is the perpendicular height of this triangle, then we can calculate the value of h by using the sine ratio:

$$\sin(A) = \frac{h}{c}$$

Transposing this equation gives $h = c \sin(A)$, which we can substitute into the rule for the area of the triangle to give the following equation.

Area of a triangle

$$\text{Area} = \frac{1}{2}bc \, \sin(A)$$

Note: We can label any sides of the triangle a, b and c, and this formula can be used as long as we have the length of two sides of a triangle and know the value of the included angle.

eles-3069

Calculate the areas of the following triangles. Give both answers correct to 2 decimal places.

a.

7 cm

63°

5 cm

b. A triangle with sides of length 8 cm and 7 cm, and an included angle of 55°

THINK

a. 1. Label the vertices of the triangle.

2. Write down the known information.

3. Substitute the known values into the formula to calculate the area of the triangle.

4. Write the answer, remembering to include the units.

b. 1. Draw a diagram to represent the triangle.

2. Write down the known information.

WRITE

a.

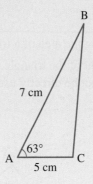

B

7 cm

63°

A   C

5 cm

$b = 5$ cm
$c = 7$ cm
$A = 63°$

$\text{Area} = \dfrac{1}{2}bc\sin(A)$

$= \dfrac{1}{2} \times 5 \times 7 \times \sin(63°)$

$= 15.592\ldots$

$= 15.59$ (to 2 decimal places)

The area of the triangle is 15.59 cm^2 correct to 2 decimal places.

b.

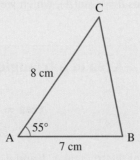

C

8 cm

55°

A   7 cm   B

$b = 8$ cm
$c = 7$ cm
$A = 55°$

3. Substitute the known values into the formula to calculate the area of the triangle.	$\text{Area} = \dfrac{1}{2} bc \sin(A)$
	$= \dfrac{1}{2} \times 8 \times 7 \times \sin(55°)$
	$= 22.936\ldots$
	$= 22.94 \text{ (to 2 decimal places)}$
4. Write the answer, remembering to include the units.	The area of the triangle is 22.94 cm^2 correct to 2 decimal places.

9.3.2 Heron's formula

Heron's formula is a way of calculating the area of the triangle if you are given all three side lengths. It is named after Hero of Alexandria, who was a Greek engineer and mathematician.

Heron's formula

Step 1: Calculate s, the value of half of the perimeter of the triangle: $s = \dfrac{a+b+c}{2}$

Step 2: Use the following formula to calculate the area of the triangle: $A = \sqrt{s(s-a)(s-b)(s-c)}$

WORKED EXAMPLE 11 Using Heron's formula to calculate the area of a triangle

Use Heron's formula to calculate the area of the triangle shown.
Give your answer correct to 1 decimal place.

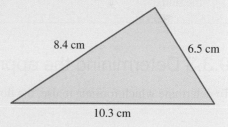

THINK	WRITE

1. Calculate the value of s.

$s = \dfrac{a+b+c}{2}$

$= \dfrac{6.5 + 8.4 + 10.3}{2}$

$= \dfrac{25.2}{2}$

$= 12.6$

2. Use Heron's formula to calculate the area of the triangle correct to 1 decimal place.

$A = \sqrt{s(s-a)(s-b)(s-c)}$

$= \sqrt{12.6(12.6-6.5)(12.6-8.4)(12.6-10.3)}$

$= \sqrt{12.6 \times 6.1 \times 4.2 \times 2.3}$

$= \sqrt{742.4676}$

≈ 27.2

3. Write the answer and give the units.

$A = 27.2 \text{ cm}^2$

Calculate the area of a triangle with sides of 4 cm, 7 cm and 9 cm, giving your answer correct to 2 decimal places.

THINK	WRITE
1. Write down the known information.	$a = 4$ cm $b = 7$ cm $c = 9$ cm
2. Calculate the value of s (the semi-perimeter).	$s = \dfrac{a+b+c}{2}$ $= \dfrac{4+7+9}{2}$ $= \dfrac{20}{2}$ $= 10$
3. Substitute the values into Heron's formula to calculate the area.	Area $= \sqrt{s(s-a)(s-b)(s-c)}$ $= \sqrt{10(10-4)(10-7)(10-9)}$ $= \sqrt{10 \times 6 \times 3 \times 1}$ $= \sqrt{180}$ $= 13.416\ldots$ ≈ 13.42
4. Write the answer, remembering to include the units.	The area of the triangle is 13.42 cm^2 correct to 2 decimal places.

9.3.3 Determining the appropriate area formula

To determine which formula to use, the flowchart below can be followed.

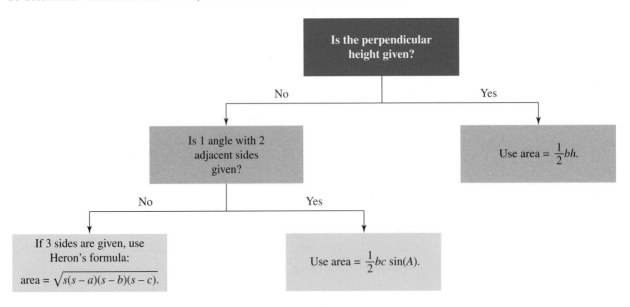

Exercise 9.3 Area of triangles

learn on

9.3 Exercise	9.3 Exam questions on

Simple familiar	Complex familiar	Complex unfamiliar
1, 2, 3, 4, 5, 6, 7, 8	9, 10	11, 12

These questions are
even better in jacPLUS!
• Receive immediate feedback
• Access sample responses
• Track results and progress

Find all this and MORE in jacPLUS ▶

Simple familiar

1. **WE10** Calculate the area of the triangle shown correct to 2 decimal places.

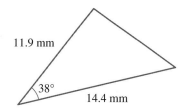

2. Calculate the area of a triangle with sides of length 14.3 mm and 6.5 mm, and an included angle of 32°. Give your answer correct to 2 decimal places.

3. The smallest two sides of a triangle are 10.2 cm and 16.2 cm respectively, and the largest angle of the same triangle is 104.5°. Calculate the area of the triangle correct to 2 decimal places.

4. **WE11** Calculate the area of the following triangles, correct to 2 decimal places where appropriate.

a.

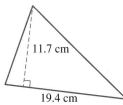

b.

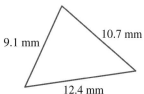

c.

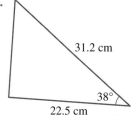

d.

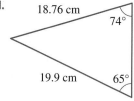

5. Calculate the areas of the following triangles, correct to 2 decimal places where appropriate.
   a. Triangle ABC, given $a = 12$ cm, $b = 15$ cm, $c = 20$ cm
   b. Triangle ABC, given $a = 10.5$ mm, $b = 11.2$ mm and $C = 40°$

6. **WE12** Calculate the area of a triangle with sides of 11 cm, 12 cm and 13 cm, giving your answer correct to 2 decimal places.

7. Calculate the area of a triangle with sides of 22.2 mm, 13.5 mm and 10.1 mm, giving your answer correct to 2 decimal places.

8. A triangular field is defined by three trees, each of which sits in one of the corners of the field, as shown in the diagram.

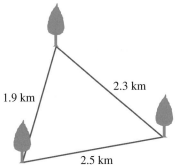

Calculate the area of the field in km^2 correct to 3 decimal places.

Complex familiar

9. A triangle has one side length of 8 cm and an adjacent angle of 45.5°. If the area of the triangle is 18.54 cm^2, calculate the length of the other side that encloses the 45.5° angle, correct to 2 decimal places.

10. Calculate the area of the following shape correct to 2 decimal places.

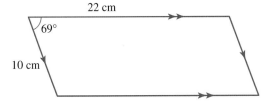

Complex unfamiliar

11. A triangle has side lengths of $3x$, $4x$ and $5x$. If the area of the triangle is 121.5 cm^2, use any appropriate method to determine the value of x.

12. A triangular-shaped piece of jewellery has two side lengths of 8 cm and an area of 31.98 cm^2. Calculate the length of the third side correct to 1 decimal place.

Fully worked solutions for this chapter are available online.

LESSON
9.4 Angles of elevation and depression

SYLLABUS LINKS

- Solve two-dimensional practical problems involving the trigonometry of right-angled and non-right-angled triangles, including problems involving angles of elevation and depression and the use of true bearings.

Source: General Mathematics Senior Syllabus 2024 © State of Queensland (QCAA) 2024; licensed under CC BY 4.0.

9.4.1 Angles of elevation and depression

An **angle of elevation** is the angle between a horizontal line from the observer to an object that is above the horizontal line.

An **angle of depression** is the angle between a horizontal line from the observer to an object that is below the horizontal line.

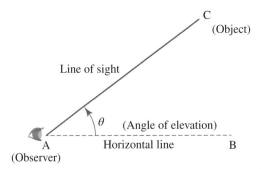

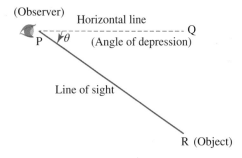

WORKED EXAMPLE 13 Using an angle of elevation to calculate the length of a side of a triangle

From a point 50 m from the foot of a building, the angle of elevation to the top of the building is measured as 40°.
Calculate the height, h, of the building, correct to the nearest metre.

THINK	WRITE
1. Label the sides of the triangle opposite, adjacent and hypotenuse.	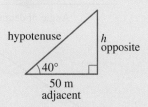
2. Choose the tangent ratio because we are determining the opposite side and have been given the adjacent side.	
3. Write the formula.	$\tan(\theta) = \dfrac{\text{opposite}}{\text{adjacent}}$
4. Substitute for θ and the adjacent side.	$\tan(40°) = \dfrac{h}{50}$
5. Make h the subject of the equation.	$h = 50\tan(40°)$
6. Simplify to calculate the value of h.	$= 42\,\text{m}$
7. Write the answer.	The height of the building is approximately 42 m.

A similar method for finding the solution is used for problems that involve an angle of depression.

eles-6394

WORKED EXAMPLE 14 Using an angle of depression to calculate the length of the side of a triangle

When an aeroplane in flight is 2 km from a runway, the angle of depression to the runway is 10°. Calculate the altitude of the aeroplane, correct to the nearest metre.

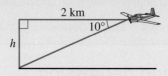

THINK	WRITE
1. Label the sides of the triangle opposite, adjacent and hypotenuse.	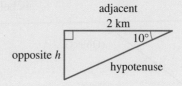
2. Choose the tangent ratio, because we are determining the opposite side given the adjacent side.	
3. Write the formula.	$\tan(\theta) = \dfrac{\text{opposite}}{\text{adjacent}}$
4. Substitute for θ and the adjacent side, converting 2 km to metres.	$\tan(10°) = \dfrac{h}{2000}$
5. Make h the subject of the equation.	$h = 2000 \tan(10°)$
6. Calculate.	$= 353 \text{ m}$
7. Write the answer.	The altitude of the aeroplane is approximately 353 m.

We use angles of elevation and depression to locate the positions of objects above or below the horizontal (reference) line. Angles of elevation and angles of depression are equal as they are alternate angles.

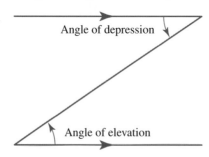

The angle of depression from a scuba diver at the water's surface to a hammerhead shark on the sea floor of the Great Barrier Reef is 40°. The depth of the water is 35 m. Calculate the horizontal distance from the scuba diver to the shark, correct to 2 decimal places.

THINK	WRITE

1. Draw a diagram to represent the information.

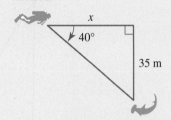

2. Label all the given information on the triangle.

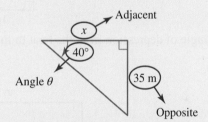

3. Since we have been given the combination of opposite, adjacent and the reference angle θ, we need to use the tangent ratio. Substitute the given values into the ratio equation.

$$\tan(\theta) = \frac{O}{A}$$

$$\tan(40°) = \frac{35}{x}$$

4. Rearrange the equation to make the unknown the subject and solve. Make sure your calculator is in degree mode.

$$x = \frac{35}{\tan(40°)}$$
$$= 41.71$$

5. Write the answer.

The horizontal distance from the scuba diver to the shark is 41.71 m.

Exercise 9.4 Angles of elevation and depression

9.4 Exercise	**9.4 Exam questions** on

Simple familiar	**Complex familiar**	**Complex unfamiliar**
1, 2, 3, 4, 5, 6, 7, 8, 9, 10	11, 12, 13	14, 15

These questions are even better in jacPLUS!
- Receive immediate feedback
- Access sample responses
- Track results and progress

Find all this and MORE in jacPLUS ▶

Simple familiar

1. **WE13** Calculate the angle of elevation of the kite from the ground, correct to 2 decimal places.

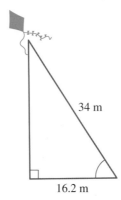

34 m

16.2 m

2. Calculate the angle of depression from the boat to the treasure at the bottom of the sea, correct to 2 decimal places.

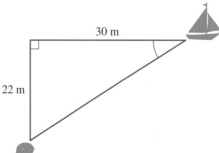

30 m

22 m

3. **WE14** A plane is 36 km from an airstrip, measured in a line from the cockpit to start of the airstrip. The angle of depression is 65°. Determine the height of the plane, correct to 2 decimal places.

4. The angle of elevation from a hammerhead shark on the sea floor of the Great Barrier Reef to a scuba diver at the water's surface is 35°. The depth of the water is 33 m.
   Calculate the horizontal distance from the shark to the scuba diver.

5. **WE15** The angle of depression from a scuba diver at the water's surface to a hammerhead shark on the sea floor of the Great Barrier Reef is 41°. The depth of the water is 32 m.
   Calculate the horizontal distance from the scuba diver to the shark.

6. A student uses a clinometer to measure an angle of elevation of 50° from the ground to the top of Uluru. If the student is standing 724 m from the base of Uluru, determine the height of Uluru correct to 2 decimal places.

7. A ski chair lift operates from the Mt Buller village and has an angle of elevation of 45° to the top of the Federation ski run. If the vertical height is 707 m, calculate the ski chair lift length, correct to 2 decimal places.

8. A tourist in Melbourne looks down from the glass floor of Eureka Tower's Skydeck to see people below on the footpath. If the angle of depression is 88° and the people are 11 m from the base of the tower, calculate how high up the tourist is standing in the glass cube.

9. A parachutist falls from a height of 5000 m to the ground while travelling over a horizontal distance of 150 m. Determine the angle of depression of the descent, correct to 2 decimal places.

10. A crocodile is fed on a 'jumping crocodile tour' on the Adelaide River. The tour guide dangles a piece of meat on a stick at an angle of elevation of 60° from the boat, horizontal to the water.
    If the stick is 2 m long and held 1 m above the water, determine the vertical distance the crocodile has to jump out of the water to get the meat, correct to 2 decimal places.

Complex familiar

11. A student uses a clinometer to measure the height of his house. The angle of elevation is 54°. He is 1.5 m tall and stands 7 m from the base of the house. Calculate the height of the house, correct to 2 decimal places.

12. A tourist 1.72 m tall is standing 50 m away from the base of the Sydney Opera House. The Opera House is 65 m tall. Calculate the angle of elevation, to the nearest degree, from the tourist to the top of the Opera House.

13. Air traffic controllers in two control towers, which are both 87 m high, spot a plane at an altitude of 500 metres. The angle of elevation from tower A to the plane is 5° and from tower B to the plane is 7°. Calculate the distance between the two control towers, correct to the nearest metre.

Complex unfamiliar

14. A surveyor on the beach finds that the angle of elevation of the top of a sand dune is 22°. On a detailed map of the area, the height of the dune is marked as 90 m.
    Determine how much further she must walk (in metres) to achieve an angle of elevation of 45°.

15. A footballer takes a set shot at goal, with the graph showing the path that the ball took as it travelled towards the goal.
    If the footballer's eye level is at 1.6 metres, calculate the angle of elevation from his eyesight to the ball, correct to 2 decimal places, after:

    a. 1 second   b. 2 seconds   c. 3 seconds
    d. 4 seconds   e. 5 seconds.

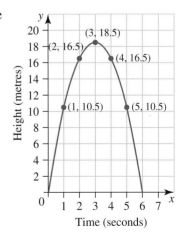

Fully worked solutions for this chapter are available online.

LESSON
9.5 The sine rule

SYLLABUS LINKS

- Solve two-dimensional problems involving a non-right-angled triangle, $\triangle ABC$, with sides, a, b, and c, and corresponding angles A, B and C.

- Sine rule: $\dfrac{a}{\sin A} = \dfrac{b}{\sin B} = \dfrac{c}{\sin C}$ (ambiguous case excluded)

Source: General Mathematics Senior Syllabus 2024 © State of Queensland (QCAA) 2024; licensed under CC BY 4.0.

9.5.1 The sine rule

The **sine rule** can be used to find the side length or angle in non-right-angled triangles.

To help us solve non-right-angled triangle problems, the labelling convention of a non-right-angled triangle, ABC, is as follows:

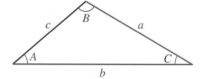

- Angle A is opposite side length a.
- Angle B is opposite side length b.
- Angle C is opposite side length c.

The largest angle will always be opposite the longest side length, and the smallest angle will always be opposite to the smallest side length.

Formulating the sine rule

We can divide an acute non-right-angled triangle into two right-angled triangles, as shown in the following diagrams.

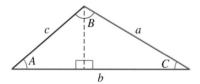 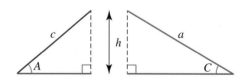

If we apply trigonometric ratios to the two right-angled triangles we get:

$$\dfrac{h}{c} = \sin(A) \qquad\qquad \dfrac{h}{a} = \sin(C)$$

$$h = c\sin(A) \qquad\qquad h = a\sin(C)$$

Equating the two expressions for h gives:

$$c\sin(A) = a\sin(C)$$

$$\dfrac{a}{\sin(A)} = \dfrac{c}{\sin(C)}$$

In a similar way, we can split the triangle into two using side a as the base, giving us:

$$\dfrac{b}{\sin(B)} = \dfrac{c}{\sin(C)}$$

This gives us the sine rule.

The sine rule

$$\frac{a}{\sin(A)} = \frac{b}{\sin(B)} = \frac{c}{\sin(C)}$$

We can apply the sine rule to determine all of the angles and side lengths of a triangle if we are given either:
- 2 side lengths and 1 non-included angle
- 1 side length and 2 angles.

WORKED EXAMPLE 16 Using the sine rule to determine the length of an unknown side of a triangle

eles-3067

Calculate the value of the unknown length *x*, correct to 2 decimal places.

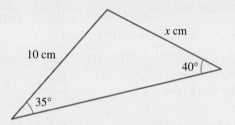

THINK	WRITE
1. Label the triangle with the given information, using the conventions for labelling. Angle *A* is opposite to side *a*. Angle *B* is opposite to side *b*.	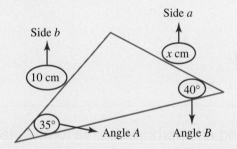
2. Substitute the known values into the sine rule.	$$\frac{a}{\sin(A)} = \frac{b}{\sin(B)}$$ $$\frac{x}{\sin(35°)} = \frac{10}{\sin(40°)}$$
3. Rearrange the equation to make *x* the subject and solve. Make sure your calculator is in degree mode.	$$\frac{x}{\sin(35°)} = \frac{10}{\sin(40°)}$$ $$x = \frac{10\sin(35°)}{\sin(40°)}$$ $$x = 8.92$$
4. Write the answer.	The unknown side length *x* is 8.92 cm.

A non-right-angled triangle has values of side $b = 12.5$, angle $A = 25.3°$ and side $a = 7.4$. Calculate the value of angle B, correct to 2 decimal places.

THINK	WRITE
1. Draw a non-right-angled triangle, labelling with the given information. Angle A is opposite to side a. Angle B is opposite to side b.	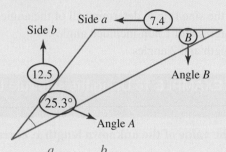
2. Substitute the known values into the sine rule.	$$\dfrac{a}{\sin(A)} = \dfrac{b}{\sin(B)}$$ $$\dfrac{7.4}{\sin(25.3°)} = \dfrac{12.5}{\sin(B°)}$$
3. Rearrange the equation to make $\sin(B)$ the subject and solve. Make sure your calculator is in degree mode.	$$\dfrac{7.4}{\sin(25.3°)} = \dfrac{12.5}{\sin(B)}$$ $$7.4 \sin(B) = 12.5 \sin(25.3°)$$ $$\sin(B) = \dfrac{12.5 \sin(25.3°)}{7.4}$$ $$B = \sin^{-1}\left(\dfrac{12.5 \sin(25.3°)}{7.4}\right)$$ $$= 46.21°$$
4. Write the answer.	Angle B is $46.21°$.

Determining an approprirate area formula to use

In some situations you may have to perform some calculations to determine either a side length or angle size before calculating the area. This may involve using the sine rule.

The following table should help if you are unsure what to do.

Given	What to do	Example
The base length and perpendicular height	Use area $= \dfrac{1}{2}bh$.	8 cm 13 cm
Two side lengths and the included angle	Use area $= \dfrac{1}{2}bc\,\sin(A)$.	5 cm 104° 9 cm

Given	What to do	Example
Three side lengths	Use Heron's formula: area $= \sqrt{s(s-a)(s-b)(s-c)}$, where $s = \dfrac{a+b+c}{2}$.	22 mm 12 mm 19 mm
Two angles and one side length	Use the sine rule to determine a second side length, and then use area $= \dfrac{1}{2} bc \sin(A)$. *Note:* The third angle may have to be calculated.	75° 29° 9 cm
Two side lengths and an angle opposite one of these lengths	Use the sine rule to calculate the other angle opposite one of these lengths, then determine the final angle before using area $= \dfrac{1}{2} bc \sin(A)$. *Note:* Check if the ambiguous case is applicable.	14 cm 72° 12 cm

on Resources

Interactivity The sine rule (int-6275)

Exercise 9.5 The sine rule

learn on

9.5 Exercise	9.5 Exam questions on

These questions are even better in jacPLUS!
- Receive immediate feedback
- Access sample responses
- Track results and progress

Find all this and MORE in jacPLUS ▶

Simple familiar	Complex familiar	Complex unfamiliar
1, 2, 3, 4, 5, 6, 7, 8, 9, 10, 11, 12, 13	14, 15, 16, 17	18, 19

Simple familiar

1. **WE16** Calculate the value of the unknown length x correct to 2 decimal places.

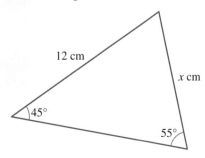

12 cm

x cm

45°

55°

2. Calculate the value of the unknown length x correct to 2 decimal places.

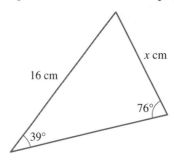

3. Calculate the value of the unknown length x correct to 2 decimal places.

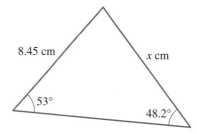

4. Calculate the area of the following triangles, correct to 2 decimal places where appropriate.
   a. Triangle DEF, given $d = 19.8$ cm, $e = 25.6$ cm and $D = 33°$
   b. Triangle PQR, given $p = 45.9$ cm, $Q = 45.5°$ and $R = 67.2°$

5. **WE17** A non-right-angled triangle has values of side $b = 10.5$, angle $A = 22.3°$ and side $a = 8.4$. Calculate the value of angle B correct to 1 decimal place.

6. A non-right-angled triangle has values of side $b = 7.63$, angle $A = 15.8°$ and side $a = 4.56$. Calculate the value of angle B correct to 1 decimal place.

7. For triangle ABC shown, determine the acute value of θ correct to 1 decimal place.

8. If triangle ABC has values $b = 19.5$, $A = 25.3°$ and $a = 11.4$, determine the acute angle value of B correct to 2 decimal places.

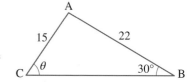

9. Calculate the values of the angles x and y shown in the diagram, correct to 1 decimal place.

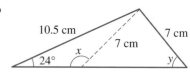

10. At a theme park, a pirate ship swings back and forth on a pendulum. The centre of the pirate ship is secured by a large metal rod that is 5.6 metres in length.
    If one of the swings covers an angle of 122°, determine the horizontal distance between the point where the rod meets the ship at both extremes of the swing. Give your answers correct to 2 decimal places.

11. A triangle ABC has values $a = 11$ cm, $b = 14$ cm and $A = 31.3°$. Answer the following correct to 2 decimal places.

    a. Calculate the size of the other two angles of the triangle.

    b. Calculate the other side length of the triangle.

    c. Calculate the area of the triangle.

12. Calculate the value of the unknown length x correct to the nearest cm.

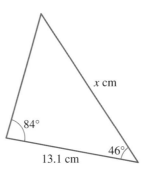

13. Determine all the side lengths, correct to 2 decimal places, for the triangle ABC, given $a = 10.5$, $B = 60°$ and $C = 72°$.

Complex familiar

14. A BMX racing track encloses two triangular sections, as shown in the diagram.

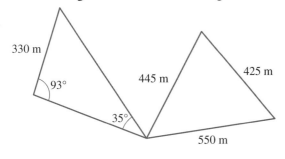

    a. Determine the total length of the BMX race track in kilometres, correct to 1 decimal place.

    b. Calculate the total area that the race track encloses to the nearest m².

15. Part of a roller-coaster track is in the shape of an isosceles triangle, ABC, as shown in the diagram. Calculate the track length AB correct to 2 decimal places.

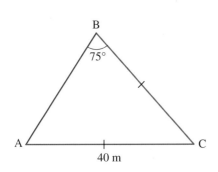

16. Andariel went for a ride on her dune buggy in the desert. She rode east for 6 km, then turned 125° to the left for the second stage of her ride. After 5 minutes riding in the same direction, she turned to the left again, and from there travelled the 5.5 km straight back to her starting position.
    Determine how far Andariel travelled in the second section of her ride, correct to 2 decimal places.

17. A dry field is in the shape of a quadrilateral, as shown in the following diagram. The longest diagonal is 164.228 m.

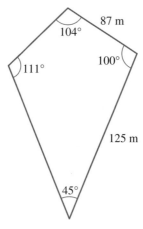

    Determine the volume of grass seed needed to cover the field in 1 mm of grass seed.
    Give your answer correct to 2 decimal places.

Complex unfamiliar

18. The shape and length of a water slide follows the path of PY and YZ in the following diagram.

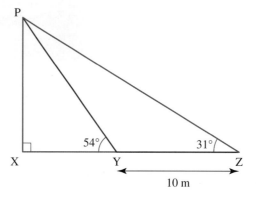

    Calculate the height of the water slide, PX, correct to 2 decimal places.

19. A shade cloth in the shape of a triangle has two sides of length 9.5 m and 13.5 m, and one angle of 40.2°.
    Calculate the maximum possible area of the shade cloth. Evaluate the reasonableness of your solution.

Fully worked solutions for this chapter are available online.

LESSON
9.6 The cosine rule

SYLLABUS LINKS

- Solve two-dimensional problems involving a non-right-angled triangle, ΔABC, with sides, a, b, and c, and corresponding angles A, B and C.
  - Cosine rule: $c^2 = a^2 + b^2 - 2ab \cos C$

Source: General Mathematics Senior Syllabus 2024 © State of Queensland (QCAA) 2024; licensed under CC BY 4.0.

9.6.1 Formulating the cosine rule

The cosine rule, like the sine rule, is used to find the length or angle in a non-right-angled triangle. We use the same labelling conventions for non-right-angled triangles as when using the sine rule.

As with the sine rule, the cosine rule is derived from a non-right-angled triangle being divided into two right-angled triangles, where the base side lengths are equal to $(b - x)$ and x.

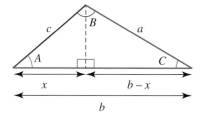
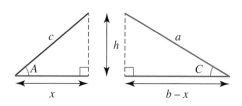

Using Pythagoras' theorem, we get:

$$c^2 = x^2 + h^2 \qquad a^2 = (b - x)^2 + h^2$$
$$h^2 = c^2 - x^2 \quad \text{and} \quad h^2 = a^2 - (b - x)^2$$

Equating the two expressions for h^2 gives:

$$c^2 - x^2 = a^2 - (b - x)^2$$
$$a^2 = (b - x)^2 + c^2 - x^2$$
$$a^2 = b^2 - 2bx + c^2$$

Substituting the trigonometric ratio $x = c \cos(A)$ from the right-angled triangle into the expression, we get:

$$a^2 = b^2 - 2b \left(c \cos(A) \right) + c^2$$
$$= b^2 + c^2 - 2bc \cos(A)$$

This is known as the cosine rule, and we can interchange the variables to get the following equations.

> **The cosine rule to determine a side length**
>
> $$a^2 = b^2 + c^2 - 2bc \cos(A)$$
> $$b^2 = a^2 + c^2 - 2ac \cos(B)$$
> $$c^2 = a^2 + b^2 - 2ab \cos(C)$$

We can apply the cosine rule to determine all of the angles and side lengths of a triangle if we are given either:
- 3 side lengths or
- 2 side lengths and the included angle.

The cosine rule can also be transposed to give the following equations.

The cosine rule to determine the size of an angle

$$\cos(A) = \frac{b^2 + c^2 - a^2}{2bc}$$

$$\cos(B) = \frac{a^2 + c^2 - b^2}{2ac}$$

$$\cos(C) = \frac{a^2 + b^2 - c^2}{2ab}$$

---

eles-3068

WORKED EXAMPLE 18 Using the cosine rule to determine a side length

Calculate the value of the unknown length x correct to 2 decimal places.

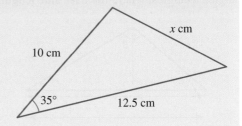

THINK	WRITE
1. Draw the non-right-angled triangle, labelling with the given information. Angle A is opposite to side a. If three sides lengths and one angle are given, always label the angle as A and the opposite side as a.	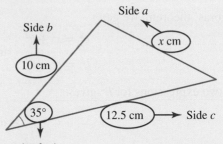
2. Substitute the known values into the cosine rule.	$a^2 = b^2 + c^2 - 2bc\,\cos(A)$ $x^2 = 10^2 + 12.5^2 - 2 \times 10 \times 12.5\,\cos(35°)$
3. Solve for x. Make sure your calculator is in degree mode.	$x^2 = 51.462$ $x = \sqrt{51.462}$ ≈ 7.17
4. Write the answer.	The unknown length x is 7.17 cm.

A non-right-angled triangle ABC has values $a = 7, b = 12$ and $c = 16$. Calculate the magnitude of angle A correct to 2 decimal places.

THINK	WRITE
1. Draw the non-right-angled triangle, labelling with the given information.	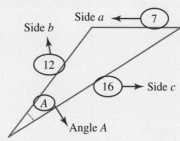
2. Substitute the known values into the cosine rule.	$a^2 = b^2 + c^2 - 2bc \cos(A)$ $7^2 = 12^2 + 16^2 - 2 \times 12 \times 16 \cos(A)$
3. Rearrange the equation to make $\cos(A)$ the subject and solve. Make sure your calculator is in degree mode.	$\cos(A) = \dfrac{12^2 + 16^2 - 7^2}{2 \times 12 \times 16}$ $A = \cos^{-1}\left(\dfrac{12^2 + 16^2 - 7^2}{2 \times 12 \times 16}\right)$ $\approx 23.93°$
4. Write the answer.	The magnitude of angle A is $23.93°$.

Note: In the example above, it would have been quicker to substitute the known values directly into the transposed cosine rule for $\cos(A)$.

Determine unknown sides or angles with given information

Knowing which rule to use for different problems will save time and help to reduce the chance for errors to appear in your working. The following table should help you determine which rule to use.

Type of triangle	What you want	What you know	What to use	Rule	Example
Right-angled	Side length	Two other sides	Pythagoras' theorem	$a^2 + b^2 = c^2$	*(right triangle with sides 8, 10, and ? at base)*
	Side length	A side length and an angle	Trigonometric ratios	$\sin(\theta) = \dfrac{O}{H}$ $\cos(\theta) = \dfrac{A}{H}$ $\tan(\theta) = \dfrac{O}{A}$	*(right triangle with side 8, ? hypotenuse, and angle 32°)*

(continued)

(continued)

Type of triangle	What you want	What you know	What to use	Rule	Example
	Angle	Two side lengths	Trigonometric ratios	$\sin(\theta) = \dfrac{O}{H}$ $\cos(\theta) = \dfrac{A}{H}$ $\tan(\theta) = \dfrac{O}{A}$	
Non-right-angled	Side length	Angle opposite unknown side and another side/angle pair	Sine rule	$\dfrac{a}{\sin(A)} = \dfrac{b}{\sin(B)}$ $= \dfrac{c}{\sin(C)}$	
	Angle	Side length opposite unknown side and another side/angle pair	Sine rule	$\dfrac{a}{\sin(A)} = \dfrac{b}{\sin(B)}$ $= \dfrac{c}{\sin(C)}$	
	Side length	Two sides and the angle between them	Cosine rule	$a^2 =$ $b^2 + c^2$ $- 2bc\cos(A)$	
	Angle	Three sides	Cosine rule	$\cos(A)$ $= \dfrac{b^2 + c^2 - a^2}{2bc}$	

on Resources

Interactivities The cosine rule (int-6276)

Solving non-right angled triangles (int-6482)

| **9.6 Exercise** | **9.6 Exam questions** on |

Simple familiar	Complex familiar	Complex unfamiliar
1, 2, 3, 4, 5, 6, 7, 8, 9, 10, 11, 12	13, 14, 15	16

Simple familiar

1. WE18 Calculate the value of the unknown length x correct to 2 decimal places.

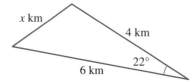

2. Calculate the value of the unknown length x correct to 2 decimal places.

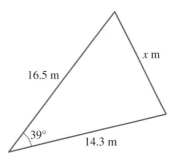

3. Calculate the value of the unknown length x correct to 1 decimal place.

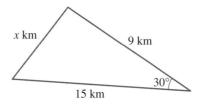

4. Calculate the value of the unknown length x correct to 2 decimal places.

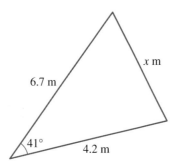

5. WE19 A non-right-angled triangle ABC has values $a = 8$, $b = 13$ and $c = 17$. Calculate the magnitude of angle A correct to 2 decimal places.

6. A non-right-angled triangle ABC has values $a = 11$, $b = 9$ and $c = 5$. Calculate the magnitude of angle A correct to 2 decimal places.

7. For triangle ABC, calculate the magnitude of angle A correct to 2 decimal places, given $a = 5$, $b = 7$ and $c = 4$.

8. For triangle ABC with $a = 12$, $B = 57°$ and $c = 8$, calculate the side length b correct to 2 decimal places.

9. Calculate the largest angle, correct to 2 decimal places, between any two legs of the sailing course shown.

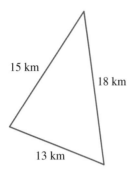

10. A triangular paddock has sides of length 40 m, 50 m and 60 m. Calculate the magnitude of the largest angle between the sides, correct to 2 decimal places.

11. A triangle has side lengths of 5 cm, 7 cm and 9 cm. Calculate the size of the smallest angle, correct to 2 decimal places.

12. Two air traffic control towers are 180 km apart. At the same time, they both detect a plane, P. The plane is at a distance 100 km from Tower A at the bearing shown in the diagram below.

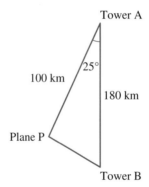

Calculate the distance of the plane from Tower B, correct to 2 decimal places.

Complex familiar

13. ABCD is a parallelogram.

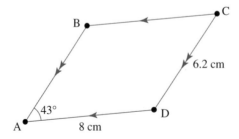

Calculate the length of the diagonal AC correct to 2 decimal places.

14. An orienteering course is shown in the following diagram.

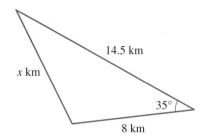

Emilia averages a walking speed of 5 km/h. Calculate the time it will take Emilia to complete the course. Give your answer to the nearest minute.

15. Britney is mapping out a new running path around her local park. She is going to run west for 2.1 km before turning 105° to the right and running another 3.3 km. From there, she will run in a straight line back to her starting position.
Determine how far Britney will run in total. Give your answer correct to the nearest metre.

Complex unfamiliar

16. A cruise boat is travelling to two destinations. To get to the first destination it travels for 4.5 hours at a speed of 48 km/h. From there, it takes a 98° turn to the left and travels for 6 hours at a speed of 54 km/h to reach the second destination.
The boat then travels directly back to the start of its journey. Determine how long this leg of the journey will take if the boat is travelling at 50 km/h. Give your answer correct to the nearest minute.

Fully worked solutions for this chapter are available online.

LESSON
9.7 True bearings

SYLLABUS LINKS

• Solve two-dimensional practical problems involving the trigonometry of right-angled and non-right-angled triangles, including problems involving angles of elevation and depression and the use of true bearings.

Source: General Mathematics Senior Syllabus 2024 © State of Queensland (QCAA) 2024; licensed under CC BY 4.0.

9.7.1 True bearings

Bearings are used to locate the positions of objects or the direction of a journey on a two-dimensional plane.

The four main directions or standard bearings of a directional compass are known as cardinal points. They are North (N), South (S), East (E) and West (W).

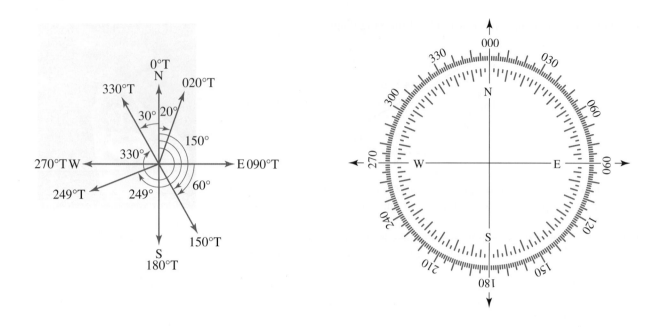

True bearings
True bearings are measured in a clockwise direction from the north line. They are written with all three digits of the angle stated.
If the angle measured is less than 100°, place a zero in front of the angle. For example, if the angle measured is 20° clockwise from the north–south line, the bearing is 020°T

Bearings from A to B

The bearing from A to B is **not** the same as the bearing from B to A.

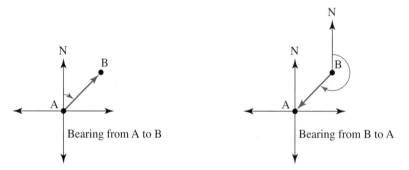

When determining a bearing from a point to another point, it is important to follow the instructions and draw a diagram. Always draw the centre of the compass at the starting point of the direction requested.

When a problem asks to find the bearing of B from A, mark in north and join a directional line to B to work out the bearing. To return to where you came from is a change in bearing of 180°.

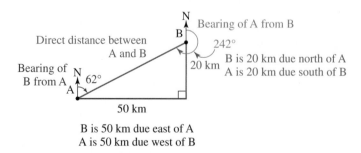

Determine the true bearing from:
a. **Town A to Town B**
b. **Town B to Town A.**

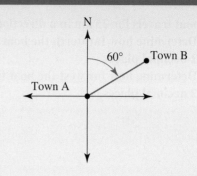

THINK	WRITE

a. 1. To determine the bearing from Town A to Town B, make sure the centre of the compass is marked at town A. The angle is measured clockwise from north to the bearing line at Town B.

a.

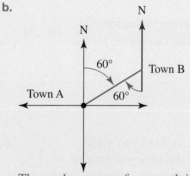

The angle measure from north is 60°.

2. Write the answer. A true bearing is written with all three digits of the angle followed by the letter T.

The true bearing from Town A to Town B is 060°T.

b. 1. To determine the bearing from Town B to Town A, make sure the centre of the compass is marked at Town B. The angle is measured clockwise from north to the bearing line at Town A.

b.

The angle measure from north is
60° + 180° = 240°.

2. Write the answer. A true bearing is written with all three digits of the angle followed by the letter T.

The true bearing from Town B to Town A is 240°T.

9.7.2 Using trigonometry in bearings problems

As the four cardinal points (N, E, S, W) are at right angles to each other, we can use trigonometry to solve problems involving bearings.

When solving a bearings problem with trigonometry, always start by drawing a diagram to represent the problem. This will help you to identify what information you already have and determine which trigonometric ratio to use.

eles-3066

WORKED EXAMPLE 21 Using trigonometry to solve bearings problems

A boat travels for 25 km in a direction of 310°T.
a. **Determine how far north the boat travels, correct to 2 decimal places.**
b. **Determine how far west the boat travels, correct to 2 decimal places.**

THINK

a. 1. Draw a diagram of the situation, remembering to label the compass points as well as all of the given information.

2. Identify the information you have in respect to the reference angle, as well as the information you need.

3. Determine which of the trigonometric ratios to use.

4. Substitute the given values into the trigonometric ratio and solve for the unknown.

5. Write the answer.

b. 1. Use your diagram from part **a** and identify the information you have in respect to the reference angle, as well as the information you need.

2. Determine which of the trigonometric ratios to use.

3. Substitute the given values into the trigonometric ratio and solve for the unknown.

4. Write the answer.

WRITE

a.

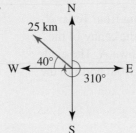

Reference angle = 40°
Hypotenuse = 25
Opposite = ?

$$\sin(\theta) = \frac{O}{H}$$

$$\sin(40°) = \frac{O}{25}$$
$$25 \sin(40°) = O$$
$$O = 16.069 \ldots$$
$$= 16.07 \text{ (to 2 decimal places)}$$

The ship travels 16.07 km north.

b. Reference angle = 40°
Hypotenuse = 25
Adjacent = ?

$$\cos(\theta) = \frac{A}{H}$$

$$\cos(40°) = \frac{A}{25}$$
$$25 \cos(40°) = A$$
$$A = 19.151 \ldots$$
$$= 19.15 \text{ (to 2 decimal places)}$$

The boat travels 19.15 km west.

Resources

Interactivity Bearings (int-6481)

Exercise 9.7 True bearings

learn on

9.7 Exercise	**9.7 Exam questions** on

Simple familiar	Complex familiar	Complex unfamiliar
1, 2, 3, 4, 5, 6, 7	8, 9	10

These questions are
even better in jacPLUS!
• Receive immediate feedback
• Access sample responses
• Track results and progress

Find all this and MORE in jacPLUS ▶

Simple familiar

1. **WE20** In the figure, determine the true bearing from:
   a. Town A to Town B
   b. Town B to Town A.

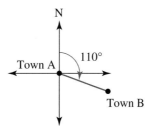

2. Consider the figure shown.

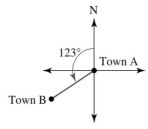

   Determine the true bearing from:
   a. Town A to Town B
   b. Town B to Town A.

3. **WE21** A boat travels for 36 km in a direction of 155°T.
   a. Determine how far south the boat travels, correct to 2 decimal places.
   b. Determine how far east the boat travels, correct to 2 decimal places.

4. A boat travels north for 6 km, west for 3 km, then south for 2 km.
   Determine the boat's true bearing from its starting point. Give your answer in decimal form to 1 decimal place.

CHAPTER 9 Applications of trigonometry **379**

5. State each of the following as a true bearing.

a.
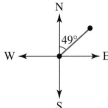

b.
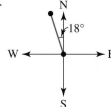

c.
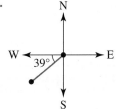

d.
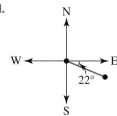

6. State each of the following as a true bearing.

a.
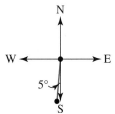

b.
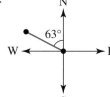

c.
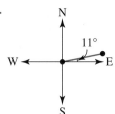

d.
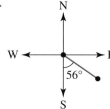

7. From the figure, determine the true bearing of:

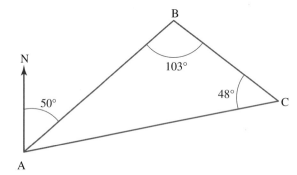

a. B from A   b. C from B   c. A from C   d. C from A.

8. A yacht race travels a triangular course. The first leg of the race, from the start to buoy 1, is 13 km due south. The second leg, from buoy 1 to buoy 2, is due west. The last leg, from buoy 2 back to the start, is 18 km.

   a. Calculate the total length of the course.
   b. Determine the bearing of the starting point from buoy 2 to the nearest degree.

9. A car travelled 5.6 km due east, then turned and travelled 800 m due south.

   a. Determine the compass bearing of the finishing point to the nearest degree.
   b. If the car could travel directly from its starting point to its finishing point, determine the difference in the distance travelled.

10. An athlete practising for a triathlon competition completes a swim, run, cycle circuit as shown in the diagram.

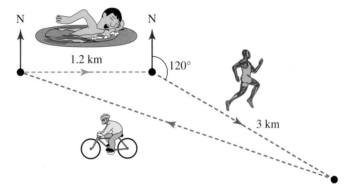

    Use the information given to calculate the bearing on which the athlete rides back to the start to the nearest degree.

Fully worked solutions for this chapter are available online.

LESSON
9.8 Review

9.8.1 Summary

doc-41492

Hey students! Now that it's time to revise this chapter, go online to:

 Access the chapter summary

 Review your results

A+ **Practise exam questions**

Find all this and MORE in jacPLUS ▶

9.8 Exercise

learn on

9.8 Exercise	9.8 Exam questions on

Simple familiar	Complex familiar	Complex unfamiliar
1, 2, 3, 4, 5, 6, 7, 8, 9, 10, 11, 12	13, 14, 15, 16	17, 18, 19, 20

These questions are even better in jacPLUS!
- Receive immediate feedback
- Access sample responses
- Track results and progress

Find all this and MORE in jacPLUS ▶

Simple familiar

1. **MC** The length x in the triangle shown can be calculated by using:

    A. $37 \cos(25°)$

    B. $37 \sin(25°)$

    C. $\dfrac{37}{\cos(25°)}$

    D. $\dfrac{37}{\sin(25°)}$

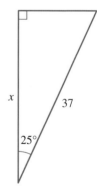

2. **MC** The length x in the triangle shown, correct to the nearest metre, is:

    A. 9

    B. 14

    C. 19

    D. 27

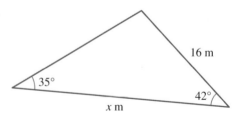

3. **MC** The magnitude of angle A in the triangle shown, correct to the nearest degree, is:

A. 22°
B. 30°
C. 34°
D. 57°

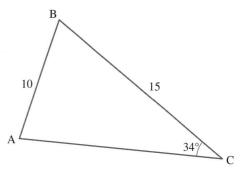

4. **MC** The largest angle in the triangle shown can be calculated by using:

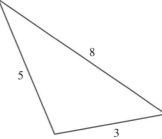

A. $\cos^{-1}\left(\dfrac{8^2 + 3^2 - 5^2}{2 \times 3 \times 8}\right)$

B. $\cos^{-1}\left(\dfrac{8^2 + 5^2 - 3^2}{2 \times 5 \times 8}\right)$

C. $\cos^{-1}\left(\dfrac{5^2 + 3^2 - 8^2}{2 \times 3 \times 5}\right)$

D. $\cos^{-1}\left(\dfrac{5^2 + 3^2 - 8^2}{2 \times 3 \times 8}\right)$

5. **MC** The acute and obtuse angles that have a sine of approximately 0.52992, correct to the nearest degree, are respectively:

A. 31° and 149°
B. 32° and 58°
C. 31° and 59°
D. 32° and 148°

6. **MC** Using Heron's formula, the area of a triangle with sides 4.2 cm, 5.1 cm and 9 cm is:

A. 5.3 cm³
B. 9.2 cm²
C. 13.7 cm²
D. 18.3 cm²

7. **MC** The locations of Jo's, Nelson's and Sammy's homes are shown in the diagram. Jo's home is due north of Nelson's home. The bearings of Jo's and Nelson's homes from Sammy's home are respectively:

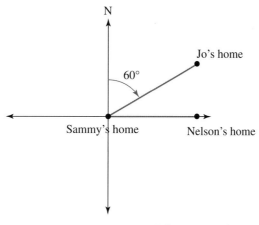

A. 030°T and 090°T
B. 060°T and 090°T
C. 030°T and 180°T
D. 060°T and 180°T

8. **MC** A boy is standing 150 m away from the base of a building. His eye level is 1.65 m above the ground. He observes a hot-air balloon hovering above the building at an angle of elevation of 30°.
If the building is 20 m high, the distance the hot air balloon is above the top of the building is closest to:

A. 64 m　　　　　　　　　　　　　　　**B.** 65 m
C. 66 m　　　　　　　　　　　　　　　**D.** 68 m

9. **MC** The area of the triangle shown, correct to 2 decimal places, is:

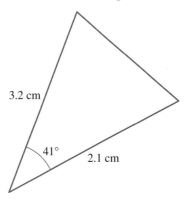

A. 2.20 cm² 　　　　**B.** 2.54 cm² 　　　　**C.** 3.36 cm² 　　　　**D.** 4.41 cm²

10. **MC** A unit of cadets walked from their camp for 7.5 km on a bearing of 064°T. They then travelled on a bearing of 148°T until they came to a signpost that indicated they were 14 km in a straight line from their camp.
The bearing from the signpost back to their camp is closest to:

A. 032°T 　　　　　　**B.** 064°T 　　　　　　**C.** 096°T 　　　　　　**D.** 296°T

11. **a.** Calculate the value of the unknown length x correct to 2 decimal places.

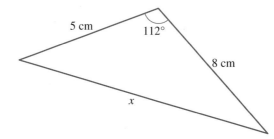

b. Calculate the value of x correct to the nearest degree.

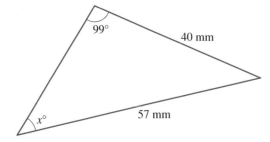

12. a. Calculate the area of the triangle shown, correct to 2 decimal places.

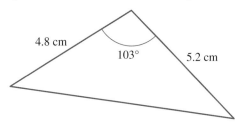

b. Calculate the area of the triangle shown, correct to 2 decimal places.

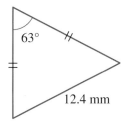

Complex familiar

13. Three treasure chests are buried on an island. Treasure chest B is 412 m on a bearing of 073°T from treasure chest A. Treasure chest C is 805 m on a bearing of 108°T from treasure chest A.

a. Show that angle *A* is 35°.

b. A treasure hunter misreads the information as 'Treasure chest B is 412 m on a bearing of 078°T from treasure chest A' rather than 'Treasure chest B is 412 m on a bearing of 073°T from treasure chest A.'

   i. Construct a diagram to represent the misread information.

   ii. Calculate the true bearing from his incorrect location of treasure chest B to the actual location of treasure chest B.

14. ABCDEF is a regular hexagon with the point B being due north of A. Calculate the true bearing of the point:

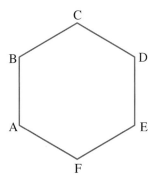

a. C from B   **b.** D from C   **c.** F from E   **d.** E from B.

15. a. Calculate the value of the unknown length x correct to 2 decimal places.

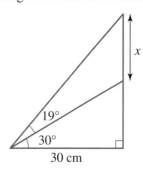

b. Calculate the values of x and y, correct to 2 decimal places where appropriate.

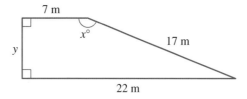

16. Three immunity idols are hidden on an island for a TV reality game. Idol B is 450 m on a bearing of 072°T from Idol A. Idol C is 885 m on a bearing of S75°E from Idol A.
Calculate the triangular area to the nearest metre between immunity idols A, B and C that contestants need to search to find all three idols.

Complex unfamiliar

17. The angle between the two sides of a parallelogram is 93°. If the longer side has length 12 cm and the longer diagonal has length 14 cm, determine the angle between the longer diagonal and the shorter side of the parallelogram, correct to the nearest degree.

18. A dog kennel is placed in the corner of a triangular garden at point C. The dog kennel is positioned 30.5 m at an angle of 32.8° from one corner of the backyard fence (A) and 20.8 m from the other corner of the backyard fence (B).
Calculate the triangular area between the dog kennel and the two corners of the backyard fence, correct to 1 decimal place.

19. A stained glass window frame consisting of five triangular sections is to be made in the shape of a regular pentagon with a side length of 30 cm.

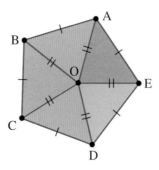

Three of the triangular panels must have coloured glass.
Calculate the total area of coloured glass required, correct to the nearest cm².

20. A triangular flag ABC has a printed design with a circle touching the sides of a flag and a 1-metre vertical line as shown in the diagram.

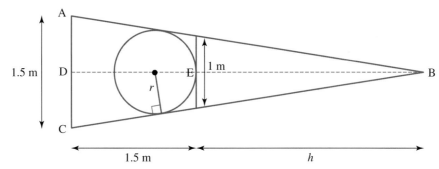

The circular printed design in the flag is coloured yellow and the rest of the flag is coloured purple. The screen printer charges a rate of $22/m^2$ for the circular design and $18/m^2$ for the rest of the flag. Determine the screen printer's cost for one flag.

Fully worked solutions for this chapter are available online.

Answers

Chapter 9 Applications of trigonometry

9.2 Trigonometric ratios

9.2 Exercise

1. $x = 2.16\,cm$
2. $x = 10.6\,mm$
3. $\theta = 46.90°$
4. $\theta = 45.48°$
5. $y = 1.99\,cm$
6. $y = 2.63\,cm$
7. $\theta = 73.97°$
8. $\theta = 35.58°$
9. $x = 13.15\,cm$
10. $y = 0.75\,cm$
11. $\theta = 53.20°$
12. $\theta = 36.38°$
13. a. 0.602   b. 51.893
    c. 6.568   d. 5.896
    e. 17.663   f. 0.657
14. a. 10°   b. 16
    c. 86°40′   d. 56°36′
15. a. 0.82   b. −0.88
    c. 0.99   d. −0.84
16. $\theta = 49.32°$
17. a. 7.4 km   b. 27.2 km
18. 36°
19. a. 47.16°   b. 1.83 m
20. $\theta = 10.29°$, $a = 79.71°$
21. 20.56°
22. 19.05 m

9.3 Area of triangles

9.3 Exercise

1. $52.75\,mm^2$
2. $24.63\,mm^2$
3. 6.50 cm
4. a. $113.49\,cm^2$ b. $47.45\,mm^2$
   c. $216.10\,cm^2$ d. $122.46\,cm^2$
5. a. $89.67\,cm^2$ b. $37.80\,mm^2$
6. $61.48\,cm^2$
7. $43.92\,mm^2$
8. $2.082\,km^2$
9. $79.99\,cm^2$
10. $205.39\,cm^2$
11. $x = 4.5\,cm$
12. 11.1 cm

9.4 Angles of elevation and depression

9.4 Exercise

1. 61.54°
2. 36.25°
3. 32.63 km
4. 47.13 m
5. 36.81 m
6. 862.83 m
7. 999.85 m
8. 315 m
9. 88.28°
10. 2.73 m
11. 11.13 m
12. 52°
13. 1357 m
14. 132 m
15. a. 83.59°   b. 82.35°   c. 79.93°
    d. 74.97°   e. 60.67°

9.5 The sine rule

9.5 Exercise

1. $x = 10.36\,cm$
2. $x = 10.38\,cm$
3. $x = 9.05\,cm$
4. a. $247.68\,cm^2$ b. $750.79\,cm^2$
5. $B = 28.3°$
6. $B = 27.1°$
7. $\theta = 47.2°$
8. $B = 46.97°$
9. $x = 142.4°$, $y = 37.6°$
10. 9.80 m
11. a. $B = 41.39°$, $C = 107.31°$
    b. $c = 20.21\,cm$
    c. $73.51\,cm^2$
12. $x = 17\,cm$
13. $b = 12.24$, $c = 13.43$
14. a. 2.8 km   b. $167\,330\,m^2$
15. 20.71 m
16. 5.91 km
17. $8.14\,m^3$
18. 10.66 m
19. $41.39\,m^2$, $61.41\,m^2$ and $59.12\,m^2$. After calculating all possible areas, the maximum area is $61.41\,m^2$.

9.6 The cosine rule

9.6 Exercise

1. $x = 2.74\,km$
2. $x = 10.49\,m$
3. $x = 8.5\,km$
4. $x = 4.48\,m$
5. $A = 26.95°$

6. $A = 99.59°$

7. $A = 44.42°$

8. $b = 10.17$

9. $79.66°$

10. $82.82°$

11. $33.56°$

12. $98.86\,\text{km}$

13. $13.23\,\text{cm}$

14. 6 hours, 20 minutes

15. $8822\,\text{m}$

16. 7 hours, 16 minutes

9.7 True bearings

9.7 Exercise

1. a. $110°\text{T}$ b. $290°\text{T}$

2. a. $237°\text{T}$ b. $057°\text{T}$

3. a. $32.63\,\text{km}$ b. $15.21\,\text{km}$

4. $323.1°\text{T}$

5. a. $049°\text{T}$ b. $342°\text{T}$
       c. $231°\text{T}$ d. $112°\text{T}$

6. a. $185°\text{T}$ b. $297°\text{T}$
       c. $79°\text{T}$ d. $124°\text{T}$

7. a. $050°\text{T}$ b. $127°\text{T}$
       c. $259°\text{T}$ d. $79°\text{T}$

8. a. $43.45\,\text{km}$ b. $044°\text{T}$

9. a. $098°\text{T}$ b. $0.74\,\text{km}$

10. a. $4.08\,\text{km}$ b. $292°\text{T}$

9.8 Review

9.8 Exercise

1. A

2. D

3. D

4. C

5. D

6. A

7. B

8. D

9. A

10. D

11. a. $10.91\,\text{cm}$ b. $44°$

12. a. $12.16\,\text{cm}^2$ b. $62.73\,\text{mm}^2$

13. a. $A = 180 - (73 + 72) = 35°$

b. i.

    ii. $345.6°\text{T}$

14. a. $060°\text{T}$ b. $120°\text{T}$
        c. $240°\text{T}$ d. $120°\text{T}$

15. a. $17.19\,\text{cm}$
        b. $x = 151.93°,\ y = 8\,\text{m}$

16. a. $563.67\,\text{m}$ b. $108\,451\,\text{m}^2$

17. $32°$

18. $316.1\,\text{m}^2$

19. $928\,\text{cm}^3$

20. \$65.20

10 Matrices and matrix arithmetic

Fully worked solutions for this chapter are available online.

EXAM PREPARATION
Access exam-style questions in every lesson, available online.

on Resources

Solutions	Solutions — Chapter 10 (sol-1251)
Exam questions	Exam question booklet — Chapter 10 (eqb-0267)
Digital documents	Learning matrix — Chapter 10 (doc-41494)
	Chapter summary — Chapter 10 (doc-41495)

LESSON
10.1 Overview

10.1.1 Introduction

Matrices are aligned with the study of solutions of linear simultaneous systems. For example, if you have two equations with two unknowns, you can use matrices to solve the two unknowns. The initial use of matrices dates back to the second century BCE; however, it was not until the end of the seventeenth century that the ideas were developed further.

Matrices are essential in solving linear simultaneous systems. Their use dates back to the Han Dynasty (200 BC–100 BCE) in China, with significant advancements by Japanese mathematician Seki in 1683.

The term 'matrix' was first used by James Joseph Sylvester in 1850, and further developed by Arthur Cayley. Their contributions laid the foundation for modern matrix theory.

During World War II, Olga Taussky-Todd applied matrices to analyse airplane vibrations. Today, matrices are widely used in fields such as quantum mechanics, computer graphics and choreography. Engineers, scientists and project managers also use matrices to help them to perform various everyday tasks.

10.1.2 Syllabus links

Lesson	Lesson title	Syllabus links
10.2	Types of matrices	○ Use matrices for storing and displaying information that can be presented in rows and columns, e.g. tables, databases, links in social or road networks. ○ Recognise different types of matrices, including row matrix, column matrix, square matrix, zero matrix and identity matrix, and determine the size of the matrix.
10.3	Operations with matrices	○ Perform matrix addition, subtraction and multiplication by a scalar.
10.4	Matrix multiplication	○ Perform matrix multiplication manually up to 3×3 matrices but not limited to square matrices. ○ Determine the power of a matrix using technology with matrix arithmetic capabilities when appropriate.
10.5	Applications of matrices	○ Use matrices, including matrix products and powers of matrices, to model and solve problems, e.g. costing or pricing problems, squaring a matrix to determine the number of ways pairs of people in a communication network can communicate with each other via a third person.

Source: General Mathematics Senior Syllabus 2024 © State of Queensland (QCAA) 2024; licensed under CC BY 4.0.

LESSON
10.2 Types of matrices

SYLLABUS LINKS

- Use matrices for storing and displaying information that can be presented in rows and columns, e.g. tables, databases, links in social or road networks.
- Recognise different types of matrices, including row matrix, column matrix, square matrix, zero matrix and identity matrix, and determine the size of the matrix.

Source: General Mathematics Senior Syllabus 2024 © State of Queensland (QCAA) 2024; licensed under CC BY 4.0.

10.2.1 Constructing matrices to display information

A **matrix** is a rectangular array of rows and columns that is used to store and display information. Matrices can be used to represent many different types of information, such as the models of cars sold in different car dealerships, the migration of people to different countries and the shopping habits of customers at different department stores. Matrices also play an important role in encryption. Before sending important information, programmers encrypt or code messages using matrices; the people receiving the information use inverse matrices as the keys to decode the messages.

A matrix is usually displayed in square brackets with no borders between the rows and columns.

Number of participants attending the dance classes

	Saturday	Sunday
Rumba	9	13
Tango	12	8
Chacha	16	14

The table shows the number of participants attending three different dance classes (rumba, waltz and chacha) over the two days of a weekend.

Matrix displaying the number of participants attending the dance classes

$$\begin{bmatrix} 9 & 13 \\ 12 & 8 \\ 16 & 14 \end{bmatrix}$$

The matrix displays the information presented in the table.

WORKED EXAMPLE 1 Constructing a matrix to display information

The table shows the number of adults and children who attended three different events over the school holidays. Construct a matrix to represent this information.

	Circus	Zoo	Show
Adults	140	58	85
Children	200	125	150

THINK

1. A matrix is like a table that stores information. What information needs to be displayed?

WRITE

The information to be displayed is the number of adults and children attending the three events: circus, zoo and show.

2. Write down how many adults and children attend each of the three events.

	Circus	Zoo	Show
Adults	140	58	85
Children	200	125	150

3. Write this information in a matrix. Remember to use square brackets.

$$\begin{bmatrix} 140 & 58 & 85 \\ 200 & 125 & 150 \end{bmatrix}$$

10.2.2 Using matrices for storing and displaying information

Matrices can also be used to display information about various types of **networks**, including road systems and social networks. The following matrix shows the links between a group of schoolmates on Facebook, with a 1 indicating that the two people are friends on Facebook and a 0 indicating that the two people aren't friends on Facebook.

$$\begin{array}{c} \\ A \\ B \\ C \\ D \end{array} \begin{array}{c} A \quad B \quad C \quad D \\ \begin{bmatrix} 0 & 1 & 1 & 0 \\ 1 & 0 & 0 & 1 \\ 1 & 0 & 0 & 1 \\ 0 & 1 & 1 & 0 \end{bmatrix} \end{array}$$

From this matrix you can see that the following people are friends with each other on Facebook:
- person A and person B
- person A and person C
- person B and person D
- person C and person D.

eles-3042

WORKED EXAMPLE 2 Interpreting information about a road network from a matrix

The distances, in kilometres, along three major roads between the Tasmanian towns Launceston (L), Hobart (H) and Devonport (D) are displayed in the matrix below.

$$\begin{array}{c} \\ H \\ D \\ L \end{array} \begin{array}{c} H \quad\ D \quad\ L \\ \begin{bmatrix} 0 & 207 & 160 \\ 207 & 0 & 75 \\ 160 & 75 & 0 \end{bmatrix} \end{array}$$

a. Determine the distance, in kilometres, between Devonport and Hobart.
b. Victor drove 75 km directly between two of the Tasmanian towns. Identify the two towns he drove between.
c. The Goldstein family would like to drive from Hobart to Launceston, and then to Devonport. Determine the total distance in kilometres that they will travel.

	THINK	**WRITE**

THINK

a. 1. Reading the matrix, locate the first city or town, i.e. Devonport (D), on the top of the matrix

2. Locate the second city or town, i.e. Hobart (H), on the side of the matrix.

3. The point where both arrows meet gives you the distance between the two towns.

b. 1. Locate the entry '75' in the matrix.

2. Locate the column and row 'titles' (L and D) for that entry.

3. Refer to the title headings in the question.

c. 1. Locate the first city or town, i.e. Hobart (H), on the top of the matrix and the second city or town, i.e. Launceston (L), on the side of the matrix.

2. Where the row and column meet gives the distance between the two towns.

3. Determine the distance between the second city or town, i.e. Launceston, and the third city or town, i.e. Devonport.

4. Where the row and column meet gives the distance between the two towns.

5. Add the two distances together.

WRITE

a.

$$\begin{array}{c c} & \begin{array}{c c c} H & D & L \end{array} \\ \begin{array}{c} H \\ D \\ L \end{array} & \left[\begin{array}{c c c} 0 & 207 & 160 \\ 207 & 0 & 75 \\ 160 & 75 & 0 \end{array}\right] \end{array}$$

$$\begin{array}{c c} & \begin{array}{c c c} H & D & L \end{array} \\ \rightarrow\begin{array}{c} H \\ D \\ L \end{array} & \left[\begin{array}{c c c} 0 & \boxed{207} & 160 \\ 207 & 0 & 75 \\ 160 & 75 & 0 \end{array}\right] \end{array}$$

207 km

b.

$$\begin{array}{c c} & \begin{array}{c c c} H & D & L \end{array} \\ \begin{array}{c} H \\ D \\ L \end{array} & \left[\begin{array}{c c c} 0 & 207 & 160 \\ 207 & 0 & \boxed{75} \\ 160 & 75 & 0 \end{array}\right] \end{array}$$

$$\begin{array}{c c} & \begin{array}{c c c} H & D & L \end{array} \\ \begin{array}{c} H \\ D \\ L \end{array} & \left[\begin{array}{c c c} 0 & 207 & 160 \\ 207 & 0 & \boxed{75} \\ 160 & 75 & 0 \end{array}\right] \end{array}$$

Victor drove between Launceston and Devonport.

c.

$$\begin{array}{c c} & \begin{array}{c c c} H & D & L \end{array} \\ \begin{array}{c} H \\ D \\ \rightarrow L \end{array} & \left[\begin{array}{c c c} 0 & 207 & 160 \\ 207 & 0 & 75 \\ \boxed{160} & 75 & 0 \end{array}\right] \end{array}$$

160 km

$$\begin{array}{c c} & \begin{array}{c c c} H & D & L \end{array} \\ \rightarrow\begin{array}{c} H \\ D \\ L \end{array} & \left[\begin{array}{c c c} 0 & 207 & 160 \\ 207 & 0 & \boxed{75} \\ 160 & 75 & 0 \end{array}\right] \end{array}$$

75 km

$160 + 75 = 235$ km

10.2.3 Different types of matrices and the size (order) of a matrix

Matrices come in different *dimensions* (sizes). We define the dimensions of a matrix by the number of rows and columns. Rows run horizontally and columns run vertically. This is known as the **order** or size of the matrix and is written as rows × columns.

In the following matrix A, there are 2 rows and 3 columns, which means it has an order of 2×3 (two by three).

$$\text{Column } 1 \quad 2 \quad 3$$
$$A = \begin{matrix} \text{Row 1} \\ \text{Row 2} \end{matrix} \begin{bmatrix} 5 & 8 & 2 \\ 1 & 3 & 0 \end{bmatrix} \quad \rightarrow \quad \textbf{rows} \times \textbf{columns} \quad \rightarrow \quad \textbf{2} \times \textbf{3} \text{ matrix}$$

Matrices are represented by capital letters. In this example, A represents a matrix.

Instead of only naming matrices by the number of rows and columns, we also use names for specific types of matrix *dimensions*.

Types of matrices		
Type of matrix	**Description**	**Example**
Column matrix	A matrix that has a single column is known as a **column matrix**.	$B = \begin{bmatrix} 1 \\ 4 \\ 2 \end{bmatrix} \quad C = \begin{bmatrix} 2 \\ 8 \\ 5 \\ 9 \end{bmatrix}$ B and C are both column matrices.
Row matrix	A matrix that has a single row is known as a **row matrix**.	$D = \begin{bmatrix} 2 & 3 \end{bmatrix} \quad E = \begin{bmatrix} 9 & 4 & 3 \end{bmatrix}$ D and E are both row matrices.
Square matrix	A matrix that has the same number of columns as rows is known as a **square matrix**.	$F = \begin{bmatrix} 8 \end{bmatrix} \quad G = \begin{bmatrix} 5 & 9 \\ -4 & 1 \end{bmatrix}$ $H = \begin{bmatrix} 7 & 6 & 4 \\ -3 & 1 & 7 \\ 4 & 9 & 8 \end{bmatrix}$ F, G and H are all square matrices.
Identity matrix	A square matrix in which all of the elements on the diagonal line from the top left to bottom right are 1s and all of the other elements are 0s is known as an **identity matrix**.	$A = \begin{bmatrix} 1 & 0 \\ 0 & 1 \end{bmatrix}$ and $B = \begin{bmatrix} 1 & 0 & 0 \\ 0 & 1 & 0 \\ 0 & 0 & 1 \end{bmatrix}$ A and B are both identity matrices.
Zero matrix	A square matrix that consists entirely of 0 elements is known as a **zero matrix**.	$A = \begin{bmatrix} 0 & 0 \\ 0 & 0 \end{bmatrix}$ A is a zero matrix.

WORKED EXAMPLE 3 Constructing a matrix and determining the order of the matrix

At High Vale College, 150 students are studying General Mathematics and 85 students are studying Mathematical Methods. Construct a column matrix to represent the number of students studying General Mathematics and Mathematical Methods, and determine the order of the matrix.

THINK	WRITE
1. Read the question and highlight the key information	150 students study General Mathematics. 85 students study Mathematical Methods.
2. Display this information in a column matrix.	$\begin{bmatrix} 150 \\ 85 \end{bmatrix}$
3. How many rows and columns are there in this matrix? Write the answer.	The order of the matrix is 2×1.

Identity matrix

An **identity matrix**, I, is a square matrix in which all of the elements on the diagonal line from the top left to bottom right are 1s and all of the other elements are 0s.

$$I_2 = \begin{bmatrix} 1 & 0 \\ 0 & 1 \end{bmatrix} \text{ and } I_3 = \begin{bmatrix} 1 & 0 & 0 \\ 0 & 1 & 0 \\ 0 & 0 & 1 \end{bmatrix} \text{ are both identity matrices.}$$

As you will see later in this chapter, identity matrices are used to determine inverse matrices, which help solve matrix equations.

Zero matrix

A **zero matrix**, 0, is a square matrix that consists entirely of '0' elements.

The matrix $\begin{bmatrix} 0 & 0 \\ 0 & 0 \end{bmatrix}$ is an example of a zero matrix.

10.2.4 Elements of matrices

The entries in a matrix are called **elements**. They are represented by a *lowercase* letter with the row and column in *subscript*. For matrix A, a_{ij} is the element in row i, column j. The position of an element is described by the corresponding row and column. For example, a_{21} means the entry in the 2nd row and 1st column of matrix A, as shown below.

$$A = \begin{bmatrix} a_{11} & a_{12} & \cdots & a_{1n} \\ \boxed{a_{21}} & a_{22} & \cdots & a_{2n} \\ \vdots & \vdots & \cdots & \vdots \\ \vdots & \vdots & \cdots & \vdots \\ a_{m1} & a_{m2} & \cdots & a_{mn} \end{bmatrix}$$

column 3

$$\text{row 1} \quad A = \begin{bmatrix} 1 & 2 & 3 \\ 4 & 5 & 6 \\ 7 & 8 & 9 \end{bmatrix}$$

a_{13} is an element of matrix A, in row 1 and column 3.

WORKED EXAMPLE 4 Identifying elements of matrices

Write the element a_{23} for the matrix $A = \begin{bmatrix} 1 & 4 & -2 \\ 3 & -6 & 5 \end{bmatrix}$.

THINK	WRITE
1. The element a_{23} means the element in the 2nd row and 3rd column. Draw lines through the 2nd row and 3rd column to help you identify this element.	$A = \begin{bmatrix} 1 & 4 & -2 \\ 3 & -6 & 5 \end{bmatrix}$
2. Identify the number that is where the lines cross over and write the answer.	$a_{23} = 5$

Exercise 10.2 Types of matrices

10.2 Exercise	10.2 Exam questions on

Simple familiar	Complex familiar	Complex unfamiliar
1, 2, 3, 4, 5, 6, 7, 8, 9, 10, 11, 12	13, 14, 15, 16	17, 18

These questions are even better in jacPLUS!
- Receive immediate feedback
- Access sample responses
- Track results and progress

Find all this and MORE in jacPLUS ▶

Simple familiar

1. **WE1** Cheap Auto sells three types of vehicles: cars, vans and motorbikes. They have two outlets at Valley Heights and Hill Vale.

   The number of vehicles in stock at each of the two outlets is shown in the table.

	Cars	Vans	Motorbikes
Valley Heights	18	12	8
Hill Vale	13	10	11

   Construct a matrix to represent this information.

2. Newton and Isaacs played a match of tennis. Newton won the match in five sets with a final score of 6–2, 4–6, 7–6, 3–6, 6–4. Construct a matrix to represent this information.

3. **WE2** The distance in kilometres between the towns Port Augusta (P), Coober Pedy (C) and Alice Springs (A) are displayed in the following matrix.

$$\begin{array}{c} \\ P \\ C \\ A \end{array} \begin{array}{ccc} P & C & A \\ \begin{bmatrix} 0 & 545 & 1225 \\ 545 & 0 & 688 \\ 1225 & 688 & 0 \end{bmatrix} \end{array}$$

   a. Determine the distance in kilometres between Port Augusta and Coober Pedy.
   b. Greg drove 688 km between two towns. Identify the two towns he travelled between.
   c. A truck driver travels from Port Augusta to Coober Pedy, then onto Alice Springs. He then drives from Alice Springs directly to Port Augusta.
      Determine the total distance in kilometres that the truck driver travelled.

4. A one-way economy train fare between Melbourne Southern Cross Station and Canberra Kingston Station is $91.13. A one-way economy train fare between Sydney Central Station and Melbourne Southern Cross Station is $110.72, and a one-way economy train fare between Sydney Central Station and Canberra Kingston Station is $48.02.
   a. Represent this information in a matrix.
   b. Drew travelled from Sydney Central to Canberra Kingston Station, and then onto Melbourne Southern Cross. Determine how much, in dollars, he paid for the train fare.

5. **WE3** An energy-saving store stocks shower water savers and energy-saving light globes. In one month they sold 45 shower water savers and 30 energy-saving light globes. Construct a column matrix to represent the number of shower water savers and energy-saving light globes sold during this month, and determine the order of the matrix.

6. Happy Greens Golf Club held a three-day competition from Friday to Sunday. Participants were grouped into three different categories: experienced, beginner and club member.
   The table shows the total entries for each type of participant on each of the days of the competition.

Category	Friday	Saturday	Sunday
Experienced	19	23	30
Beginner	12	17	18
Club member	25	33	36

   a. Determine how many entries were received for the competition on Friday.
   b. Calculate the total number of entries for the three-day competition.
   c. Construct a row matrix to represent the number of beginners participating in the competition for each of the three days.

7. Write the order of matrices A, B and C.

$$A = [3], \quad B = \begin{bmatrix} 2 \\ 5 \\ 6 \end{bmatrix}, \quad C = \begin{bmatrix} 4 & -2 \end{bmatrix}$$

8. Identify which of the following represent matrices. Justify your answers.

   a. $\begin{bmatrix} 3 \\ 5 \end{bmatrix}$
   b. $\begin{bmatrix} 4 & 0 \\ & 3 \end{bmatrix}$
   c. $\begin{bmatrix} & 5 \\ 4 & \\ & 7 \end{bmatrix}$
   d. $\begin{bmatrix} a & c & e & g \\ b & d & f & h \end{bmatrix}$

9. **WE4** Write down the values of the following elements for matrix D.

$$D = \begin{bmatrix} 4 & 5 & 0 \\ 2 & -1 & -3 \\ 1 & -2 & 6 \\ 0 & 3 & 7 \end{bmatrix}$$

   a. d_{12}
   b. d_{33}
   c. d_{43}

10. a. The following matrix represents an incomplete 3×3 identity matrix. Complete the matrix.

$$\begin{bmatrix} 1 & 0 & 0 \\ 0 & & 0 \\ & 0 & \end{bmatrix}$$

   b. Construct a 2×2 zero matrix.

11. Matrices D and E are shown. Determine the values of the following elements.

$$D = \begin{bmatrix} 5 & 0 & 2 & -1 \\ 8 & 1 & 3 & 6 \end{bmatrix}$$

$$E = \begin{bmatrix} 0.5 & 0.3 \\ 1.2 & 1.1 \\ 0.4 & 0.9 \end{bmatrix}$$

a. d_{23} b. d_{14} c. d_{22} d. e_{11} e. e_{32}

12. Consider the matrix $E = \begin{bmatrix} \dfrac{2}{3} & 0 & \dfrac{1}{4} \\ -1 & -\dfrac{1}{2} & -3 \end{bmatrix}$.

    a. Explain why the element e_{24} does not exist.
    b. Identify the element that has a value of -3.
    c. Nadia was asked to write down the value of element e_{12} and wrote -1. Explain Nadia's mistake and write down the correct value of element e_{12}.
    Explain Nadia's mistake and write down the correct value of element e_{12}.

Complex familiar

13. The elements in matrix H are shown below.
    $h_{12} = 3$
    $h_{11} = 4$
    $h_{21} = -1$
    $h_{31} = -4$
    $h_{32} = 6$
    $h_{22} = 7$
    Construct matrix H and write down its order.

14. The land area and population of each Australian state and territory were recorded and summarised in the table shown. (The table lists populations at 30 September, 2023.)

State/territory	Land area (km^2)	Population (millions)
Australian Capital Territory	2 358	0.5
Queensland	1 727 200	5.5
New South Wales	801 428	8.4
Northern Territory	1 346 200	0.3
South Australia	984 000	1.9
Western Australia	2 529 875	2.9
Tasmanian	68 330	0.6
Victoria	227 600	6.9

Population density is a measure of how many people live in each square kilometre.
Construct a column matrix of the population density of the states and territories listed.

15. The estimated number of Aboriginal and Torres Strait Islander Australians living in each state and territory in Australia in 2023 is shown in the following table.

State and territory	Number of Aboriginal and Torres Strait Islander persons	% of population that is from an Aboriginal or Torres Strait Islander background
New South Wales	303 186	4
Victoria	68 693	1
Queensland	258 411	5
South Australia	48 338	3
Western Australia	114 312	4
Tasmania	31 849	6
Northern Territory	79 695	27
Australian Capital Territory	9 113	2

Construct an appropriate matrix of this information and determine the total number of Aboriginal and Torres Strait Islander persons who were estimated to be living in Australia in 2023.

16. AeroWings is a budget airline specialising in flights between four mining towns: Olympic Dam (O), Broken Hill (B), Dampier (D) and Mount Isa (M).

The cost of airfares (in dollars) to fly from the towns in the top row to the towns in the first column is shown in the matrix below.

$$\text{To} \begin{array}{c} \\ O \\ B \\ D \\ M \end{array} \overset{\begin{array}{ccc} \text{From} \\ O \quad\; B \quad\;\; D \quad\;\; M \end{array}}{\begin{bmatrix} 0 & 70 & 150 & 190 \\ 89 & 0 & 85 & 75 \\ 175 & 205 & 0 & 285 \\ 307 & 90 & 101 & 0 \end{bmatrix}}$$

a. In the context of this problem, explain the meaning of the zero entries.
b. Calculate the cost, in dollars, to fly from Olympic Dam to Dampier.
c. Yen paid $101 for his airfare with AeroWings. Identify the town at which he arrived.
d. AeroWings offers a 25% discount for passengers flying between Dampier and Mount Isa, and a 15% discount for passengers flying from Broken Hill to Olympic Dam.
   Construct another matrix that includes the discounted airfares (in dollars) between the four mining towns.

17. The matrix below displays the number of roads connecting five towns: Ross (R), Stanley (S), Thomastown (T), Edenhope (E) and Fairhaven (F).

$$N = \begin{array}{c} \\ \\ \\ \\ \\ \end{array} \begin{array}{ccccc} R & S & T & E & F \\ \begin{bmatrix} 0 & 0 & 1 & 1 & 2 \\ 0 & 1 & 0 & 0 & 1 \\ 1 & 0 & 0 & 2 & 0 \\ 1 & 0 & 2 & 0 & 1 \\ 2 & 1 & 0 & 1 & 0 \end{bmatrix} & \begin{array}{c} R \\ S \\ T \\ E \\ F \end{array} \end{array}$$

A major flood washes away part of the road connecting Ross and Thomastown. Identify the elements in matrix N that will need to be changed to reflect the new road conditions between the towns.

18. Mackenzie is sitting a Mathematics multiple choice test with ten questions. There are five possible responses for each question: A, B, C, D and E.

She selects A for the first question and then determines the answers to the remaining questions using the following matrix.

$$\begin{array}{c} \text{Answer} \\ \text{Next question} \end{array} \begin{array}{c} \\ \\ \\ \\ \\ \end{array} \begin{array}{c} \quad\quad \text{Answer} \\ \text{This question} \\ \begin{array}{ccccc} A & B & C & D & E \end{array} \\ \begin{array}{c} A \\ B \\ C \\ D \\ E \end{array} \begin{bmatrix} 0 & 0 & 0 & 0 & 0 \\ 0 & 0 & 1 & 1 & 0 \\ 0 & 0 & 0 & 0 & 0 \\ 1 & 0 & 0 & 0 & 1 \\ 0 & 1 & 0 & 0 & 0 \end{bmatrix} \end{array}$$

a. Explain using mathematical reasoning why it is impossible for Mackenzie to have more than one answer with response A. Evaluate the reasonableness of your response.

b. Mackenzie used another matrix to help her answer the multiple choice test. Her responses using this matrix are shown in this grid.

Question	1	2	3	4	5	6	7	8	9	10
Response	A	D	C	B	E	A	D	C	B	E

Complete the matrix that Mackenzie used for the test by determining the values of the missing elements.

$$\text{Next question} \begin{array}{c} \\ \\ \\ \\ \\ \end{array} \begin{array}{c} \quad\quad \text{This question} \\ \begin{array}{ccccc} A & B & C & D & E \end{array} \\ \begin{array}{c} A \\ B \\ C \\ D \\ E \end{array} \begin{bmatrix} 0 & 0 & 0 & 0 & \\ 0 & & & 0 & 0 \\ 0 & & & & 0 \\ 1 & 0 & 0 & & 0 \\ 0 & & 0 & 0 & \end{bmatrix} \end{array}$$

Fully worked solutions for this chapter are available online.

LESSON
10.3 Operations with matrices

SYLLABUS LINKS

- Perform matrix addition, subtraction and multiplication by a scalar.

Source: General Mathematics Senior Syllabus 2024 © State of Queensland (QCAA) 2024; licensed under CC BY 4.0.

10.3.1 Matrix addition

Matrices can be added using the same rules as in regular arithmetic. However, matrices can only be added if they are the same order (that is, if they have the same number of rows and columns).

To add matrices, you need to add the corresponding elements of each matrix together (that is, the numbers in the same position).

eles-3044

WORKED EXAMPLE 5 Adding matrices

If $A = \begin{bmatrix} 4 & 2 \\ 3 & -2 \end{bmatrix}$ and $B = \begin{bmatrix} 1 & 0 \\ 5 & 3 \end{bmatrix}$, calculate the value of $A + B$.

THINK	WRITE
1. Write down the two matrices in a sum.	$\begin{bmatrix} 4 & 2 \\ 3 & -2 \end{bmatrix} + \begin{bmatrix} 1 & 0 \\ 5 & 3 \end{bmatrix}$
2. Identify the elements in the same position. For example, 4 and 1 are both in the first row and first column. Add the elements in the same positions together.	$\begin{bmatrix} 4+1 & 2+0 \\ 3+5 & -2+3 \end{bmatrix}$
3. Work out the sums and write the answer.	$\begin{bmatrix} 5 & 2 \\ 8 & 1 \end{bmatrix}$

10.3.2 Matrix subtraction

As for matrices addition, matrices subtraction can also be performed by using the same rules as in regular arithmetic, but only if they are the same order.

To subtract matrices, you need to subtract the corresponding elements in the same order as presented in the question.

eles-6395

WORKED EXAMPLE 6 Subtracting matrices

If $A = \begin{bmatrix} 6 & 0 \\ 2 & -2 \end{bmatrix}$ and $B = \begin{bmatrix} 4 & 2 \\ -2 & 3 \end{bmatrix}$, calculate the value of $A - B$.

THINK	WRITE
1. Write the two matrices.	$\begin{bmatrix} 6 & 0 \\ 2 & -2 \end{bmatrix} - \begin{bmatrix} 4 & 2 \\ -2 & 3 \end{bmatrix}$

2. Subtract the elements in the same position together.

$$\begin{bmatrix} 6-4 & 0-2 \\ 2-2 & -2-3 \end{bmatrix}$$

3. Work out the subtractions and write the answer.

$$\begin{bmatrix} 2 & -2 \\ 4 & -5 \end{bmatrix}$$

10.3.3 Scalar multiplication

If $A = \begin{bmatrix} 3 & 2 \\ 5 & 1 \\ 0 & 7 \end{bmatrix}$, then $A + A$ can be found by multiplying each element in matrix A by the scalar number 2, because $A + A = 2A$.

$$A + A = \begin{bmatrix} 3 & 2 \\ 5 & 1 \\ 0 & 7 \end{bmatrix} + \begin{bmatrix} 3 & 2 \\ 5 & 1 \\ 0 & 7 \end{bmatrix} = \begin{bmatrix} 6 & 4 \\ 10 & 2 \\ 0 & 14 \end{bmatrix}$$

$$2A = \begin{bmatrix} 2 \times 3 & 2 \times 2 \\ 2 \times 5 & 2 \times 1 \\ 2 \times 0 & 2 \times 7 \end{bmatrix} = \begin{bmatrix} 6 & 4 \\ 10 & 2 \\ 0 & 14 \end{bmatrix}$$

The number 2 is known as a scalar quantity, and the matrix $2A$ represents a **scalar multiplication**. Any matrix can be multiplied by any scalar quantity and the order of the matrix will remain the same. A scalar quantity can be any real number, such as negative or positive numbers, fractions or decimal numbers.

eles-3045

WORKED EXAMPLE 7 Multiplying a matrix by a scalar

Consider the matrix $A = \begin{bmatrix} 120 & 90 \\ 80 & 60 \end{bmatrix}$.

Evaluate the following.

a. $2A$ b. $-10A$ c. $\dfrac{1}{4}A$ d. $0.1A$

THINK	WRITE
a. 1. Identify the scalar for the matrix. In this case it is 2, which means that each element in A is multiplied by 2.	**a.** $2\begin{bmatrix} 120 & 90 \\ 80 & 60 \end{bmatrix}$
2. Multiply each element in A by the scalar.	$\begin{bmatrix} 2 \times 120 & 2 \times 90 \\ 2 \times 80 & 2 \times 60 \end{bmatrix}$
3. Write the answer.	$\begin{bmatrix} 240 & 180 \\ 160 & 120 \end{bmatrix}$
b. 1. Identify the scalar for the matrix. In this case it is -10, which means that each element in A is multiplied by -10.	**b.** $-10\begin{bmatrix} 120 & 90 \\ 80 & 60 \end{bmatrix}$
2. Multiply each element in A by the scalar.	$\begin{bmatrix} -10 \times 120 & -10 \times 90 \\ -10 \times 80 & -10 \times 60 \end{bmatrix}$

3. Write the answer.

$$\begin{bmatrix} -1200 & -900 \\ -800 & -600 \end{bmatrix}$$

c. 1. Identify the scalar for the matrix. In this case it is $\dfrac{1}{4}$, which means that each element in A is multiplied by $\dfrac{1}{4}$ (or divided by 4).

c. $\dfrac{1}{4}\begin{bmatrix} 120 & 90 \\ 80 & 60 \end{bmatrix}$

2. Multiply each element in A by the scalar.

$$\begin{bmatrix} \dfrac{1}{4} \times 120 & \dfrac{1}{4} \times 90 \\ \dfrac{1}{4} \times 80 & \dfrac{1}{4} \times 60 \end{bmatrix}$$

3. Simplify each multiplication by determining common factors and write the answer.

$$\begin{bmatrix} \cancel{120}^{30} \times \dfrac{1}{\cancel{4}^{1}} & \cancel{90}^{45} \times \dfrac{1}{\cancel{4}^{2}} \\ \cancel{80}^{20} \times \dfrac{1}{\cancel{4}^{1}} & \cancel{60}^{15} \times \dfrac{1}{\cancel{4}^{1}} \end{bmatrix} = \begin{bmatrix} 30 & \dfrac{45}{2} \\ 20 & 15 \end{bmatrix}$$

d. 1. Identify the scalar. In this case it is 0.1, which means that each element in A is multiplied by 0.1 (or divided by 10).

d. $0.1\begin{bmatrix} 120 & 90 \\ 80 & 60 \end{bmatrix}$

2. Multiply each element in A by the scalar.

$$\begin{bmatrix} 0.1 \times 120 & 0.1 \times 90 \\ 0.1 \times 80 & 0.1 \times 60 \end{bmatrix}$$

3. Calculate the values for each element and write the answer.

$$\begin{bmatrix} 12 & 9 \\ 8 & 6 \end{bmatrix}$$

Exercise 10.3 Operations with matrices

learn on

Simple familiar

1. a. **WE5** If $A = \begin{bmatrix} 2 & -3 \\ -1 & -8 \end{bmatrix}$ and $B = \begin{bmatrix} -1 & 9 \\ 0 & 11 \end{bmatrix}$, determine the value of $A + B$.

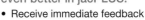

b. If $A = \begin{bmatrix} 0.5 \\ 0.1 \\ 1.2 \end{bmatrix}$, $B = \begin{bmatrix} -0.5 \\ 2.2 \\ 0.9 \end{bmatrix}$ and $C = \begin{bmatrix} -0.1 \\ -0.8 \\ 2.1 \end{bmatrix}$, determine the matrix sum $A + B + C$.

2. **WE6** If $A = \begin{bmatrix} 1 & 1 \\ -3 & -1 \end{bmatrix}$ and $B = \begin{bmatrix} 2 & 3 \\ -2 & 4 \end{bmatrix}$, calculate the value of $A - B$.

3. If $A = \begin{bmatrix} 5 \\ 4 \\ -2 \end{bmatrix}$, $B = \begin{bmatrix} -1 \\ 0 \\ 4 \end{bmatrix}$ and $C = \begin{bmatrix} -4 \\ 3 \\ 2 \end{bmatrix}$, calculate the following.

   a. $A + C$ **b.** $B + C$ **c.** $A - B$ **d.** $A + B - C$

4. Consider the matrices $C = \begin{bmatrix} 1 & -3 \\ 7 & 5 \\ b & 8 \end{bmatrix}$ and $D = \begin{bmatrix} 0 & a \\ -5 & -4 \\ 2 & -9 \end{bmatrix}$.

   If $C + D = \begin{bmatrix} 1 & 1 \\ 2 & 1 \\ -4 & -1 \end{bmatrix}$, determine the values of a and b.

5. Consider the following.

$$B - A = \begin{bmatrix} 4 \\ 0 \\ 2 \end{bmatrix}, \quad A + B = \begin{bmatrix} 4 \\ 2 \\ 8 \end{bmatrix} \text{ and } A = \begin{bmatrix} 0 \\ 1 \\ 3 \end{bmatrix}.$$

   a. Explain why matrix B must have an order of 3×1.
   b. Determine matrix B.

6. **WE7** Consider the matrix $C = \begin{bmatrix} 2 & 3 & 7 \\ 1 & 4 & 6 \end{bmatrix}$. Evaluate the following.

   a. $4C$ **b.** $\dfrac{1}{5}C$ **c.** $0.3C$ **d.** $-C$

7. **MC** Consider the matrix $M = \begin{bmatrix} 12 & 9 & 15 \\ 36 & 6 & 21 \end{bmatrix}$. Identify which of the following matrices is equal to the matrix M.

   A. $0.1 \begin{bmatrix} 1.2 & 0.9 & 1.5 \\ 3.6 & 0.6 & 2.1 \end{bmatrix}$ **B.** $3 \begin{bmatrix} 3 & 3 & 5 \\ 9 & 2 & 7 \end{bmatrix}$ **C.** $3 \begin{bmatrix} 4 & 3 & 5 \\ 12 & 2 & 7 \end{bmatrix}$ **D.** $3 \begin{bmatrix} 36 & 27 & 45 \\ 108 & 18 & 63 \end{bmatrix}$

8. Evaluate the following.

   a. $[0.5 \quad 0.25 \quad 1.2] - [0.75 \quad 1.2 \quad 0.9]$

   b. $\begin{bmatrix} 1 & 0 \\ 3 & 1 \end{bmatrix} + \begin{bmatrix} 2 & -1 \\ 6 & 0 \end{bmatrix}$

   c. $\begin{bmatrix} 12 & 17 & 10 \\ 35 & 20 & 25 \\ 28 & 32 & 29 \end{bmatrix} - \begin{bmatrix} 13 & 12 & 9 \\ 31 & 22 & 22 \\ 25 & 35 & 31 \end{bmatrix}$

   d. $\begin{bmatrix} 11 & 6 & 9 \\ 7 & 12 & -1 \end{bmatrix} + \begin{bmatrix} 2 & 8 & 8 \\ 6 & 7 & 6 \end{bmatrix} - \begin{bmatrix} -2 & -1 & 10 \\ 4 & 9 & -3 \end{bmatrix}$

Complex familiar

9. Matrix D was multiplied by the scalar quantity x. If $3D = \begin{bmatrix} 15 & 0 \\ 21 & 12 \\ 33 & 9 \end{bmatrix}$ and $xD = \begin{bmatrix} 12.5 & 0 \\ 17.5 & 10 \\ 27.5 & 7.5 \end{bmatrix}$, calculate the value of x.

10. If $\begin{bmatrix} 3 & 0 \\ 5 & a \end{bmatrix} + \begin{bmatrix} 2 & 2 \\ -b & 1 \end{bmatrix} = \begin{bmatrix} c & 2 \\ 3 & -4 \end{bmatrix}$, calculate the values of a, b and c.

11. If $\begin{bmatrix} 12 & 10 \\ 25 & 13 \\ 20 & a \end{bmatrix} - \begin{bmatrix} 9 & 11 \\ 26 & c \\ b & 9 \end{bmatrix} = \begin{bmatrix} 3 & -1 \\ -1 & 8 \\ 21 & -3 \end{bmatrix}$, calculate the values of a, b and c.

12. By calculating the order of each of the following matrices, identify which of the matrices can be added and/or subtracted to each other and explain why.

$$A = \begin{bmatrix} 1 & -5 \end{bmatrix} \quad B = \begin{bmatrix} 2 \\ -8 \end{bmatrix} \quad C = \begin{bmatrix} -1 \\ -9 \end{bmatrix} \quad D = \begin{bmatrix} -4 \end{bmatrix} \quad E = \begin{bmatrix} -3 & 6 \end{bmatrix}$$

13. Hard Eggs sells both free-range and barn-laid eggs in three different egg sizes (small, medium and large) to two shops, Appleton and Barntown. The number of cartons ordered for the Appleton shop is shown in the table below.

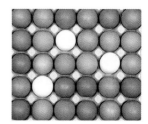

Eggs	Small	Medium	Large
Free range	2	3	5
Barn laid	4	6	3

The total orders for both shops are shown in the table below.

Eggs	Small	Medium	Large
Free range	3	4	8
Barn laid	6	8	5

a. Set up a matrix difference that would determine the order for the Barntown shop.

b. Use the matrix difference from part **a** to determine the order for the Barntown shop. Show the order in a table.

14. Marco was asked to complete the matrix sum $\begin{bmatrix} 8 & 126 & 59 \\ 17 & 102 & -13 \end{bmatrix} + \begin{bmatrix} 22 & 18 & 38 \\ 16 & 27 & 45 \end{bmatrix}$.

He gave $\begin{bmatrix} 271 \\ 194 \end{bmatrix}$ as his answer.

a. By referring to the order of matrices, explain why Marco's answer must be incorrect.

b. By explaining how to add matrices, write simple steps for Marco to follow so that he is able to add and subtract any matrices. Use the terms 'order of matrices' and 'elements' in your explanation.

15. There are three types of fish in a pond: speckles, googly eyes and fantails. At the beginning of the month there were 12 speckles, 9 googly eyes and 8 fantails in the pond. By the end of the month there were 9 speckles, 6 googly eyes and 8 fantails in the pond.

a. Construct a matrix sum to represent this information.

b. After six months, there were 12 speckles, 4 googly eyes and 10 fantails in the pond. Starting from the end of the first month, construct another matrix sum to represent this information.

16. Using technology or otherwise, evaluate the matrix sum

$$\begin{bmatrix} \frac{1}{2} & \frac{3}{4} & \frac{5}{6} \\ \frac{3}{5} & \frac{2}{7} & \frac{1}{3} \\ \frac{2}{3} & \frac{1}{4} & \frac{2}{9} \end{bmatrix} + \begin{bmatrix} \frac{1}{4} & \frac{3}{8} & \frac{1}{3} \\ \frac{1}{10} & \frac{3}{14} & \frac{4}{9} \\ \frac{1}{6} & \frac{1}{2} & \frac{2}{3} \end{bmatrix} - \begin{bmatrix} \frac{1}{8} & \frac{1}{2} & \frac{7}{6} \\ \frac{2}{15} & \frac{8}{21} & \frac{4}{3} \\ \frac{2}{9} & \frac{5}{8} & \frac{10}{9} \end{bmatrix}.$$

17. Frederick, Harold, Mia and Petra are machinists who work for Stitch in Time. The table below shows the hours worked by each of the four employees and the number of garments completed each week for the last three weeks.

| Employee | Week 1 | | Week 2 | | Week 3 | |
	Hours worked	Number of garments	Hours worked	Number of garments	Hours worked	Number of garments
Frederick	35	150	32	145	38	166
Harold	41	165	36	152	35	155
Mia	38	155	35	135	35	156
Petra	25	80	30	95	32	110

a. Construct a 4×1 matrix to represent the number of garments each employee made in week 1.

b. i. Create a matrix sum that would determine the total number of garments each employee made over the three weeks.

ii. Using your matrix sum from part **bi**, determine the total number of garments each employee made over the three weeks.

c. Nula is the manager of Stitch in Time. She uses the following matrix sum to determine the total number of hours worked by each of the four employees over the three weeks.

$$\begin{bmatrix} 35 \\ 41 \\ 38 \\ 25 \end{bmatrix} + \begin{bmatrix} 32 \\ 36 \\ 35 \\ 30 \end{bmatrix} + \begin{bmatrix} 38 \\ 35 \\ 35 \\ 32 \end{bmatrix} = \begin{bmatrix} \\ \\ \\ \end{bmatrix}$$

Complete the matrix sum by filling in the missing values.

Complex unfamiliar

18. Consider the following matrix sum: $A - C + B = D$. Matrix D has an order of 3×2.

A has elements $a_{11} = x$, $a_{21} = 20$, $a_{31} = 3c_{31}$, $a_{12} = 7$, $a_{22} = y$ and $a_{32} = -8$.

B has elements $b_{11} = x$, $b_{21} = 2x$, $b_{31} = 3x$, $b_{12} = y$, $b_{22} = 5$ and $b_{32} = 6$.

C has elements $c_{11} = 12$, $c_{21} = \frac{1}{2}a_{21}$, $c_{31} = 5$, $c_{12} = 9$, $c_{22} = 2y$ and $c_{32} = 3x$.

If $D = \begin{bmatrix} -8 & 1 \\ 14 & 2 \\ 16 & -8 \end{bmatrix}$, show that $x = 2$ and $y = 3$.

19. Consider the matrices A and B.

$$A = \begin{bmatrix} 21 & 10 & 9 \\ 18 & 7 & 12 \end{bmatrix} \qquad B = \begin{bmatrix} -10 & 19 & 11 \\ 36 & -2 & 15 \end{bmatrix}$$

The matrix sum $A + B$ was performed using a spreadsheet. The elements for A were entered into a spreadsheet in the following cells: a_{11} was entered in cell A1, a_{21} into cell A2, a_{12} in cell B1, a_{22} in cell B2, a_{13} in cell C2 and a_{23} in cell C3.

If the respective elements for B were entered into cells D1, D2, E1, E2, F1 and F2, write the formulas required to determine the matrix sum $A + B$.

20. The inverse of a 2×2 matrix, $\begin{bmatrix} a & b \\ c & d \end{bmatrix}$, is determined by $\dfrac{1}{ad-bc} \times \begin{bmatrix} d & -b \\ -c & a \end{bmatrix}$.

The inverse of $\begin{bmatrix} 4 & x \\ 2 & 6 \end{bmatrix}$ is $\begin{bmatrix} 0.6 & -0.7 \\ -0.2 & 0.4 \end{bmatrix}$.

Calculate the value of x.

Fully worked solutions for this chapter are available online.

LESSON
10.4 Matrix multiplication

SYLLABUS LINKS

- Perform matrix multiplication manually up to 3×3 matrices but not limited to square matrices.
- Determine the power of a matrix using technology with matrix arithmetic capabilities when appropriate.

Source: General Mathematics Senior Syllabus 2024 © State of Queensland (QCAA) 2024; licensed under CC BY 4.0.

10.4.1 Rules for multiplying matrices

Multiplying a matrix by another matrix is not as straightforward as multiplying a matrix by a scalar.

For example, a local bakery sells wholemeal loaves of bread for \$5, white loaves of bread for \$4 and sourdough loaves of bread for \$6.50.

The bakery sales for 5 days of one week can be summarised in the following table.

	Mon	Tues	Wed	Thurs	Frid
Wholemeal (\$5)	20	15	10	18	25
White (\$4)	22	28	12	15	30
Sourdough (\$6.50)	18	14	8	15	20

Total sales:

Monday: $5 \times 20 + 4 \times 22 + 6.5 \times 18 = \$305$

Tuesday: $5 \times 15 + 4 \times 28 + 6.5 \times 14 = \$278$

Wednesday: $5 \times 10 + 4 \times 12 + 6.5 \times 8 = \$150$

Thursday: $5 \times 18 + 4 \times 15 + 6.5 \times 15 = \$247.50$

Friday: $5 \times 25 + 4 \times 30 + 6.5 \times 20 = \$375$

The price is matched to how many sold, and the values are then added together to give the sales for each day.

Writing this calculation as the product of two matrices gives:

$$\begin{bmatrix} 5 & 4 & 6.5 \end{bmatrix} \begin{bmatrix} 20 & 15 & 10 & 18 & 25 \\ 22 & 28 & 12 & 15 & 30 \\ 18 & 14 & 8 & 15 & 20 \end{bmatrix} = \begin{bmatrix} 305 & 278 & 150 & 247.50 & 375 \end{bmatrix}$$

Note that the prices in the first row of the first matrix are multiplied by the quantities in the first column of the second matrix to get the total sales for Monday and so on.

The resulting matrix gives the total sales for the five days of the week.

The product of these two matrices can only be found if the number of columns in the first matrix is equal to the number of rows in the second matrix.

> **Rules for multiplying matrices**
>
> **For matrices to be able to be multiplied together, the *number of columns* in the first matrix must equal *the number of rows* in the second matrix.**

For example, consider matrices A and B.	$A = \begin{bmatrix} 7 & 5 \\ 1 & 3 \end{bmatrix} \qquad B = \begin{bmatrix} 8 \\ 6 \end{bmatrix}$
Consider the order of each matrix to determine whether the product matrix is defined. Because the number of columns in matrix A and the number of rows in matrix B are the same, the product matrix is defined. The order of the resultant matrix is the remaining two numbers (number of rows in A and the number of columns in B).	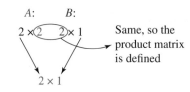

WORKED EXAMPLE 8 Determining the order of a product matrix

If $A = \begin{bmatrix} 3 \\ 2 \end{bmatrix}$ and $B = \begin{bmatrix} 1 & 2 \end{bmatrix}$, show that the product matrix AB exists and hence write down the order of AB.

THINK	WRITE
1. Write the order of each matrix.	$A: 2 \times 1$ $B: 1 \times 2$
2. Write the orders next to each other.	$2 \times 1 \quad 1 \times 2$
3. Circle the two middle numbers.	$2 \times \boxed{1 \quad 1} \times 2$
4. If the two numbers are the same, then the product matrix exists.	Number of columns in $A =$ number of rows in B; therefore, the product matrix AB exists.
5. The order of the resultant matrix (the product) will be the first and last number.	$\textcircled{2} \times 1 \quad 1 \times \textcircled{2}$
6. Write the answer.	The order of AB is 2×2.

10.4.2 Matrix multiplication

To multiply matrices together, use the following steps.

Step 1: Confirm that the product matrix exists (that is, the number of columns in the first matrix equals the number of rows in the second matrix).

Step 2: Multiply the elements of each row of the first matrix by the elements of each column of the second matrix.

Step 3: Sum the products in each element of the product matrix.

Consider the matrices C and D.	$C = \begin{bmatrix} 1 & 3 & 2 \\ 4 & -4 & 5 \end{bmatrix}$ and $D = \begin{bmatrix} -6 & 3 \\ 2 & 1 \\ 4 & -1 \end{bmatrix}$
Check the orders to see if the product matrix is defined. The remaining numbers make up the order of the product matrix.	Orders: $2 \times \boxed{3}$ and $\boxed{3} \times 2$ Same, so the product matrix is defined $2 \times 2 \longrightarrow$ Order of product matrix
Multiply the elements in the first row of matrix C with the elements in the first column of matrix D, then sum their products into the corresponding element of the product matrix.	$\begin{bmatrix} 1 & 3 & 2 \\ 4 & -4 & 5 \end{bmatrix} \times \begin{bmatrix} -6 & 3 \\ 2 & 1 \\ 4 & -1 \end{bmatrix} = \begin{bmatrix} 1 \times (-6) + 3 \times 2 + 2 \times 4 & \\ & \end{bmatrix}$
Repeat as above with the first row of matrix C and the second column of matrix D.	$\begin{bmatrix} 1 & 3 & 2 \\ 4 & -4 & 5 \end{bmatrix} \times \begin{bmatrix} -6 & 3 \\ 2 & 1 \\ 4 & -1 \end{bmatrix} = \begin{bmatrix} 8 & 1 \times 3 + 3 \times 1 + 2 \times (-1) \\ & \end{bmatrix}$
Repeat as above with the second row of matrix C and the first column of matrix D.	$\begin{bmatrix} 1 & 3 & 2 \\ 4 & -4 & 5 \end{bmatrix} \times \begin{bmatrix} -6 & 3 \\ 2 & 1 \\ 4 & -1 \end{bmatrix} = \begin{bmatrix} 8 & 4 \\ 4 \times (-6) + (-4) \times 2 + 5 \times 4 & \end{bmatrix}$
Repeat as above with the second row of matrix C and the second column of matrix D.	$\begin{bmatrix} 1 & 3 & 2 \\ 4 & -4 & 5 \end{bmatrix} \times \begin{bmatrix} -6 & 3 \\ 2 & 1 \\ 4 & -1 \end{bmatrix} = \begin{bmatrix} 8 & 4 \\ -12 & 4 \times 3 + (-4) \times 1 + 5 \times (-1) \end{bmatrix}$
We now have the product matrix.	$CD = \begin{bmatrix} 1 & 3 & 2 \\ 4 & -4 & 5 \end{bmatrix} \times \begin{bmatrix} -6 & 3 \\ 2 & 1 \\ 4 & -1 \end{bmatrix}$ $= \begin{bmatrix} 8 & 4 \\ -12 & 3 \end{bmatrix}$

TIP

The element a_{nm} of a product matrix is obtained by multiplying the elements in row n of the first matrix with the elements in column m of matrix the second matrix, then summing their products. For instance, the term a_{23} of the product matrix DC is obtained by multiplying the elements in row 2 of matrix D by the elements in column 3 of matrix C, then summing their products. It can be useful to write the matrices and their product as shown below to better visualise the rows and columns involved at any step.

$$\begin{bmatrix} 1 & 2 & 2 \\ 4 & -4 & 5 \end{bmatrix} \Big\} C$$

$$D \Big\{ \begin{bmatrix} -6 & 3 \\ 2 & 1 \\ 4 & -1 \end{bmatrix} \begin{bmatrix} R_1 \times C_1 & R_1 \times C_2 & R_1 \times C_3 \\ R_2 \times C_1 & R_2 \times C_2 & R_2 \times C_3 \\ R_3 \times C_1 & R_3 \times C_2 & R_3 \times C_3 \end{bmatrix} \Big\} DC$$

Unlike the multiplication of real numbers, the multiplication of matrices is not commutative. This means that, in most cases, $AB \neq BA$. So, it's very important to keep in mind the order of matrix multiplication.

Furthermore, the fact that the product matrix exists for AB does not mean necessarily that it exists for BA.

Try to test this out using matrices C and D from earlier. You can verify that in this case the product matrix DC is also defined:

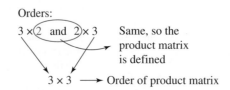

Orders:

3×2 and 2×3 Same, so the product matrix is defined

3×3 ⟶ Order of product matrix

If you try to perform the matrix multiplication $D \times C$, the resultant product matrix will be a 3×3 matrix, so it won't be the same as CE, which is a 2×2 matrix. You can check this by hand or by using your calculator.

WORKED EXAMPLE 9 Determining the product matrix

If $A = \begin{bmatrix} 3 & 5 \end{bmatrix}$ and $B = \begin{bmatrix} 2 \\ 6 \end{bmatrix}$, determine the product matrix AB.

THINK

1. Set up the product matrix.

2. Determine the order of the product matrix AB by writing the order of each matrix A and B.

3. Multiply each element in the first row by the corresponding element in the first column, then calculate the sum of the results.

4. Write the answer as a matrix.

WRITE

$\begin{bmatrix} 3 & 5 \end{bmatrix} \times \begin{bmatrix} 2 \\ 6 \end{bmatrix}$

$A \times B$
$① \times 2$ and $2 \times ①$
AB has an order of 1×1.

3×2 and $5 \times 6 = 36$

$[36]$

WORKED EXAMPLE 10 Determining the product matrix

Determine the product matrix MN if $M = \begin{bmatrix} 3 & 6 \\ 5 & 2 \end{bmatrix}$ and $N = \begin{bmatrix} 1 & 8 \\ 5 & 4 \end{bmatrix}$.

THINK

1. Set up the product matrix.

2. Determine the order of the product matrix MN by writing the order of each matrix M and N.

3. To determine the element mn_{11}, multiply the corresponding elements in the first row and first column and calculate the sum of the results.

4. To determine the element mn_{12}, multiply the corresponding elements in the first row and second column and calculate the sum of the results.

WRITE

$\begin{bmatrix} 3 & 6 \\ 5 & 2 \end{bmatrix} \times \begin{bmatrix} 1 & 8 \\ 5 & 4 \end{bmatrix}$

$M \times N$
$② \times 2$ and $2 \times ②$
MN has an order of 2×2.

$\begin{bmatrix} 3 & 6 \\ 5 & 2 \end{bmatrix} \times \begin{bmatrix} 1 & 8 \\ 5 & 4 \end{bmatrix}$

$3 \times 1 + 6 \times 5 = 33$

$\begin{bmatrix} 3 & 6 \\ 5 & 2 \end{bmatrix} \times \begin{bmatrix} 1 & 8 \\ 5 & 4 \end{bmatrix}$

$3 \times 8 + 6 \times 4 = 48$

5. To determine the element mn_{21}, multiply the corresponding elements in the second row and first column and calculate the sum of the results.

$$\begin{bmatrix} 3 & 6 \\ \boxed{5} & \boxed{2} \end{bmatrix} \times \begin{bmatrix} \boxed{1} & 8 \\ \boxed{5} & 4 \end{bmatrix}$$

$5 \times 1 + 2 \times 5 = 15$

6. To determine the element mn_{22}, multiply the corresponding elements in the second row and second column and calculate the sum of the results.

$$\begin{bmatrix} 3 & 6 \\ \boxed{5} & \boxed{2} \end{bmatrix} \times \begin{bmatrix} 1 & \boxed{8} \\ 5 & \boxed{4} \end{bmatrix}$$

$5 \times 8 + 2 \times 4 = 48$

7. Construct the matrix MN by writing in each of the elements.

$$\begin{bmatrix} 33 & 48 \\ 15 & 48 \end{bmatrix}$$

eles-6397

WORKED EXAMPLE 11 Multiplying matrices

Determine the product matrix PQ if $P = \begin{bmatrix} 2 & -1 \\ 3 & 1 \\ 4 & -2 \end{bmatrix}$ and $Q = \begin{bmatrix} -1 & 3 & 5 \\ 1 & 2 & -3 \end{bmatrix}$.

THINK

WRITE

1. Set up the product matrix.

$$\begin{bmatrix} 2 & -1 \\ 3 & 1 \\ 4 & -2 \end{bmatrix} \times \begin{bmatrix} -1 & 3 & 5 \\ 1 & 2 & -3 \end{bmatrix}$$

2. Determine the order of product matrix PQ by writing the order of each matrix P and Q.

$P \times Q$

$\boxed{3} \times 2 \times 2 \times \boxed{3}$

PQ has an order of 3×3.

3. To determine PQ_{11}, multiply the corresponding elements in the first row and first column and calculate the sum of the results.

$$\begin{bmatrix} \boxed{2 \,\, -1} \\ 3 & 1 \\ 4 & -2 \end{bmatrix} \times \begin{bmatrix} \boxed{-1} & 3 & 5 \\ \boxed{1} & 2 & -3 \end{bmatrix}$$

$PQ_{11} = (2 \times -1) + (-1 \times 1) = -3$

4. To determine PQ_{12}, multiply the corresponding elements in the second row and first column and calculate the sum of the results.

$$\begin{bmatrix} \boxed{2 \,\, -1} \\ 3 & 1 \\ 4 & -2 \end{bmatrix} \times \begin{bmatrix} -1 & \boxed{3} & 5 \\ 1 & \boxed{2} & -3 \end{bmatrix}$$

$PQ_{12} = (2 \times 3) + (-1 \times 2) = 4$

5. To determine PQ_{13}, multiply the corresponding elements in the third row and first column and calculate the sum of the results.

$$\begin{bmatrix} \boxed{2 \,\, -1} \\ 3 & 1 \\ 4 & -2 \end{bmatrix} \times \begin{bmatrix} -1 & 3 & \boxed{5} \\ 1 & 2 & \boxed{-3} \end{bmatrix}$$

$PQ_{13} = (2 \times 5) + (-1 \times -3) = 13$

6. To determine PQ_{21}, multiply the corresponding elements in the second row and first column and calculate the sum of the results.

$$\begin{bmatrix} 2 & -1 \\ \boxed{3 \,\, 1} \\ 4 & -2 \end{bmatrix} \times \begin{bmatrix} \boxed{-1} & 3 & 5 \\ \boxed{1} & 2 & -3 \end{bmatrix}$$

$PQ_{21} = (3 \times -1) + (1 \times 1) = -2$

7. To determine PQ_{22}, multiply the corresponding elements in the second row and second column and calculate the sum of the results.

$$\begin{bmatrix} 2 & -1 \\ \boxed{3 \,\, 1} \\ 4 & -2 \end{bmatrix} \times \begin{bmatrix} -1 & \boxed{3} & 5 \\ 1 & \boxed{2} & -3 \end{bmatrix}$$

$PQ_{22} = (3 \times 3) + (1 \times 2) = 11$

8. To determine PQ_{23}, multiply the corresponding elements in the second row and third column and calculate the sum of the results.

$$\begin{bmatrix} 2 & -1 \\ 3 & 1 \\ 4 & -2 \end{bmatrix} \times \begin{bmatrix} -1 & 3 & 5 \\ 1 & 2 & -3 \end{bmatrix}$$

$$PQ_{23} = (3 \times 5) + (1 \times -3) = 12$$

9. To determine PQ_{31}, multiply the corresponding elements in the third row and first column and calculate the sum of the results.

$$\begin{bmatrix} 2 & -1 \\ 3 & 1 \\ 4 & -2 \end{bmatrix} \times \begin{bmatrix} -1 & 3 & 5 \\ 1 & 2 & -3 \end{bmatrix}$$

$$PQ_{31} = (4 \times -1) + (-2 \times 1) = -6$$

10. To determine PQ_{32}, multiply the corresponding elements in the third row and second column and calculate the sum of the results.

$$\begin{bmatrix} 2 & -1 \\ 3 & 1 \\ 4 & -2 \end{bmatrix} \times \begin{bmatrix} -1 & 3 & 5 \\ 1 & 2 & -3 \end{bmatrix}$$

$$PQ_{32} = (4 \times 3) + (-2 \times 2) = 8$$

11. To determine PQ_{33}, multiply the corresponding elements in the third row and third column and calculate the sum of the results.

$$\begin{bmatrix} 2 & -1 \\ 3 & 1 \\ 4 & -2 \end{bmatrix} \times \begin{bmatrix} -1 & 3 & 5 \\ 1 & 2 & -3 \end{bmatrix}$$

$$PQ_{33} = (4 \times 5) + (-2 \times -3) = 26$$

12. Construct the matrix PQ by writing in each of the elements.

$$\begin{bmatrix} -3 & 4 & 13 \\ -2 & 11 & 12 \\ -6 & 8 & 26 \end{bmatrix}$$

10.4.3 Multiplying by the identity matrix

The identity matrix, I, plays a similar role to the number 1 in the real number system. When a matrix is multiplied by an identity matrix, the result is the same as if a real number is multiplied by 1. The matrix will remain the same.

So, for a matrix A, the following will apply:

$$AI = IA = A$$

For example, if the matrix $A = \begin{bmatrix} 2 & 4 & 6 \\ 3 & 5 & 7 \end{bmatrix}$ is multiplied by an identity matrix on its left (that is IA), it will be multiplied by a 2×2 identity matrix (because A has 2 rows). If A is multiplied by an identity matrix on its right (that is AI), it will be multiplied by a 3×3 identity matrix (because A has 3 columns).

$$
\underset{A}{\begin{bmatrix} 2 & 4 & 6 \\ 3 & 5 & 7 \end{bmatrix}} \underset{I}{\begin{bmatrix} 1 & 0 & 0 \\ 0 & 1 & 0 \\ 0 & 0 & 1 \end{bmatrix}} = \underset{I}{\begin{bmatrix} 1 & 0 \\ 0 & 1 \end{bmatrix}} \underset{A}{\begin{bmatrix} 2 & 4 & 6 \\ 3 & 5 & 7 \end{bmatrix}} = \underset{A}{\begin{bmatrix} 2 & 4 & 6 \\ 3 & 5 & 7 \end{bmatrix}}
$$

$$
\begin{bmatrix} 2 \times 1 + 4 \times 0 + 6 \times 0 & 2 \times 0 + 4 \times 1 + 6 \times 0 & 2 \times 0 + 4 \times 0 + 6 \times 1 \\ 3 \times 1 + 5 \times 0 + 7 \times 0 & 3 \times 0 + 5 \times 1 + 7 \times 0 & 3 \times 0 + 5 \times 0 + 7 \times 1 \end{bmatrix}
$$

$$
= \begin{bmatrix} 1 \times 2 + 0 \times 3 & 1 \times 4 + 0 \times 5 & 1 \times 6 + 0 \times 7 \\ 0 \times 2 + 1 \times 3 & 0 \times 4 + 1 \times 5 & 0 \times 6 + 1 \times 7 \end{bmatrix} = \begin{bmatrix} 2 & 4 & 6 \\ 3 & 5 & 7 \end{bmatrix}
$$

Multiplication with the identity matrix is an exception to the rule that most matrix multiplication isn't commutative.

Multiplying by the identity matrix

For a matrix A, the following will apply:

$$AI = IA = A$$

where I is an identity matrix with 1s in the top left to bottom right diagonal and 0s for all other elements.

For example, $[1]$, $\begin{bmatrix} 1 & 0 \\ 0 & 1 \end{bmatrix}$ and $\begin{bmatrix} 1 & 0 & 0 \\ 0 & 1 & 0 \\ 0 & 0 & 1 \end{bmatrix}$ are all identity matrices.

WORKED EXAMPLE 12 Multiplying a matrix by the identity matrix

Show that $AI = A$ by evaluating the following matrix product.

$$\begin{bmatrix} 6 & 2 \\ 8 & 3 \end{bmatrix} \times \begin{bmatrix} 1 & 0 \\ 0 & 1 \end{bmatrix}$$

THINK

WRITE

1. Multiply the elements of the first row of the first matrix with the elements of the first column of the second matrix. Sum the products.

$$\begin{bmatrix} 6 & 2 \\ 8 & 3 \end{bmatrix} \times \begin{bmatrix} 1 & 0 \\ 0 & 1 \end{bmatrix} = \begin{bmatrix} 6 \times 1 + 2 \times 0 & - \\ - & - \end{bmatrix}$$

2. Repeat step 1 with the first row of the first matrix and the second column of the second matrix

$$\begin{bmatrix} 6 & 2 \\ 8 & 3 \end{bmatrix} \times \begin{bmatrix} 1 & 0 \\ 0 & 1 \end{bmatrix} = \begin{bmatrix} 6 & 6 \times 0 + 2 \times 1 \\ - & - \end{bmatrix}$$

3. Repeat step 1 with the second row of the first matrix and the first column of the second matrix.

$$\begin{bmatrix} 6 & 2 \\ 8 & 3 \end{bmatrix} \times \begin{bmatrix} 1 & 0 \\ 0 & 1 \end{bmatrix} = \begin{bmatrix} 6 & 2 \\ 8 \times 1 + 3 \times 0 & - \end{bmatrix}$$

4. Repeat step 1 with the second row of the first matrix and the second column of the second matrix.

$$\begin{bmatrix} 6 & 2 \\ 8 & 3 \end{bmatrix} \times \begin{bmatrix} 1 & 0 \\ 0 & 1 \end{bmatrix} = \begin{bmatrix} 6 & 2 \\ 8 & 8 \times 0 + 3 \times 1 \end{bmatrix}$$

$$= \begin{bmatrix} 6 & 2 \\ 8 & 3 \end{bmatrix}$$

Note that you can similarly show that $IA = A$.

10.4.4 Powers of square matrices

When a square matrix is multiplied by itself, the order of the resultant matrix is equal to the order of the original square matrix. Because of this fact, whole-number powers of square matrices always exist.

You can use a calculator to quickly determine powers of square matrices.

WORKED EXAMPLE 13 Calculating powers of a square matrix

If $A = \begin{bmatrix} 3 & 5 \\ 5 & 1 \end{bmatrix}$, calculate the value of A^3.

THINK	WRITE
1. Write the matrix multiplication in full.	$A^3 = AAA$ $= \begin{bmatrix} 3 & 5 \\ 5 & 1 \end{bmatrix} \begin{bmatrix} 3 & 5 \\ 5 & 1 \end{bmatrix} \begin{bmatrix} 3 & 5 \\ 5 & 1 \end{bmatrix}$
2. Calculate the first matrix multiplication (AA).	$AA = \begin{bmatrix} 3 & 5 \\ 5 & 1 \end{bmatrix} \begin{bmatrix} 3 & 5 \\ 5 & 1 \end{bmatrix}$ $= \begin{bmatrix} 3 \times 3 + 5 \times 5 & 3 \times 5 + 5 \times 1 \\ 5 \times 3 + 1 \times 5 & 5 \times 5 + 1 \times 1 \end{bmatrix}$ $= \begin{bmatrix} 34 & 20 \\ 20 & 26 \end{bmatrix}$
3. Rewrite the full matrix multiplication, substituting the answer found in the previous part. Calculate the second matrix multiplication (AAA).	$A^3 = AAA$ $= \begin{bmatrix} 34 & 20 \\ 20 & 26 \end{bmatrix} \begin{bmatrix} 3 & 5 \\ 5 & 1 \end{bmatrix}$ $= \begin{bmatrix} 34 \times 3 + 20 \times 5 & 34 \times 5 + 20 \times 1 \\ 20 \times 3 + 26 \times 5 & 20 \times 5 + 26 \times 1 \end{bmatrix}$ $= \begin{bmatrix} 202 & 190 \\ 190 & 126 \end{bmatrix}$
4. Write the answer.	$A^3 = \begin{bmatrix} 202 & 190 \\ 190 & 126 \end{bmatrix}$

on Resources

Interactivity Matrix multiplication (int-6464)

Exercise 10.4 Matrix multiplication

10.4 Exercise	10.4 Exam questions on	These questions are even better in jacPLUS!

Simple familiar	Complex familiar	Complex unfamiliar
1, 2, 3, 4, 5, 6, 7, 8, 9, 10	11, 12, 13, 14, 15, 16, 17	18

- Receive immediate feedback
- Access sample responses
- Track results and progress

Find all this and MORE in jacPLUS ▶

Simple familiar

1. **a.** **WE8** If $X = \begin{bmatrix} 3 & 5 \end{bmatrix}$ and $Y = \begin{bmatrix} 4 \\ 2 \end{bmatrix}$, show that the product matrix XY exists and write down the order of XY.

   b. Determine which of the following matrices can be multiplied together and write down the order of any product matrices that exist.

$$D = \begin{bmatrix} 7 & 4 \\ 3 & 5 \\ 1 & 2 \end{bmatrix}, C = \begin{bmatrix} 5 & 7 \\ 8 & 9 \end{bmatrix} \text{ and } E = \begin{bmatrix} 4 & 1 & 2 \\ 6 & 2 & 6 \end{bmatrix}$$

2. The product matrix ST has an order of 3×4. If matrix S has 2 columns, write down the order of matrices S and T.

3. Determine which of the following matrices can be multiplied together. Justify your answers by determining the order of the product matrices.

$$D = \begin{bmatrix} 3 \\ 7 \\ 8 \end{bmatrix}, E = \begin{bmatrix} 5 & 8 \\ 7 & 1 \\ 9 & 3 \end{bmatrix}, F = \begin{bmatrix} 12 & 7 & 3 \\ 15 & 8 & 4 \end{bmatrix}, G = \begin{bmatrix} 13 & 15 \end{bmatrix}$$

4. **a.** **WE9** If $M = \begin{bmatrix} 4 \\ 3 \end{bmatrix}$ and $N = \begin{bmatrix} 7 & 12 \end{bmatrix}$, determine the product matrix MN.

   b. Determine whether the product matrix NM exists. Justify your answer by calculating the product matrix NM and stating its order.

5. **WE10** Determine the product matrix PQ if $P = \begin{bmatrix} 3 & 7 \\ 8 & 4 \end{bmatrix}$ and $Q = \begin{bmatrix} 2 & 1 \\ 5 & 6 \end{bmatrix}$.

6. **WE11** Determine the product matrices when the following pairs of matrices are multiplied together.

   a. $\begin{bmatrix} 7 \\ 2 \\ 9 \end{bmatrix}$ and $\begin{bmatrix} 10 & 15 \end{bmatrix}$

   b. $\begin{bmatrix} 6 & 5 \\ 8 & 3 \end{bmatrix}$ and $\begin{bmatrix} 2 \\ 9 \end{bmatrix}$

   c. $\begin{bmatrix} 4 & 6 \\ 2 & 3 \\ 3 & 1 \end{bmatrix} \begin{bmatrix} 3 & 5 & 4 \\ 1 & 2 & 5 \end{bmatrix}$

   d. $\begin{bmatrix} 5 & 7 & 1 \\ 6 & 5 & 2 \end{bmatrix} \begin{bmatrix} 4 & 2 \\ 1 & 7 \\ 3 & 1 \end{bmatrix}$

   e. $\begin{bmatrix} 1 & 3 & -1 \\ 2 & -3 & 0 \\ 1 & -2 & 1 \end{bmatrix} \times \begin{bmatrix} -2 & 0 & 1 \\ 3 & -1 & 4 \\ 0 & 1 & -2 \end{bmatrix}$

   f. $\begin{bmatrix} -2 & 1 & 0 \\ 0 & -1 & 0 \\ 2 & 0 & 2 \end{bmatrix} \times \begin{bmatrix} -1 & 1 & -1 \\ 2 & 1 & 2 \\ 0 & 1 & 0 \end{bmatrix}$

7. Evaluate the following matrix multiplications.

   a. $\begin{bmatrix} 4 & 6 \\ 2 & 3 \end{bmatrix} \begin{bmatrix} 1 & 0 \\ 0 & 1 \end{bmatrix}$

   b. $\begin{bmatrix} 1 & 0 \\ 0 & 1 \end{bmatrix} \begin{bmatrix} 4 & 6 \\ 2 & 3 \end{bmatrix}$

   c. Using your results from parts **a** and **b**, determine when AB will be equal to BA.

   d. If A and B are not of the same order, explain whether it is possible for AB to be equal to BA.

8. **WE12** The 3×3 identity matrix, $I = \begin{bmatrix} 1 & 0 & 0 \\ 0 & 1 & 0 \\ 0 & 0 & 1 \end{bmatrix}$.

  a. Calculate the value of I^2.
  b. Calculate the value of I^3.
  c. Calculate the value of I^4.
  d. Comment on your answers to parts a–c.

9. **WE13** If $P = \begin{bmatrix} 8 & 2 \\ 4 & 7 \end{bmatrix}$, calculate the value of P^2.

10. If $T = \begin{bmatrix} 3 & 5 \\ 0 & 6 \end{bmatrix}$, calculate the value of T^3.

Complex familiar

11. By using technology or otherwise, calculate the following powers of square matrices.

  a. $\begin{bmatrix} 4 & 8 \\ 7 & 2 \end{bmatrix}^4$

  b. $\begin{bmatrix} 3 & 1 & 7 \\ 4 & 2 & 8 \\ 5 & 6 & 9 \end{bmatrix}^3$

12. Matrix $S = \begin{bmatrix} 1 & 4 & 3 \end{bmatrix}$, matrix $T = \begin{bmatrix} 2 \\ 3 \\ t \end{bmatrix}$ and the product matrix $ST = [5]$. Calculate the value of t.

13. In an AFL game of football, 6 points are awarded for a goal and 1 point is awarded for a behind. St Kilda and Collingwood played two games, with the two results given by the following matrix multiplication.

$$\begin{bmatrix} 9 & 14 \\ 10 & 8 \\ 16 & 12 \\ 7 & 10 \end{bmatrix} \begin{bmatrix} 6 \\ 1 \end{bmatrix} = \begin{bmatrix} C_1 \\ S_1 \\ C_2 \\ S_2 \end{bmatrix}$$

  Complete the matrix multiplication to determine the scores in the two games.

14. The table below shows the percentage of students who are expected to be awarded grades A–E on their final examinations for Mathematics and Physics.

Grade	A	B	C	D	E
Percentage of students	5	18	45	25	7

  The number of students studying Mathematics and Physics is 250 and 185 respectively.

  a. Construct a 1×5 matrix, A, to represent the percentage of students expected to receive each grade, expressing each element in decimal form.
  b. Construct a column matrix, S, to represent the number of students studying Mathematics and Physics.
  c. i. In the context of this problem, explain what the product matrix SA represents.
     ii. Determine the product matrix SA. Write your answers correct to the nearest whole numbers.
  d. In the context of this problem, explain what element SA_{12} represents.

15. Dodgy Bros sell vans, utes and sedans. The average selling price for each type of vehicle is shown in the first table.

Type of vehicle	Selling price ($)
Vans	$4 000
Utes	$12 500
Sedans	$8 500

The second table shows the total number of vans, utes and sedans sold at Dodgy Bros in one month.

Type of vehicle	Number of sales
Vans	5
Utes	8
Sedans	4

Stan is the owner of Dodgy Bros and wants to determine the total amount of monthly sales.

a. Explain how matrices could be used to help Stan determine the total amount, in dollars, of monthly sales.
b. Perform a matrix multiplication that finds the total amount of monthly sales.
c. Brian is Stan's brother and the accountant for Dodgy Bros. In finding the total amount of monthly sales, he performs the following matrix multiplication.

$$\begin{bmatrix} 5 \\ 8 \\ 4 \end{bmatrix} \begin{bmatrix} 4000 & 12\,500 & 8500 \end{bmatrix}$$

Explain using mathematical reasoning why this matrix multiplication is not valid for this problem.

16. For a concert, three different types of tickets can be purchased; adult, senior and child. The cost of each type of ticket is $12.50, $8.50 and $6.00 respectively. The number of people attending the concert is shown in the following table.

Ticket type	Number of people
Adult	65
Senior	40
Child	85

a. Construct a column matrix to represent the cost of the three different tickets in the order adult, senior and child.

If the number of people attending the concert is written as a row matrix, a matrix multiplication can be performed to determine the total amount in ticket sales for the concert.

b. By determining the orders of each matrix and then the product matrix, explain why this is the case.
c. By completing the matrix multiplication from part b, determine the total amount (in dollars) in ticket sales for the concert.

17. The number of adults, children and seniors attending the zoo over Friday, Saturday and Sunday is shown in the table.

Day	Adults	Children	Seniors
Friday	125	245	89
Saturday	350	456	128
Sunday	421	523	102

Entry prices for adults, children and seniors are $35, $25 and $$20 respectively.

a. Perform a matrix multiplication that will find the entry fee collected for each of the three days.

b. Write the calculation that finds the entry fee collected for Saturday.

c. Determine if it is possible to perform a matrix multiplication that would find the total for each type of entry fee (adults, children and seniors) over the three days. Explain your answer using mathematical reasoning.

Complex unfamiliar

18. A product matrix, $N = MPR$, has order 3×4. Matrix M has m rows and n columns, matrix P has order $1 \times q$, and matrix R has order $2 \times s$. Determine the values of m, n, s and q.

Fully worked solutions for this chapter are available online.

LESSON
10.5 Applications of matrices

SYLLABUS LINKS

- Use matrices, including matrix products and powers of matrices, to model and solve problems, e.g. costing or pricing problems, squaring a matrix to determine the number of ways pairs of people in a communication network can communicate with each other via a third person.

Source: General Mathematics Senior Syllabus 2024 © State of Queensland (QCAA) 2024; licensed under CC BY 4.0.

10.5.1 Solving problems with matrices

In this section we use matrices to solve application questions.

eles-6399

WORKED EXAMPLE 14 Using a matrix to solve a finance problem

Three tuckshop suppliers, Ace Catering (*A*), Best Buddies (*B*) and Café Co-op (*C*), can supply juice, sushi, burgers and wraps to the local schools.

The following matrix shows the prices ($) of each item supplied by the individual suppliers.

$$
\begin{array}{c} \\ A \\ B \\ C \end{array}
\begin{array}{cccc}
\text{Juice} & \text{Sushi} & \text{Burgers} & \text{Wraps} \\
\left[\begin{array}{cccc}
3.50 & 4.80 & 10.20 & 9.00 \\
2.20 & 4.20 & 11.50 & 10.50 \\
2.90 & 3.90 & 12.00 & 11.00
\end{array}\right]
\end{array}
$$

a. Identify how much Ace Catering is charging for sushi.

b. Identify the supplier that has the cheapest juice.

c. A school needs 50 each of sushi, burgers and wraps, and 200 orders of juice. If the school wishes to purchase all the items from one supplier, use matrices to determine which supplier will be the cheapest and the cost of the order.

THINK	WRITE
a. Consider the element a_{12}.	**a.** Ace Catering is charging $4.80 for each sushi item.
b. Consider the juice column. The cheapest juice is $2.20, supplied by Best Buddies.	**b.** Best Buddies is supplying the cheapest juice at $2.20.
c. 1. A matrix product needs to be calculated. The matrix given is a 3×4 matrix, so the order matrix needs to be 4×1.	**c.** Order matrix: $$\begin{bmatrix} 200 \\ 50 \\ 50 \\ 50 \end{bmatrix}$$
2. Multiply the matrices.	$$\begin{bmatrix} 3.50 & 4.80 & 10.20 & 9.00 \\ 2.20 & 4.20 & 11.50 & 10.50 \\ 2.90 & 3.90 & 12.00 & 11.00 \end{bmatrix} \begin{bmatrix} 200 \\ 50 \\ 50 \\ 50 \end{bmatrix} = \begin{bmatrix} 1900 \\ 1750 \\ 1925 \end{bmatrix}$$
3. Check the cheapest total price.	The cheapest total is $1750.
4. Write the answer.	Best Buddies provides the cheapest order at $1750.

10.5.2 Adjacency matrices

When matrices are used to analyse a graph, the graph must first be converted into a square matrix. This matrix is called the **adjacency matrix**, and it is used to represent the information in a graph.

The following are the general steps that need to be taken to create an adjacency matrix:

Step 1: Start with an $n \times n$ matrix where n is the number of objects/places.

Step 2: Input the number of connections between objects in the respective element in the adjacency matrix.

Take, for example, the following graph.

WORKED EXAMPLE 15 Constructing an adjacency matrix

The diagram shows the number of roads connecting four towns, A, B, C and D.
Construct an adjacency matrix to represent this information.

		THINK			WRITE			

THINK

1. Since there are four connecting towns, a 4×4 adjacency matrix needs to be constructed. Label the row and columns with the relevant towns A, B, C and D.

WRITE

$$\begin{array}{c} \\ A \\ B \\ C \\ D \end{array} \begin{array}{cccc} A & B & C & D \\ \begin{bmatrix} - & - & - & - \\ - & - & - & - \\ - & - & - & - \\ - & - & - & - \end{bmatrix} \end{array}$$

2. There is one road connecting town A to town B, so enter 1 in the cell from A to B.

$$\begin{array}{c} \\ A \\ B \\ C \\ D \end{array} \begin{array}{cccc} A & B & C & D \\ \begin{bmatrix} - & - & - & - \\ ① & - & - & - \\ - & - & - & - \\ - & - & - & - \end{bmatrix} \end{array}$$

3. There is also only one road between town A and towns C and D; therefore, enter 1 in the appropriate matrix positions. There are no loops at town A (i.e. a road connecting A to A); therefore, enter 0 in this position.

$$\begin{array}{c} \\ A \\ B \\ C \\ D \end{array} \begin{array}{cccc} A & B & C & D \\ \begin{bmatrix} 0 & - & - & - \\ 1 & - & - & - \\ 1 & - & - & - \\ 1 & - & - & - \end{bmatrix} \end{array}$$

4. Repeat this process for towns B, C and D. Note that there are two roads connecting towns B and D, and that town C only connects to town A.

$$\begin{array}{c} \\ A \\ B \\ C \\ D \end{array} \begin{array}{cccc} A & B & C & D \\ \begin{bmatrix} 0 & 1 & 1 & 1 \\ 1 & 0 & 0 & 2 \\ 1 & 0 & 0 & 0 \\ 1 & 2 & 0 & 0 \end{bmatrix} \end{array}$$

10.5.3 Determining the number of connections between objects

An adjacency matrix allows us to determine the number of connections either directly between or via objects. If a direct connection between two objects is denoted as one 'step', 'two steps' means a connection between two objects via a third object, for example the number of ways a person can travel between towns A and D via another town.

> ### Determining the number of connections between objects
>
> **You can determine the number of connections of differing 'steps' by raising the adjacency matrix to the power that reflects the number of steps in the connection.**

For example, the following diagram shows the number of roads connecting five towns, A, B, C, D and E. There are a number of ways to travel between towns A and D. There is one direct path between the towns; this is a one-step path.

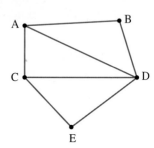

A one-step path matrix for the roads connecting the five towns can be shown as:

$$
\begin{array}{c}
\\
\\
\text{FROM}
\end{array}
\begin{array}{c}
\\
\text{A} \\
\text{B} \\
\text{C} \\
\text{D} \\
\text{E}
\end{array}
\overset{\displaystyle \text{TO}}{
\overset{\text{A\quad B\quad C\quad D\quad E}}{
\begin{bmatrix}
0 & 1 & 1 & 1 & 0 \\
1 & 0 & 0 & 1 & 0 \\
1 & 0 & 0 & 1 & 1 \\
1 & 1 & 1 & 0 & 1 \\
0 & 0 & 1 & 1 & 0
\end{bmatrix}}}
$$

However, you can also travel between towns A and D via town C or B. These are considered two-step paths as there are two links (or roads) in these paths. This is referred to as the link length. The power on the adjacency matrix would therefore be 2 in this case.

WORKED EXAMPLE 16 Using an adjacency matrix to determine the number of two-step paths

The following adjacency matrix shows the number of pathways between three attractions at the zoo: lions (L), seals (S) and monkeys (M).

$$
\begin{array}{c}
\text{L} \\
\text{S} \\
\text{M}
\end{array}
\overset{\text{L\quad S\quad M}}{
\begin{bmatrix}
0 & 1 & 1 \\
1 & 0 & 2 \\
1 & 2 & 0
\end{bmatrix}}
$$

Determine how many ways a family can travel from the lions to the monkeys via the seals.

THINK	WRITE
1. Determine the link length.	The required path is between two attractions via a third attraction, so the link length is 2.
2. As the link length is 2, this will raise the matrix to a power of 2. Evaluate the matrix using technology such as graphing or a calculator.	$\begin{bmatrix} 0 & 1 & 1 \\ 1 & 0 & 2 \\ 1 & 2 & 0 \end{bmatrix}^2 = \begin{bmatrix} 2 & 2 & 2 \\ 2 & 5 & 1 \\ 2 & 1 & 5 \end{bmatrix}$
3. Interpret the information in the matrix and locate the required value.	$\begin{bmatrix} 2 & ② & 2 \\ 2 & 5 & 1 \\ 2 & 1 & 5 \end{bmatrix}$
4. Write the answer.	There are 2 ways in which a family can travel from the lions to the monkeys via the seals.

---

on Resources

Interactivity The adjacency matrix (int-6466)

Exercise 10.5 Applications of matrices

10.5 Exercise	10.5 Exam questions on

Simple familiar	Complex familiar	Complex unfamiliar
1, 2, 3, 4	5, 6, 7, 8, 9, 10, 11	12

These questions are even better in jacPLUS!
- Receive immediate feedback
- Access sample responses
- Track results and progress

Find all this and MORE in jacPLUS ⊳

Simple familiar

1. **WE15** The diagram shows the network cable between five main computers (A, B, C D and E) in an office building.

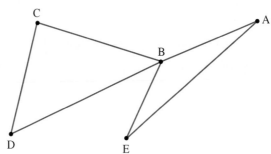

Construct an adjacency matrix to represent this information.

2. Given the following communication matrix, answer the following questions.

$$
\begin{array}{c}
 \\
A \\
B \\
C \\
D
\end{array}
\begin{array}{cccc}
A & B & C & D \\
\begin{bmatrix}
0 & 1 & 0 & 1 \\
1 & 0 & 0 & 1 \\
1 & 1 & 0 & 1 \\
1 & 1 & 1 & 0
\end{bmatrix}
\end{array}
$$

a. Identify who C can talk to.
b. Identify who B can receive calls from.
c. Explain why the main diagonal is all zeros.
d. Identify who B cannot call.

3. **WE14** Three wholesale electrical retailers, Calypso, Supersaver and Triumph, supply discounted fridges, dishwashers, ovens and microwaves.
The following matrix shows the prices ($) of each item supplied by the individual suppliers.

$$
\begin{array}{c}
 \\
C \\
S \\
T
\end{array}
\begin{array}{cccc}
\text{Fridge} & \text{Dish} & \text{Ovens} & \text{Micro} \\
\begin{bmatrix}
650 & 670 & 990 & 400 \\
800 & 560 & 690 & 380 \\
780 & 720 & 830 & 250
\end{bmatrix}
\end{array}
$$

a. Identify how much Supersaver is charging for ovens.
b. Identify the retailer that has the most expensive fridges.
c. A retail store needs 15 fridges, 20 ovens, 30 dishwashers and 10 microwaves. If the retailer wishes to purchase all the items from one wholesaler, use matrices to determine which wholesaler will be the cheapest and the cost of the order.

4. The adjacency matrix below shows the number of text messages sent between three friends, Stacey (S), Ruth (R) and Toiya (T), immediately after school one day.

$$
\begin{array}{c c}
 & \begin{array}{c c c} S & R & T \end{array} \\
\begin{array}{c} S \\ R \\ T \end{array} &
\left[\begin{array}{c c c}
0 & 3 & 2 \\
3 & 0 & 1 \\
2 & 1 & 0
\end{array}\right]
\end{array}
$$

a. Determine the number of text messages sent between Stacey and Ruth.

b. Determine the total number of text messages sent between all three friends.

Complex familiar

5. The number of Google Home and Google Home Minis sold by four stores is shown in the table.

	Google Home	Google Home Mini
Store A	25	8
Store B	12	9
Store C	20	12
Store D	15	10

If the Google Homes were priced at $155 each and the Google Home Minis were priced at $60 each, use matrix operations to determine:

a. the total sales figures of Google Homes at each store

b. the total sales figures for each store

c. the store that had the highest sales figures for:

   i. Google Homes
   ii. Google Home Minis.

6. There are five friends on a social media site: Peta, Seth, Tran, Ned and Wen. The number of communications made between these friends in the last 24 hours is shown in the adjacency matrix below.

$$
\begin{array}{c c}
 & \begin{array}{c c c c c} P & S & T & N & W \end{array} \\
\begin{array}{c} P \\ S \\ T \\ N \\ W \end{array} &
\left[\begin{array}{c c c c c}
0 & 1 & 3 & 1 & 0 \\
1 & 0 & 0 & 0 & 4 \\
3 & 0 & 0 & 2 & 1 \\
1 & 0 & 2 & 0 & 0 \\
0 & 4 & 1 & 0 & 0
\end{array}\right]
\end{array}
$$

a. Determine how many times Peta and Tran communicated over the last 24 hours.

b. Determine whether Seth communicated with Ned at any time during the last 24 hours.

c. In the context of this problem, explain the existence of the zeros along the diagonal.

d. Using the adjacency matrix, construct a diagram that shows the number of communications between the five friends.

7. **WE16** The adjacency matrix below shows the number of roads between three country towns, Gladstone (G), Rockhampton (R) and Bundaberg (B).

$$
\begin{array}{c c}
 & \begin{array}{c c c} G & R & B \end{array} \\
\begin{array}{c} G \\ R \\ B \end{array} &
\left[\begin{array}{c c c}
0 & 1 & 1 \\
1 & 0 & 2 \\
1 & 2 & 0
\end{array}\right]
\end{array}
$$

Using a technology of your choice, determine the number of ways a person can travel from Gladstone to Rockhampton via Bundaberg.

8. The direct Cape Air flights between five cities, Boston (B), Hyannis (H), Martha's Vineyard (M), Nantucket (N) and Providence (P), are shown in the adjacency matrix.

$$\begin{array}{c} \\ B \\ H \\ M \\ N \\ P \end{array} \begin{array}{c} \begin{array}{ccccc} B & H & M & N & P \end{array} \\ \begin{bmatrix} 0 & 1 & 1 & 1 & 0 \\ 1 & 0 & 1 & 1 & 0 \\ 1 & 1 & 0 & 1 & 1 \\ 1 & 1 & 1 & 0 & 1 \\ 0 & 0 & 1 & 1 & 0 \end{bmatrix} \end{array}$$

a. Construct a diagram to represent the direct flights between the five cities.
b. Construct a matrix that determines the number of ways a person can fly between two cities via another city.
c. Explain how you would determine the number of ways a person can fly between two cities via two other cities.
d. Determine whether it is possible to fly from Boston and stop at every other city. Explain how you would answer this question.

9. Consider the directed network shown.

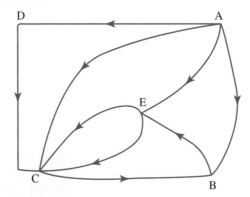

a. Determine the number and name of:

  i. the one-stage pathways to get to C   ii. the two-stage pathways to get to C.
b. Represent the one-stage and two-stage pathways of the directed network in matrix form.

10. Construct the communication matrix from the following communication network.

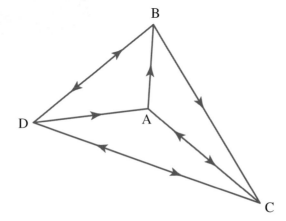

11. Airlink flies charter flights in the Cape Lancaster region. The direct flights between Williamton, Cowal, Hugh River, Kokialah and Archer are shown in the diagram.

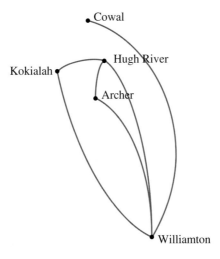

a. Determine how many ways a person can travel between Williamton and Kokialah via another town.
b. Determine whether it is possible to fly between Cowal and Archer and stop over at two other towns. Justify your answer.

Complex unfamiliar

12. The senior school manager developed a matrix formula to determine the number of school jackets to order for Years 11 and 12 students. The column matrix, J_0, shows the number of jackets ordered last year.

$$J_0 = \begin{bmatrix} 250 \\ 295 \end{bmatrix}$$

J_1 is the column matrix that lists the number of Year 11 and 12 jackets to be ordered this year. J_1 is given by the matrix formula

$$J_1 = AJ_0 + B,$$

where $A = \begin{bmatrix} 0.65 & 0 \\ 0 & 0.82 \end{bmatrix}$ and $B = \begin{bmatrix} 13 \\ 19 \end{bmatrix}$.

Determine J_1 and the jacket order for the next year.
Write your answer for the jacket order to the nearest whole number.

Fully worked solutions for this chapter are available online.

LESSON
10.6 Review

10.6.1 Summary

doc-
41495

10.6 Exercise

 learnon

10.6 Exercise	10.6 Exam practice on

Simple familiar	Complex familiar	Complex unfamiliar
1, 2, 3, 4, 5, 6, 7, 8, 9, 10, 11, 12	13, 14, 15, 16	17, 18, 19, 20

Simple familiar

1. **MC** Consider the matrix equation $\begin{bmatrix} 6 & a \\ 3 & 2 \end{bmatrix} - \begin{bmatrix} 2 & -1 \\ 0 & b \end{bmatrix} = \begin{bmatrix} 4 & 2 \\ 3 & 1 \end{bmatrix}$.

   The values of a and b respectively are:

   A. $a = 1$, $b = 1$ **B.** $a = 1$, $b = -1$ **C.** $a = 2$, $b = 1$ **D.** $a = 2$, $b = -1$

2. **MC** The order of the matrix $\begin{bmatrix} 6 & 1 & 2 \\ 7 & 3 & 5 \\ 9 & 5 & 0 \\ 0 & 7 & 2 \end{bmatrix}$ is:

   A. 3×4 **B.** 4×2 **C.** 4×3 **D.** 8

The following information relates to questions 3 and 4.

The following matrix shows the airline ticket price between four cities: Perth, Melbourne, Hobart and Sydney.

$$
\begin{array}{c}
 \\
\text{To} \\
\end{array}
\begin{array}{cc}
 & \text{From} \\
\begin{array}{c} \\ P \\ M \\ H \\ S \end{array} &
\begin{array}{cccc}
P & M & H & S \\
\begin{bmatrix}
0 & 140 & 450 & 190 \\
180 & 0 & 90 & 50 \\
350 & 80 & 0 & 80 \\
240 & 60 & 110 & 0
\end{bmatrix}
\end{array}
\end{array}
$$

3. **MC** The cost to fly from Sydney to Hobart is:

   A. $50   **B.** $80   **C.** $90   **D.** $110

4. **MC** A passenger flew from one city and then on to another, with the total cost of the flight being $320. The order in which the plane flew between the three cities was:

   A. Sydney to Melbourne to Perth   **B.** Hobart to Melbourne to Perth

   C. Perth to Melbourne to Sydney   **D.** Perth to Sydney to Hobart

5. **MC** Sarah purchased 2 apples and 3 bananas from her local market for $3.80. Later that week she went back to the market and purchased 4 more apples and 4 more bananas for $6.20.

   A matrix equation that could be set up to determine the price of a single apple and banana is:

   A. $\begin{bmatrix} 2 & 3 \\ 4 & 4 \end{bmatrix}\begin{bmatrix} x \\ y \end{bmatrix} = \begin{bmatrix} 3.8 \\ 6.2 \end{bmatrix}$ **B.** $\begin{bmatrix} x \\ y \end{bmatrix}\begin{bmatrix} 2 & 3 \\ 4 & 4 \end{bmatrix} = \begin{bmatrix} 3.8 \\ 6.2 \end{bmatrix}$ **C.** $\begin{bmatrix} 2 & 4 \\ 3 & 4 \end{bmatrix}\begin{bmatrix} x \\ y \end{bmatrix} = \begin{bmatrix} 3.8 \\ 6.2 \end{bmatrix}$ **D.** $\begin{bmatrix} 2 & 3 \\ 4 & 4 \end{bmatrix}\begin{bmatrix} 3.8 \\ 6.2 \end{bmatrix} = \begin{bmatrix} x \\ y \end{bmatrix}$

6. **MC** The matrix that represents the product $\begin{bmatrix} 5 & 1 & 2 \\ 3 & 7 & 8 \end{bmatrix}\begin{bmatrix} 12 \\ 10 \\ 9 \end{bmatrix}$ is:

   A. $\begin{bmatrix} 88 \\ 178 \end{bmatrix}$ **B.** $\begin{bmatrix} 60 & 10 & 18 \\ 36 & 63 & 72 \end{bmatrix}$ **C.** $\begin{bmatrix} 88 & 178 \end{bmatrix}$ **D.** $\begin{bmatrix} 48 \\ 49 \end{bmatrix}$

7. **MC** The population age structure (in percentages) in 2010 for selected countries is shown in the following table.

Country	Percentage of population between the age groups		
	0–14 years	15–64 years	Over 65 years
Australia	18.9	67.6	13.5
China	16.4	69.5	14.1
Indonesia	27.0	67.4	5.6

   A 3×1 matrix that could be used to represent the percentage of population across the three age groups for Indonesia is:

   A. $\begin{bmatrix} 18.9 \\ 16.4 \\ 27.0 \end{bmatrix}$ **B.** $\begin{bmatrix} 27.0 & 67.4 & 5.6 \end{bmatrix}$ **C.** $\begin{bmatrix} 18.9 & 67.6 & 13.5 \\ 16.4 & 69.5 & 14.1 \\ 27.0 & 67.4 & 5.6 \end{bmatrix}$ **D.** $\begin{bmatrix} 27.0 \\ 67.4 \\ 5.6 \end{bmatrix}$

8. **MC** Matrix A has an order of 3×2. Matrix B has an order of 1×3. Matrix C has an order of 2×1. Identify which one of the following matrix multiplications is not possible.

   A. AC **B.** BA **C.** BC **D.** CB

9. Running paths through a park are shown in the diagram. Construct an adjacency matrix for this communication network.

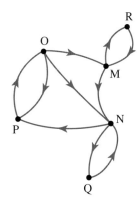

10. Given matrices $A = \begin{bmatrix} 2 & 3 \\ 1 & 4 \end{bmatrix}$, $B = \begin{bmatrix} 1 \\ 0 \end{bmatrix}$ and $C = \begin{bmatrix} 3 \\ 2 \end{bmatrix}$, evaluate the following.

    a. $C + B$ b. $B - 2C$ c. $AB + C$ d. $1.5A$

11. Matrix D has an order of 3×2, matrix E has an order of $1 \times p$ and matrix F has an order of 2×2.

    a. Determine the value of p for which the product matrix would ED exist.
    b. If the product matrix H exists and $H = EDF$, determine the order of H.

12. a. For each of the following pairs of matrices, determine the order of the matrices, and hence determine the order of the product matrix.

      i. $\begin{bmatrix} 3 & 5 \end{bmatrix} \begin{bmatrix} 2 \\ 6 \end{bmatrix}$

      ii. $\begin{bmatrix} 4 & 6 \end{bmatrix} \begin{bmatrix} 2 & 3 \\ 4 & 7 \end{bmatrix}$

      iii. $\begin{bmatrix} -1 & 9 \\ 10 & 5 \end{bmatrix} \begin{bmatrix} 3 & -2 \\ 5 & 11 \end{bmatrix}$

      iv. $\begin{bmatrix} 2 \\ 7 \\ 8 \end{bmatrix} \begin{bmatrix} 5 & 3 & 4 \end{bmatrix}$

    b. Determine the product matrices of the matrix multiplications given in part a.

Complex familiar

13. **MC** To help him to answer his ten multiple choice questions, Trei is using the following matrix.

		Answers This question				
		A	B	C	D	E
Answers Next question	A	0	0	0	1	0
	B	0	0	0	0	0
	C	0	1	0	0	0
	D	0	0	1	0	0
	E	1	0	0	0	1

Trei answered B to question 1 and then used the matrix to answer the remaining nine questions. Determine Trei's answer to question 6.

    A. C   **B.** D   **C.** E   **D.** A

14. The table shows the three different ticket prices, in dollars, and the number of tickets sold for a school concert.

Ticket type	Ticket price	Number of tickets sold
Adult	$12.50	140
Child/student	$6.00	225
Teacher	$10.00	90

    a. Construct a column matrix to represent the ticket prices for adults, children/students and teachers respectively.
    b. Perform a matrix multiplication to determine the total amount of ticket sales in dollars.

15. Rhonda was asked to perform the following matrix multiplication to determine the product matrix *GH*.

$$GH = \begin{bmatrix} 6 & 5 \\ 3 & 8 \\ 5 & 9 \end{bmatrix} \begin{bmatrix} 10 \\ 13 \end{bmatrix}$$

Rhonda's answer was $\begin{bmatrix} 60 & 65 \\ 30 & 104 \\ 50 & 117 \end{bmatrix}$.

Determine the product matrix *GH* and explain why Rhonda's method of multiplying matrices is an incorrect method.

16. The number of adults, children and seniors attending the aquarium over Friday, Saturday and Sunday is shown in the table.

Day	Adults	Children	Seniors
Friday	90	155	95
Saturday	260	306	120
Sunday	310	420	40

Entry prices for adults, children and seniors are $30, $20, $15 respectively.

Using technology or otherwise, determine the total entry fee collected over the three days. Explain your answer using mathematical reasoning.

Complex unfamiliar

17. The energy content and amounts of fat and protein contained in each slice of bread and cheese and one teaspoon of margarine is shown in the table below.

Food	Energy content (kilojoules)	Fat (grams)	Protein (grams)
Bread	410	0.95	3.7
Cheese	292	5.5	1.6
Margarine	120	3.3	0.5

Pedro made toasted cheese sandwiches for himself and his friends for lunch. The total amount of fat and protein (in grams) for each of the three foods — bread, cheese and margarine — in the prepared lunch were recorded in the following matrix.

$$\begin{bmatrix} 7.6 & 29.6 \\ 44.0 & x \\ 13.2 & 2 \end{bmatrix}$$

a. Determine how many bread slices Pedro used.
b. If each sandwich used two pieces of bread, determine how many cheese sandwiches Pedro made.
c. i. Show that each sandwich had two slices of cheese.
   ii. Hence, calculate the exact value of *x*.
d. Construct a 1×3 matrix to represent the number of slices of bread and cheese and servings of margarine for each sandwich.

18. Tootin' Travel Agents sell three different types of train travel packages on the Midnight Express: Platinum, Gold and Red class. The price for each travel package is shown in the table.

Class	Price
Platinum	$3890
Gold	$2178
Red	$868

In one month, 62 Platinum packages, 125 Gold packages and 270 Red packages were sold.

a. Determine the total amount in dollars for train travel packages in the month. Explain using mathematical reasoning.

b. Travellers who book in a year in advance receive a 10% discount. For a specific month, 30, 45 and 60 customers have booked Platinum, Gold and Red packages respectively.
Determine the total amount in dollars corresponding to the discounted packages sold for that month.

19. TruSport owns two stores at the LeisureLand and SportLand shopping centres. The number of tennis racquets, baseball bats and soccer balls sold in the last week at the two stores is shown in the table below.

Store	Tennis racquets	Baseball bats	Soccer balls
LeisureLand	10	8	9
SportLand	9	12	11

The selling price of each item is shown in the table below.

	Tennis racquet	Baseball bat	Soccer ball
Selling price	$45.95	$25.50	$18.60

Determine the takings for each store.

20. There are 472 students studying History and 424 studying Economics at a university. At the end of the academic year, 25% of students will be awarded a Pass grade, 38% will be awarded a Credit grade, 19% will be awarded a Distinction grade, 8% will be awarded a High Distinction grade and the remaining students will not pass.

The column matrix $N = \begin{bmatrix} 472 \\ 424 \end{bmatrix}$ represents the number of students studying History and Economics.

a. Determine, using matrices, the number of students studying economics who will receive a distinction grade.

b. The cost for textbooks for a student studying History is $125; for Economics, the textbooks costs $235. Calculate, using matrices, the total cost for the textbooks.

Fully worked solutions for this chapter are available online.

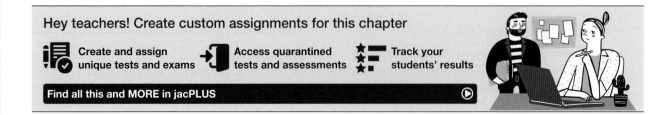

Hey teachers! Create custom assignments for this chapter

Create and assign unique tests and exams

Access quarantined tests and assessments

Track your students' results

Find all this and MORE in jacPLUS

Answers

Chapter 10 Matrices and matrix arithmetic

10.2 Types of matrices

10.2 Exercise

1. $\begin{bmatrix} 18 & 12 & 8 \\ 13 & 10 & 11 \end{bmatrix}$

2. a.
$$\begin{array}{c} \\ M \\ C \\ S \end{array} \begin{array}{ccc} M & C & S \end{array}$$
$$\begin{array}{c} M \\ C \\ S \end{array} \begin{bmatrix} 0 & 91.13 & 110.72 \\ 91.13 & 0 & 48.02 \\ 110.72 & 48.02 & 0 \end{bmatrix}$$

   b. $139.15

3. a. 545 km

   b. Coober Pedy and Alice Springs

   c. 2458 km

4. $\begin{bmatrix} 6 & 4 & 7 & 3 & 6 \\ 2 & 6 & 6 & 6 & 4 \end{bmatrix}$

5. $\begin{bmatrix} 45 \\ 30 \end{bmatrix}$, order 2×1

6. a. 56 b. 213 c. $\begin{bmatrix} 12 & 17 & 18 \end{bmatrix}$

7. A: 1×1, B: 3×1, C: 1×2

8. **a** and **d** are matrices with orders of 2×1 and 2×4 respectively. The matrix shown in **b** is incomplete, and the matrix shown in **c** has a different number of rows in each column.

9. a. 5 b. 6 c. 7

10. a. $\begin{bmatrix} 1 & 0 & 0 \\ 0 & 1 & 0 \\ 0 & 0 & 1 \end{bmatrix}$ b. $\begin{bmatrix} 0 & 0 \\ 0 & 0 \end{bmatrix}$

11. a. 3 b. -1 c. 1 d. 0.5 e. 0.9

12. a. There is no 4th column.

    b. e_{23}

    c. Nadia thought that e_{12} was read as 1st column, 2nd row. The correct value is 0.

13. $H = \begin{bmatrix} 4 & 3 \\ -1 & 7 \\ -4 & 6 \end{bmatrix}$; the order is 3×2.

14. $\begin{bmatrix} 212 \\ 0.318 \\ 10.48 \\ 0.22 \\ 1.93 \\ 1.15 \\ 8.78 \\ 30.32 \end{bmatrix}$

15. $\begin{bmatrix} 303\,186 & 4 \\ 68\,693 & 1 \\ 258\,411 & 5 \\ 48\,338 & 3 \\ 114\,312 & 4 \\ 31\,849 & 6 \\ 79\,695 & 27 \\ 9\,113 & 2 \end{bmatrix}$

    The total number of Aboriginal and Torres Strait Islander persons living in Australia is 913 597.

16. a. The zeros mean they don't fly from one place back to the same place.

    b. $175

    c. Mount Isa

    d.
$$\begin{array}{c} \\ O \\ B \\ D \\ M \end{array} \begin{array}{cccc} O & B & D & M \end{array}$$
$$\begin{array}{c} O \\ B \\ D \\ M \end{array} \begin{bmatrix} 0 & 59.50 & 150 & 190 \\ 89 & 0 & 85 & 75 \\ 175 & 205 & 0 & 213.75 \\ 307 & 90 & 75.75 & 0 \end{bmatrix}$$

17. N_{31} and N_{13}

18. a. There are no 1s in row A, just 0s.

    b.

		A	B	C	D	E
This question						
	A	0	0	0	0	1
	B	0	0	1	0	0
Next question	C	0	0	0	1	0
	D	1	0	0	0	0
	E	0	1	0	0	0

10.3 Operations with matrices

10.3 Exercise

1. a. $\begin{bmatrix} 1 & 6 \\ -1 & 3 \end{bmatrix}$ b. $\begin{bmatrix} -0.1 \\ 1.5 \\ 4.2 \end{bmatrix}$

2. $a = 4$, $b = -6$

3. a. $\begin{bmatrix} 1 \\ 7 \\ 0 \end{bmatrix}$ b. $\begin{bmatrix} -5 \\ 3 \\ 6 \end{bmatrix}$ c. $\begin{bmatrix} 6 \\ 4 \\ -6 \end{bmatrix}$ d. $\begin{bmatrix} 8 \\ 1 \\ 0 \end{bmatrix}$

4. $\begin{bmatrix} -1 & -2 \\ -1 & -5 \end{bmatrix}$

5. a. Both matrices must be of the same order for it to be possible to add and subtract them.

   b. $B = \begin{bmatrix} 4 \\ 1 \\ 5 \end{bmatrix}$

6. a. $\begin{bmatrix} 8 & 12 & 28 \\ 4 & 16 & 24 \end{bmatrix}$ b. $\begin{bmatrix} \frac{2}{5} & \frac{3}{5} & \frac{7}{5} \\ \frac{1}{5} & \frac{4}{5} & \frac{6}{5} \end{bmatrix}$

   c. $\begin{bmatrix} 0.6 & 0.9 & 2.1 \\ 0.3 & 1.2 & 1.8 \end{bmatrix}$ d. $\begin{bmatrix} -2 & -3 & -7 \\ -1 & -4 & -6 \end{bmatrix}$

7. C

8. a. $\begin{bmatrix} -0.25 & -0.95 & 0.3 \end{bmatrix}$

b. $\begin{bmatrix} 3 & -1 \\ 9 & 1 \end{bmatrix}$

c. $\begin{bmatrix} -1 & 5 & 1 \\ 4 & -2 & 3 \\ 3 & -3 & -2 \end{bmatrix}$

d. $\begin{bmatrix} 15 & 15 & 7 \\ 9 & 10 & 8 \end{bmatrix}$

9. $x = 2.5$

10. $a = -5, \ b = 2, \ c = 5$

11. $a = 6, \ b = -1, \ c = 5$

12. A and E have the same order, 1×2.
B and C have the same order, 2×1.

13. a. $\begin{bmatrix} 3 & 4 & 8 \\ 6 & 8 & 5 \end{bmatrix} - \begin{bmatrix} 2 & 3 & 5 \\ 4 & 6 & 3 \end{bmatrix}$

b. $\begin{bmatrix} 1 & 1 & 3 \\ 2 & 2 & 2 \end{bmatrix}$

Eggs	Small	Medium	Large
Free range	1	1	3
Barn laid	2	2	2

14. a. Both matrices are of the order 2×3; therefore, the answer matrix must also be of the order 2×3. Marco's answer matrix is of the order 2×1, which is incorrect.

b. Sample responses can be found in the worked solutions in the online resources. A possible response is:
Step 1: Check that all matrices are the same order.
Step 2: Add or subtract the corresponding elements.

15. a. $\begin{bmatrix} 12 \\ 9 \\ 8 \end{bmatrix} + \begin{bmatrix} -3 \\ -3 \\ 0 \end{bmatrix} = \begin{bmatrix} 9 \\ 6 \\ 8 \end{bmatrix}$ or $\begin{bmatrix} 12 \\ 9 \\ 8 \end{bmatrix} - \begin{bmatrix} 9 \\ 6 \\ 8 \end{bmatrix} = \begin{bmatrix} 3 \\ 3 \\ 0 \end{bmatrix}$

b. $\begin{bmatrix} 9 \\ 6 \\ 8 \end{bmatrix} + \begin{bmatrix} 3 \\ -2 \\ 2 \end{bmatrix} = \begin{bmatrix} 12 \\ 4 \\ 10 \end{bmatrix}$ or $\begin{bmatrix} 12 \\ 4 \\ 10 \end{bmatrix} - \begin{bmatrix} 9 \\ 6 \\ 8 \end{bmatrix} = \begin{bmatrix} 3 \\ -2 \\ 2 \end{bmatrix}$

16. $\begin{bmatrix} \dfrac{5}{8} & \dfrac{5}{8} & 0 \\ \dfrac{17}{30} & \dfrac{5}{42} & \dfrac{-5}{9} \\ \dfrac{11}{18} & \dfrac{1}{8} & \dfrac{-2}{9} \end{bmatrix}$

17. a. $\begin{bmatrix} 150 \\ 165 \\ 155 \\ 80 \end{bmatrix}$

b. i. $\begin{bmatrix} 150 \\ 165 \\ 155 \\ 80 \end{bmatrix} + \begin{bmatrix} 145 \\ 152 \\ 135 \\ 95 \end{bmatrix} + \begin{bmatrix} 166 \\ 155 \\ 156 \\ 110 \end{bmatrix}$

ii. $\begin{bmatrix} 461 \\ 472 \\ 446 \\ 285 \end{bmatrix}$

c. $\begin{bmatrix} 35 \\ 41 \\ 38 \\ 25 \end{bmatrix} + \begin{bmatrix} 32 \\ 36 \\ 35 \\ 30 \end{bmatrix} + \begin{bmatrix} 38 \\ 35 \\ 35 \\ 32 \end{bmatrix} = \begin{bmatrix} 105 \\ 112 \\ 108 \\ 87 \end{bmatrix}$

18. Sample responses can be found in the worked solutions in the online resources.

19. Sample responses can be found in the worked solutions in the online resources. Possible answer:

$\begin{bmatrix} = A_1 + D_1 = B_1 + E_1 = C_1 + F_1 \\ = A_2 + D_2 = B_2 + E_2 = C_2 + F_2 \end{bmatrix}$

20. $x = 7$

10.4 Matrix multiplication

10.4 Exercise

1. a. $1 \times \underbrace{2} \quad \underbrace{2} \times 1$

Number of columns = number of rows; therefore, XY exists and is of order 1×1.

b. $DE: 3 \times 3, \ DC: 3 \times 2, \ ED: 2 \times 2, \ CE: 2 \times 3$

2. $S: 3 \times 2, \ T: 2 \times 4$

3. $DG: 3 \times 2, \ FD: 2 \times 1, \ FE: 2 \times 2, \ EF: 3 \times 3, \ GF: 1 \times 3$

4. a. $MN = \begin{bmatrix} 28 & 48 \\ 21 & 36 \end{bmatrix}$

b. Yes, $[64]$ is of the order 1×1.

5. $PQ = \begin{bmatrix} 41 & 45 \\ 36 & 32 \end{bmatrix}$

6. a. $\begin{bmatrix} 70 & 105 \\ 20 & 30 \\ 90 & 135 \end{bmatrix}$ **b.** $\begin{bmatrix} 57 \\ 43 \end{bmatrix}$

c. $\begin{bmatrix} 18 & 32 & 46 \\ 9 & 16 & 23 \\ 10 & 17 & 17 \end{bmatrix}$ **d.** $\begin{bmatrix} 30 & 60 \\ 35 & 49 \end{bmatrix}$

e. $\begin{bmatrix} 7 & -4 & 15 \\ -13 & 3 & -10 \\ -8 & 3 & -9 \end{bmatrix}$ **f.** $\begin{bmatrix} 4 & -1 & 4 \\ -2 & -1 & -2 \\ -2 & 4 & -2 \end{bmatrix}$

7. a. $\begin{bmatrix} 4 & 6 \\ 2 & 3 \end{bmatrix}$ **b.** $\begin{bmatrix} 4 & 6 \\ 2 & 3 \end{bmatrix}$

c. When either A or B is the identity matrix

d. No. Consider matrix A with order of $m \times n$ and matrix B with order of $p \times q$, where $m \neq p$ and $n \neq q$. If AB exists, then it has order $m \times q$ and $n = p$. If BA exists, then it has order $p \times n$ and $q = m$. Therefore, $AB \neq BA$, unless $m = p$ and $n = q$, which is not possible since they are of different orders.

8. a. $I^2 = \begin{bmatrix} 1 & 0 & 0 \\ 0 & 1 & 0 \\ 0 & 0 & 1 \end{bmatrix}$

b. $I^3 = \begin{bmatrix} 1 & 0 & 0 \\ 0 & 1 & 0 \\ 0 & 0 & 1 \end{bmatrix}$

c. $I^4 = \begin{bmatrix} 1 & 0 & 0 \\ 0 & 1 & 0 \\ 0 & 0 & 1 \end{bmatrix}$

d. Whatever power you raise I to, the matrix stays the same.

9. $\begin{bmatrix} 72 & 30 \\ 60 & 57 \end{bmatrix}$

10. $\begin{bmatrix} 27 & 315 \\ 0 & 216 \end{bmatrix}$

11. a. $\begin{bmatrix} 4 & 8 \\ 7 & 2 \end{bmatrix}^4 = \begin{bmatrix} 7200 & 6336 \\ 5544 & 5616 \end{bmatrix}$

b. $\begin{bmatrix} 3 & 1 & 7 \\ 4 & 2 & 8 \\ 5 & 6 & 9 \end{bmatrix}^3 = \begin{bmatrix} 792 & 694 & 1540 \\ 984 & 868 & 1912 \\ 1356 & 1210 & 2632 \end{bmatrix}$

12. $t = -3$

13. $C_1 = 68$, $S_1 = 68$, $C_2 = 108$, $S_2 = 52$. The two results were $68 - 68$ and $108 - 52$.

14. a. $\begin{bmatrix} 0.05 & 0.18 & 0.45 & 0.25 & 0.07 \end{bmatrix}$

b. $\begin{bmatrix} 250 \\ 185 \end{bmatrix}$

c. i. The number of expected grades (A–E) for students studying Mathematics and Physics

ii. $\begin{bmatrix} 13 & 45 & 113 & 63 & 18 \\ 9 & 33 & 83 & 46 & 13 \end{bmatrix}$

d. 45 students studying maths are expected to be awarded a B grade.

15. a. Possible answer:
Represent the number of vehicles in a row matrix and the cost for each vehicle in a column matrix, then multiply the two matrices together. The product matrix will have an order of 1×1.

b. $\begin{bmatrix} 154\,000 \end{bmatrix}$ or $154\,000$

c. Possible answer:
In this multiplication, each vehicle is multiplied by the price of each type of vehicle, which is incorrect. For example, the ute is valued at $12\,500, but in this multiplication the eight utes sold are multiplied by $4000, $12\,500 and $8500 respectively.

16. a. $\begin{bmatrix} 12.50 \\ 8.50 \\ 6.00 \end{bmatrix}$

b. Total tickets requires an order of 1×1, and the order of the ticket price is 3×1. The number of people must be of order 1×3 to result in a product matrix of order 1×1. Therefore, the answer must be a row matrix.

c. $1662.50

17. a. $\begin{bmatrix} 12\,280 \\ 26\,210 \\ 29\,850 \end{bmatrix}$;

Friday $12\,280, Saturday $26\,210, Sunday $29\,850

b. $350 \times 35 + 456 \times 25 + 128 \times 20$

c. No, because you cannot multiply the entry price (3×1) by the number of people (3×3).

18. $n = 1$, $m = 3$, $s = 4$, $q = 2$

10.5 Applications of matrices

10.5 Exercise

1.
	A	B	C	D	E
A	0	1	0	0	1
B	1	0	1	1	1
C	0	1	0	1	0
D	0	1	1	0	0
E	1	1	0	0	0

2. a. C can talk to A, B and D.
b. B can receive calls from A, C and D.
c. Since they don't communicate with themselves
d. B cannot call C.

3. a. $690
b. Supersaver
c. Supersaver is the cheapest at a total of $46\,400.

4. a. 3
b. 6

5. a. Store A: $3875
Store B: $1860
Store C: $3100
Store D: $2325

b. Sales figures:
Store A: $4355
Store B: $2400
Store C: $3820
Store D: $2925

c. i. Store A: $3875
ii. Store C: $720

6. a. 3
b. No
c. They did not communicate with themselves.

d.

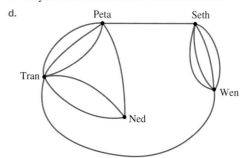

7. 2

8. a.

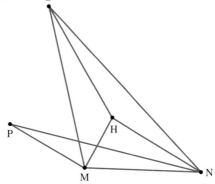

b.

	B	H	M	N	P
B	3	2	2	2	2
H	2	3	2	2	2
M	2	2	4	3	1
N	2	2	3	4	1
P	2	2	1	1	2

c. Raise the matrix to a power of 2.

d. Yes. Raise the matrix to a power of 4, as there are five cities in total.

9. a.i. A to C   ii. A to D to C
     D to C   A to E to C
     E to C

b. $A = \begin{bmatrix} 0 & 1 & 1 & 1 & 1 \\ 0 & 0 & 0 & 0 & 1 \\ 0 & 1 & 0 & 0 & 0 \\ 0 & 0 & 1 & 0 & 0 \\ 0 & 0 & 2 & 0 & 0 \end{bmatrix}$ $A^2 = \begin{bmatrix} 0 & 1 & 3 & 0 & 1 \\ 0 & 0 & 2 & 0 & 0 \\ 0 & 0 & 0 & 0 & 1 \\ 0 & 1 & 0 & 0 & 0 \\ 0 & 2 & 0 & 0 & 0 \end{bmatrix}$

10. $\begin{bmatrix} 0 & 1 & 1 & 0 \\ 0 & 0 & 1 & 1 \\ 1 & 0 & 0 & 1 \\ 1 & 1 & 1 & 0 \end{bmatrix}$

11. a. 1

b. Yes. The matrix raised to the power of 3 will provide the number of ways possible.

12. $J_1 = \begin{bmatrix} 175.5 \\ 260.9 \end{bmatrix}$; 127 Year 11 jackets and 233 Year 12 jackets

10.6 Review

10.6 Exercise

1. A

2. C

3. B

4. D

5. A

6. A

7. D

8. C

9. $\begin{bmatrix} 0 & 1 & 0 & 0 & 0 & 1 \\ 0 & 0 & 0 & 1 & 1 & 0 \\ 1 & 1 & 0 & 1 & 0 & 0 \\ 0 & 0 & 1 & 0 & 0 & 0 \\ 0 & 1 & 0 & 0 & 0 & 0 \\ 1 & 0 & 0 & 0 & 0 & 0 \end{bmatrix}$

10. a. $\begin{bmatrix} 4 \\ 2 \end{bmatrix}$ b. $\begin{bmatrix} -5 \\ -4 \end{bmatrix}$

     c. $\begin{bmatrix} 5 \\ 3 \end{bmatrix}$ d. $\begin{bmatrix} 3 & 4.5 \\ 1.5 & 6 \end{bmatrix}$

11. a. 3   b. 1×2

12. a. i. 1×2 and 2×1; product matrix: 1×1
     ii. 1×2 and 2×2; product matrix: 1×2
     iii. 2×2 and 2×2; product matrix: 2×2
     iv. 3×1 and 1×3; product matrix: 3×3

b. i. $\begin{bmatrix} 36 \end{bmatrix}$ ii. $\begin{bmatrix} 32 & 54 \end{bmatrix}$

     iii. $\begin{bmatrix} 42 & 101 \\ 55 & 35 \end{bmatrix}$ iv. $\begin{bmatrix} 10 & 6 & 8 \\ 35 & 21 & 28 \\ 40 & 24 & 32 \end{bmatrix}$

13. C

14. a. $\begin{bmatrix} 12.5 \\ 6 \\ 10 \end{bmatrix}$ b. $4000

15. $\begin{bmatrix} 125 \\ 134 \\ 167 \end{bmatrix}$

Possible answer:
Rhonda multiplied the first column with the first row, and then the second column with the second row.

16. $41 245

17. a. 8 slices of bread

b. 4 sandwiches

c. i. There are 4 sandwiches. A total of 44.0 g of fat means:
$5.5 \times x = 44.0 \therefore x = 8$
So 2 slices per sandwich.

     ii. 12.8

d. $\begin{bmatrix} 2 & 2 & 1 \end{bmatrix}$

18. a. $747 790

b. $240 111

19. LeisureLand: $830.90, SportLand: $924.15

20. a. 81 students studying Economics will receive Distinction grades.

b. $158 640

11 Univariate data analysis

LEARNING SEQUENCE

Fully worked solutions for this chapter are available online.

EXAM PREPARATION

Access exam-style questions in every lesson, available online.

on Resources

Solutions	Solutions — Chapter 11 (sol-1252)
Exam questions	Exam question booklet — Chapter 11 (eqb-0268)
Digital documents	Learning matrix — Chapter 11 (doc-41497)
	Chapter summary — Chapter 11 (doc-41498)

LESSON
11.1 Overview

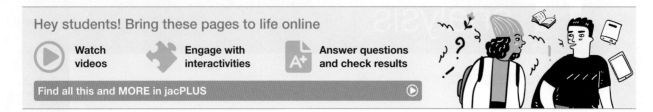
11.1.1 Introduction

The study and analysis of univariate data informs decision making in all areas of society: polls to determine voting patterns, the expected size of babies in the first year to determine good health, traffic data to determine a need for new roads and many more.

Data is published in the media and on government and business websites to explain or justify more or less spending. Scientists use data to decide whether to proceed with or abandon new drugs or technologies.

All of us must be able to work with and analyse sets of data to inform our decisions both personally and professionally.

The ability to analyse univariate data empowers individuals and organisations to make evidence-based decisions. This analytical skill is essential for driving progress and improvement in society, ensuring that choices are grounded in factual understanding and statistical evidence. As data becomes increasingly integral to all facets of life, developing proficiency in univariate analysis is crucial for informed personal and professional decision-making.

11.1.2 Syllabus links

Lesson	Lesson title	Syllabus links
11.2	**Classifying data and displaying categorical data**	◯ Understand the meaning of univariate data.
		◯ Classify a statistical variable as categorical or numerical.
		◯ Classify a categorical variable as ordinal or nominal and use tables and pie, bar and column charts to organise and display the data, e.g. ordinal: income level (high, medium, low); nominal: place of birth (Australia, overseas).
		◯ Classify a numerical variable as discrete or continuous, e.g. discrete: the number of people in a room; continuous: the temperature in degrees Celsius.
11.3	**Dot plots, stem-and-leaf plots, column charts and histograms**	◯ Select, construct and justify an appropriate graphical display to describe the distribution of a numerical dataset, including dot plot, stem-and-leaf plot, column chart and histogram.
11.4	**Measures of centre**	◯ Describe a graphical display in terms of the number of modes, shape (symmetric versus positively or negatively skewed), measures of centre and spread, and outliers, and interpret this information in the context of the data.
		◯ Understand and calculate the mean, median, mode, range and interquartile range (IQR) of a dataset, with and without technology. • Mean: $\bar{x} = \dfrac{\sum x}{n}$ • Median: $\left(\dfrac{n+1}{2} \right)^{\text{th}}$ data value
11.5	**Measures of spread**	◯ Understand and calculate the (sample) standard deviation, s_x, of a dataset, using technology only.
		◯ Use statistics as measures of centre and spread of a data distribution, being aware of their limitations.
		◯ Compare datasets in terms of mean, median, range, IQR and standard deviation, interpret the differences observed in the context of the data, and report the findings in a systematic and concise manner.

Source: General Mathematics Senior Syllabus 2024 © State of Queensland (QCAA) 2024; licensed under CC BY 4.0.

LESSON
11.2 Classifying data and displaying categorical data

SYLLABUS LINKS

- Understand the meaning of univariate data.
- Classify a statistical variable as categorical or numerical.
- Classify a categorical variable as ordinal or nominal and use tables and pie, bar and column charts to organise and display the data, e.g. ordinal: income level (high, medium, low); nominal: place of birth (Australia, overseas).
- Classify a numerical variable as discrete or continuous, e.g. discrete: the number of people in a room; continuous: the temperature in degrees Celsius.

Source: General Mathematics Senior Syllabus 2024 © State of Queensland (QCAA) 2024; licensed under CC BY 4.0.

11.2.1 Data types

Univariate data is used to describe a set of data that consists of observations or measurements on only a single characteristic or attribute of that data. For example, the number of attendees at each home game of the Brisbane Lions in one season or the fastest times recorded for the Gold Coast Marathon for the last 10 years.

Univariate data can be split into two major groups: **categorical data** and **numerical data**. Both of these can be further divided into two subgroups. The information in the flowchart shown can be used to determine the type of data being considered.

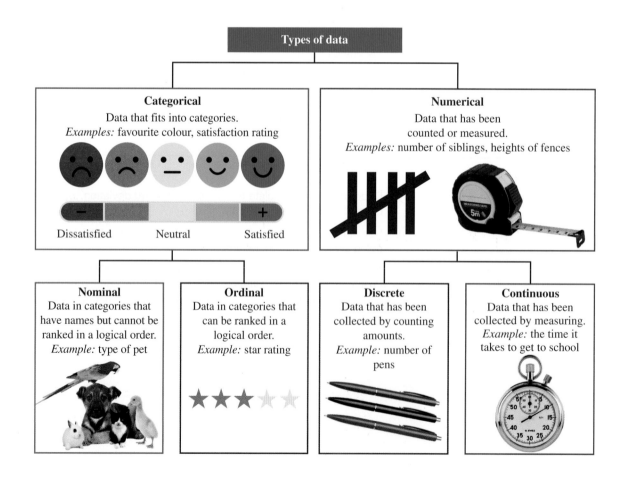

Categorical data

Data that can be organised into groups or categories is known as categorical data. Categorical data is often an 'object', 'thing' or 'idea', with examples including brand names, colours, general sizes and opinions. Categorical data can be classified as either ordinal or nominal.

Nominal data is data that can be grouped according to a particular characteristic, where the subgroups have no particular order or ranking. Examples include eye colour and make of car.

Ordinal data is data that can be grouped **and** ordered according to a particular characteristic. The subgroups can be placed into a natural order or ranking. Examples include fitness levels such as high, medium or low.

Numerical data

Data that can be counted or measured is known as numerical data, with examples including weights, number of children in a family, heights and number of cars. Numerical data can be classified as either discrete or continuous.

Discrete data is data that is counted as exact values, with the values usually being whole numbers, for example the number of people in a queue or the number of bedrooms in a house.

Continuous data is data that can be measured and have an infinite number of values, with an additional value always possible between any two given values, for example maximum daily temperature data or marathon finishing times.

WORKED EXAMPLE 1 Classifying a statistical variable as categorical or numerical

Classify the following statistical variables as categorical or numerical.

a. **Favourite colour of Year 11 students**
b. **Number of children in a family**
c. **Height of players in a netball team**
d. **Number of songs on a Spotify playlist**
e. **Varieties of fish caught in Moreton Bay**

THINK	WRITE
a. Favourite colour of Year 11 students would be a list of colours, so this is categorical data.	a. Categorical
b. The number of children in a family needs to be counted, so this is numerical data.	b. Numerical
c. The height of players in a netball team needs to be measured, so this is numerical data.	c. Numerical
d. The number of songs on a Spotify playlist needs to be counted, so this is numerical data.	d. Numerical
e. Varieties of fish caught in Moreton Bay would be a list of the varieties, so this is categorical data.	e. Categorical

WORKED EXAMPLE 2 Determining types of categorical data

Data on the different brands of cars on display in a car yard is collected. Classify the categorical data as ordinal or nominal.

THINK	WRITE
1. Identify the type of data.	The data collected is the brands of cars, so this is categorical data.
2. Identify whether there is a natural order or ranking of the subgroups.	When assessing the types of different cars, there is no natural order or ranking to the subgroups, so this is nominal data.
3. Write the answer.	The data collected is nominal data.

WORKED EXAMPLE 3 Determining types of numerical data

Data on the number of people attending matches at sporting venues is collected. Classify the numerical data as either discrete or continuous.

THINK	WRITE
1. Identify the type of data.	The data collected is the number of people in sporting venues, so this is numerical data because the attendees were counted.
2. Does the data have a restricted or infinite set of possible values?	The data involves counting people, so only whole number values are possible.
3. Write the answer.	The data collected is discrete data.

11.2.2 Displaying categorical data

Once raw data has been collected, it is helpful to summarise the information into a table or display. Categorical data is usually displayed in either **frequency tables**, **bar (column) charts** or **pie charts**. Both of these display the frequency (number of times) that a piece of data occurs in the collected data.

Frequency tables

Frequency tables split the collected data into defined categories and register the frequency of each category in a separate column. A tally column is often included to help count the frequency.

For example, if we have collected the following data about people's favourite colours, we could display it in a frequency table.

Red, Blue, Yellow, Red, Purple, Blue, Red, Yellow, Blue, Red

Favourite colour	Tally	Frequency
Red	||||	4
Blue	|||	3
Yellow	||	2
Purple	|	1

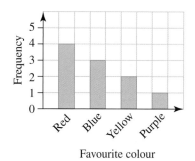

Bar or column charts

Bar charts (column charts) display the categories of data on the horizontal axis and the frequency of the data on the vertical axis. As the categories are distinct, there should be a space between all of the bars in the chart.

The bar chart above displays the previous data about favourite colours.

eles-3080

WORKED EXAMPLE 4 Displaying data in a bar chart

The number of students from a particular school who participate in organised sport on weekends is shown in the frequency table.
Display the data in a bar chart.

Sport	Frequency
Tennis	40
Swimming	30
Cricket	60
Basketball	50
No sport	70

THINK

1. Choose an appropriate scale for the bar chart. As the frequencies go up to 70 and all of the values are multiples of 10, we will mark our intervals in 10s. Display the different categories along the horizontal axis.

WRITE

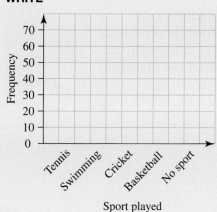

2. Draw bars to represent the frequency of each category, making sure there are spaces between the bars.

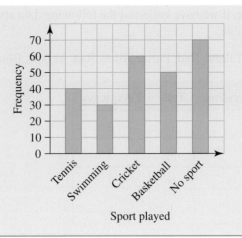

11.2.3 Pie charts

A pie chart is used when we want the graph to display a comparison of quantities. It shows the relative sizes of each quantity. The amount of each quantity is represented by a slice of the pie, calculated using a fraction of 360°.

eles-6400

WORKED EXAMPLE 5 Drawing a pie chart

The table shows the results of a survey on favourite sports. Draw a pie chart of the results.

Sport	Frequency
AFL	6
Basketball	2
Cricket	7
Netball	2
Rugby League	3
Rugby Union	1
Soccer	2
Tennis	1

THINK

1. Calculate each angle as a fraction of 360° by dividing the frequency of each sport by the total frequency and multiplying by 360°.

WRITE

$$AFL = \frac{6}{24} \times 360°$$
$$= 90°$$

$$Basketball = \frac{6}{24} \times 360°$$
$$= 30°$$

$$Cricket = \frac{7}{24} \times 360°$$
$$= 105°$$

$$Netball = \frac{2}{24} \times 360°$$
$$= 30°$$

$$Rugby\ League = \frac{3}{24} \times 360°$$
$$= 45°$$

$$Rugby\ Union = \frac{1}{24} \times 360°$$
$$= 15°$$

$$Soccer = \frac{2}{24} \times 360°$$
$$= 30°$$

$$Tennis = \frac{1}{24} \times 360°$$
$$= 15°$$

2. Construct the pie chart, labelling each section of the chart or providing a legend.

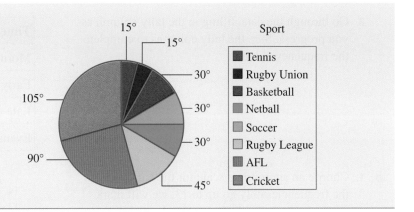

11.2.4 Analysing data — the mode

For categorical data, the **mode** is the category that has the highest frequency. When displaying categorical data in a bar chart, the modal category is the highest bar.

Identifying the mode allows us to know which category is the most common or most popular, which can be particularly useful when analysing data.

In some instances there may be either no modal category or more than one modal category. If the data has no modal category, then there is no mode; if it has 2 modal categories, then it is bimodal; and if it has 3 modal categories, it is trimodal.

WORKED EXAMPLE 6 Displaying and interpreting categorical data

Thirty students were asked to pick their favourite time of the day between the following categories: Morning (M), Early afternoon (A), Late afternoon (L), Evening (E)
The following data was collected:

A, E, L, E, M, L, E, A, E, M, E, L, E, A, L, M, E, E, L, M, E, A, E, M, L, L, E, E, A, E

a. **Represent the data in a frequency table.**
b. **Construct a bar chart to represent the data.**
c. **Determine which time of day is the most popular.**

THINK	WRITE
a. 1. Create a frequency table to capture the data.	a.

a.

Time of day	Tally	Frequency
Morning		
Early afternoon		
Late afternoon		
Evening		

▶

2. Go through the data, filling in the tally column as you progress. Sum the tally columns to complete the frequency column.

Time of day	Tally	Frequency
Morning	卌	5
Early afternoon	卌	5
Late afternoon	卌 ||	7
Evening	卌 卌 || |	13

b. 1. Choose an appropriate scale for the bar chart. As the frequencies only go up to 13, we will mark our intervals in single digits. Display the different categories along the horizontal axis.

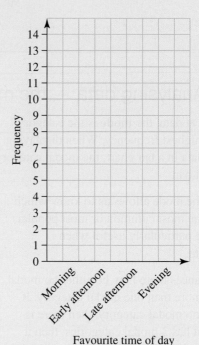

2. Draw bars to represent the frequency of each category, making sure there are spaces between the bars.

b.

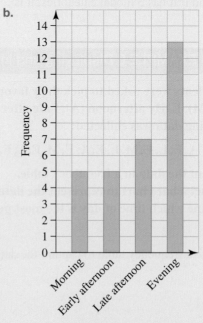

c. 1. The highest bar is the modal category. This is the most popular category. Write the answer.

c. Evening is the most popular time of day among the students.

Exercise 11.2 Classifying data and displaying categorical data **learn**on

11.2 Exercise	**11.2 Exam questions** on

Simple familiar	Complex familiar	Complex unfamiliar
1, 2, 3, 4, 5, 6, 7, 8, 9, 10, 11, 12, 13, 14	15, 16, 17, 18, 19, 20	N/A

Simple familiar

1. **WE1** Classify the following statistical variables as categorical or numerical.

   a. The weights of Brisbane Broncos players
   b. The brands of shoes worn by Olympic runners
   c. The types of plants on display in the botanical gardens
   d. The amount of soft drink consumed by teenagers each week
   e. The amount of time spent gaming each day

2. **WE2** Data on the different brands of cereal on supermarket shelves is collected.
   Classify the categorical data as either ordinal or nominal.

3. Data on the rating of hotels from 'one star' to 'five star' is collected.
   Classify the categorical data as either ordinal or nominal.

4. **WE3** Identify whether the following are categorical or numerical data and whether any numerical data is discrete or continuous.

   a. The amount of daily rainfall in Geelong
   b. The heights of players in the National Basketball League
   c. The number of children in families
   d. The type of pet owned by families

5. Identify whether the following categorical or numerical data is nominal, ordinal, discrete or continuous.

   a. The times taken for the place getters in the Olympic 100 metres sprinting final
   b. The number of gold medals won by countries competing at the Olympic Games
   c. The type of medals won by a country at the Olympic Games
   d. The countries that won at least one gold medal in any Olympics Games

6. **WE4** The preferred movie genre of 100 students is shown in the following frequency table.

Favourite movie genre	Frequency
Action	32
Comedy	19
Romance	13
Drama	15
Horror	7
Musical	4
Animation	10

Construct a bar chart to represent the data.

7. **WE5** Display the data in question 6 in a pie chart.

8. The favourite pizza type of 60 students is shown in the bar chart.

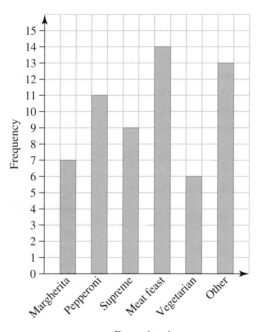

Favourite pizza

Construct a frequency table to represent the data.

9. A group of students at a university were surveyed about their usual method of travel, with the results shown in the following table.

Student	Transport method	Student	Transport method
A	Bus	N	Car
B	Walk	O	Bus
C	Train	P	Car
D	Bus	Q	Bus
E	Car	R	Bicycle
F	Bus	S	Car
G	Walk	T	Train
H	Bicycle	U	Bus
I	Bus	V	Walk
J	Car	W	Car
K	Car	X	Train
L	Train	Y	Bus
M	Bicycle	Z	Bus

a. Identify the type of data that is being collected.
b. Organise the data into a frequency table.
c. Construct a bar chart to represent the data.

10. Complete the following table by indicating the type of data:

Data	Type	
Example: The types of meat displayed in a butcher shop.	Categorical	Nominal
a. Wines rated as high, medium or low quality		
b. The number of downloads from a website		
c. Electricity usage over a three-month period		
d. The volume of petrol sold by a petrol station per day		

11. **WE6** Twenty-five students were asked to pick their favourite type of animal to keep as a pet. The following data was collected:

Dog, Cat, Cat, Rabbit, Dog,
Guinea pig, Dog, Cat, Cat, Rat,
Rabbit, Ferret, Dog, Guinea pig, Cat,
Rabbit, Rat, Dog, Dog, Rabbit,
Cat, Cat, Guinea pig, Cat, Dog

a. Construct a frequency table to represent the data.
b. Construct a bar chart to represent the data.
c. Identify which animal is the most popular.

12. The different types of coffee sold at a café in one hour are displayed in the bar chart.

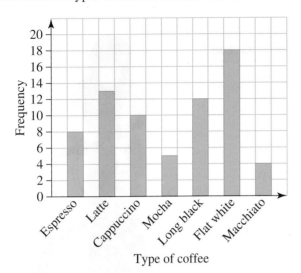

Type of coffee

a. Determine the modal category of the coffees sold.

b. Determine how many coffees were sold in that hour.

13. Exam results for a group of students are shown in the following table.

Student	Result	Student	Result	Student	Result	Student	Result
1	A	6	C	11	B	16	C
2	B	7	C	12	C	17	A
3	D	8	C	13	C	18	C
4	E	9	E	14	C	19	D
5	A	10	D	15	D	20	E

a. Construct a frequency table to represent the exam result data.

b. Construct a bar chart to represent the data.

c. Identify the type of data that has been collected.

14. The number of properties sold in the capital cities of Australia for a particular time period is shown in the following table.

City	Number of bedrooms			
	2	**3**	**4**	**5**
Adelaide	8	12	5	4
Brisbane	15	11	8	6
Canberra	8	12	9	2
Hobart	3	9	5	1
Melbourne	16	18	12	11
Sydney	23	19	15	9
Perth	7	9	12	3

Use the given information to construct a bar graph that represents the number of bedrooms of properties sold in the capital cities during this time period.

15. Thirty students were asked to pick their favourite type of music from the following categories: Pop (P), Rock (R), Classical (C), Folk (F), Electronic (E).
    The following data was collected:

    E, R, R, P, P, E, F, E, E, P, R, C, E, P, E,
    P, C, R, P, F, E, P, P, E, R, R, E, F, P, R

    a. Display the data in a suitable graph. Justify your choice.
    b. Calculate the percentage of students who chose the most popular type of music.

16. The results of an opinion survey are displayed in the bar chart.

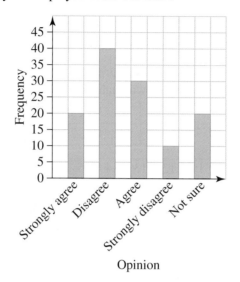

    a. Identify the type of data that is being displayed.
    b. Explain what is wrong with the current data display.
    c. Redraw the bar chart to display the data correctly.

17. The following frequency table displays the different categories of purchases in a shopping basket.

Category	Frequency
Fruit	6
Vegetables	8
Frozen goods	5
Packaged goods	11
Toiletries	3
Other	7

Calculate what percentage of the total purchases were fruit.

18. The birthplaces of 200 Australian citizens were recorded and are shown in the following frequency table.

Birthplace	Frequency
Australia	128
United Kingdom	14
India	10
China	9
Ireland	6
Other	33

Determine what percentage of the respondents were born in Australia.

19. The maximum daily temperatures (°C) in Adelaide during a 15-day period in February are listed in the following table.

Day	1	2	3	4	5	6	7	8
Temperature (°C)	31	32	40	42	32	34	41	29

Day	9	10	11	12	13	14	15
Temperature (°C)	25	33	34	24	22	24	30

Temperatures greater than or equal to 39 °C are considered above average and those less than 25 °C are considered below average.

a. Organise the data into three categories and display the results in a frequency table.
b. Identify the type of data in your frequency table and choose and construct a suitable display for the data.

20. The bar chart shown represents the ages of attendees at a local sporting event.

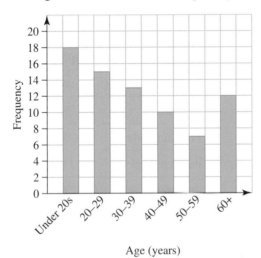

Age (years)

a. The age groups are changed to Under '20', '20–39', '40–59' and '60+'. Redraw the bar chart with these new categories.
b. Explain whether this changes the modal category.

Fully worked solutions for this chapter are available online.

LESSON
11.3 Dot plots, stem-and-leaf plots, column charts and histograms

SYLLABUS LINKS

- Select, construct and justify an appropriate graphical display to describe the distribution of a numerical dataset, including dot plot, stem-and-leaf plot, column chart and histogram.

Source: General Mathematics Senior Syllabus 2024 © State of Queensland (QCAA) 2024; licensed under CC BY 4.0.

11.3.1 Dot plots

Discrete numerical data can also be displayed as a **dot plot**. In these plots every data value is represented by a dot. The most common values can then be clearly identified. You can also use dot plots to represent categorical data.

When drawing a dot plot, be careful to make sure that the dots are evenly and consistently spaced.

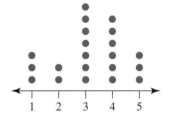

WORKED EXAMPLE 7 Constructing a dot plot

The frequency table shows the number of floors in apartment buildings in a particular area. Construct a dot plot to represent the data.

Number of floors	Frequency
2	2
3	5
4	3
5	0
6	4
7	2

THINK

1. Draw a horizontal scale using the discrete data values shown.

2. Place one dot directly above the number on the scale for each discrete data value present, making sure to keep corresponding dots at the same level.

WRITE

The discrete data values are given by the number of floors.

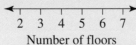
Number of floors

Number of floors

11.3.2 Stem-and-leaf plots

Stem-and-leaf plots can be used to display both discrete and continuous numerical data. The data is grouped according to its numerical place value (the 'stem') and then displayed horizontally as a single digit (the 'leaf'). In an ordered stem-and-leaf plot, the values are placed in numerical order with the smallest values closest to the stem.

If there are 4 or fewer different place values in your data, it may be preferable to make the stems of the plot represent a class interval or class size of 5 instead of a class interval or class size of 10. This can be done by inserting an asterisk (*) after the second of the stems with the same number, as shown in the following example.

Note that the data has been presented in neat vertical columns, making it easy to read.

Always remember to include a key with your stem-and-leaf plot to indicate what the stem and the leaf represent when put together.

Key: 1 |4 = 14
1*|7 = 17

Stem	Leaf
1	4
1*	7 7 9
2	2 3 4 4
2*	6 8
3	0 1 3
3*	5 6 9
4	0

eles-3169

WORKED EXAMPLE 8 Constructing a stem-and-leaf plot

The following data set (of 31 values) shows the maximum daily temperature (in °C) during the month of January in a particular area.

26, 22, 24, 26, 28, 28, 27, 42, 25, 25, 29, 31, 23, 33, 34, 27,

39, 44, 35, 34, 27, 30, 36, 30, 30, 28, 33, 23, 24, 34, 37

Construct a stem-and-leaf plot to represent the data.

THINK	WRITE
1. Identify the place values for the data. If there are 4 or fewer different place values, split each into two.	The temperature data has values in the 20s, 30s and 40s. Stem | Leaf 2 | 2* | 3 | 3* | 4 |
2. Write the units for each stem place value in numerical order, with the smallest values closest to the stem. Make sure to keep consecutive numbers level as they move away from the stem. Remember to add a key to your plot.	Key: 2 |2 = 22 °C Stem | Leaf 2 | 2 3 3 4 4 2* | 5 5 6 6 7 7 7 8 8 8 9 3 | 0 0 0 1 3 3 4 4 4 3* | 5 6 7 9 4 | 2 4

Note: We should not use a stem plot to represent a data if the range of values in the data set is large, or if the data values have a high number of units in them (ignoring decimal places), as these stem plots can become unwieldy and difficult to use.

11.3.3 Column charts

The types of display we choose to represent numerical data depend on whether that data is discrete or continuous. Representations of discrete data show distinct points that are not connected. On the other hand, continuous data displays have no gaps, as all possible values between the listed values are possible.

In section 11.2.2, column charts were used to display categorical data. Column charts can also be used to display discrete numerical data.

WORKED EXAMPLE 9 Constructing a column chart for discrete numerical data

The table below shows the results of the survey on the number of times a song is played on the radio over a weekend.
Construct a column chart to represent this information.

Song	Frequency
2	6
3	2
4	7
5	2
6	3
7	1
8	2
9	1

THINK

1. Draw the horizontal axis showing each song.
2. Draw a vertical axis to show frequencies up to 7.
3. Use a ruler to draw the columns all the same width with gaps between.
4. Label the axes.
5. Give the graph a title.

WRITE

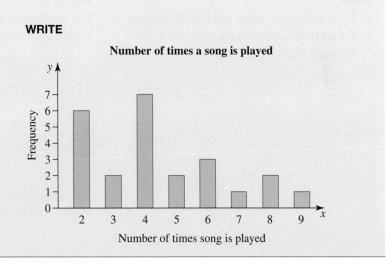

11.3.4 Histograms

We can represent continuous numerical data using a **histogram**, which is very similar to a column chart with a few essential differences.

In a histogram, the width of each column represents all values up to but not including the last value. The next column includes the last value of the previous column. The heights represent their frequencies.

For example, in the following histogram the first column represents the frequency of data values that are greater than or equal to 10 but less than 20 ($10 \leq x < 20$).

Data values	Frequency
10–< 20	10
20–< 30	20
30–< 40	30
40–< 50	35
50–< 60	45
60–< 70	20
70–< 80	15
80–< 90	5

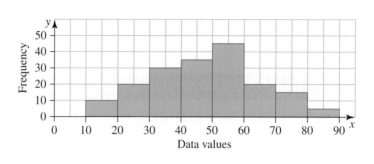

WORKED EXAMPLE 10 Constructing a histogram

eles-3081

The following frequency table represents the heights of players in a basketball squad.

Height (cm)	175–<180	180–<185	185–<190	190–<195	195–<200	200–<205
Frequency	1	3	6	3	1	1

Construct a histogram to represent this data.

THINK

1. Look at the data range and use the starting values from each interval in the table for the scale of the horizontal axis.

WRITE

The height data in the table has intervals starting from 175 cm and increasing by 5 cm.

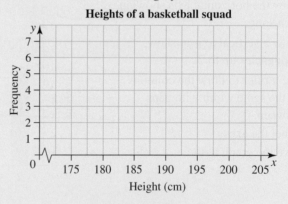

2. Draw rectangles for each interval to the height of the frequency indicated by the data in the table. (Remember to label the axes and give the graph a title.)

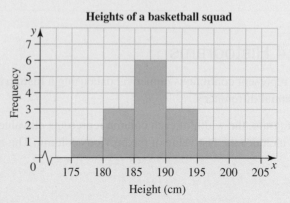

Exercise 11.3 Dot plots, stem-and-leaf plots, column charts and histograms

learnon

11.3 Exercise	**11.3 Exam questions** on

Simple familiar	Complex familiar	Complex unfamiliar
1, 2, 3, 4, 5, 6, 7, 8, 9	10, 11	12

These questions are even better in jacPLUS!

• Receive immediate feedback
• Access sample responses
• Track results and progress

Find all this and MORE in jacPLUS ▶

Simple familiar

1. **WE7** Construct a dot plot to represent each of the following collections of data.

a. The number of wickets per game taken by a bowler in a cricket season.

Number of wickets	Frequency
0	4
1	6
2	4
3	2
4	1
5	1

b. The number of hours per week spent checking emails by a group of workers at a particular company.

Hours checking emails	Frequency
1	1
2	1
3	2
4	4
5	8
6	4

2. Construct dot plots to represent the following collections of data.

a. The scores per round of a golfer over a particular time period (40 values):

73, 77, 74, 77, 73, 74, 72, 75, 72, 76, 77, 75, 74, 73, 75, 77, 78, 73, 77, 74,
72, 77, 73, 70, 72, 75, 73, 70, 77, 75, 77, 76, 70, 73, 75, 76, 78, 77, 74, 75

b. The scores out of 10 in a multiple choice test for a group of students (30 values):

6, 7, 4, 7, 3, 7, 7, 5, 7, 6, 7, 5, 1, 3, 5, 7, 8, 3, 7, 4, 9, 5, 4, 6, 7, 9, 10, 5, 7, 4

3. **WE8** Construct stem plots for the following sets of data.

a. The dollars spent per day on lunch by a group of 15 people:

22, 21, 22, 24, 19, 22, 24, 21, 22, 23, 25, 26, 22, 23, 22

b. The number of hours spent per week playing computer games by a group of 20 students at a particular school:

14, 21, 25, 7, 25, 20, 21, 14, 21, 20,
6, 23, 26, 23, 17, 13, 9, 24, 17, 24

4. **WE9** The following data represents the mark out of 10 achieved by 25 students on a quick quiz.

9, 3, 5, 6, 9, 2, 4, 5, 2, 7, 8, 9, 2, 4, 4, 5, 6, 8, 9, 2, 5, 6, 5, 9, 4

Construct a column chart to display this data.

5. The marks scored on a Maths exam, out of 100, by 25 Year 11 students are shown below.

87, 44, 95, 66, 78, 69, 66, 92, 78, 54, 60, 66, 69,
66, 77, 79, 66, 71, 71, 83, 74, 81, 69, 70, 57

a. Copy and complete the table.

Mark	Tally	Frequency
40–49		
50–59		
60–69		
70–79		
80–89		
90–99		

b. Construct a column graph to display the data from part **a**.

6. **WE10** The following frequency table represents the cholesterol levels measured for a group of people. Construct a histogram to represent this data.

Cholesterol level (mmol/L)	1–<2.5	2.5–<4.0	4.0–<5.5	5.5–<7.0	7.0–<8.5
Frequency	2	8	12	14	10

7. The following frequency table represents the distances travelled to school by a group of students. Construct a histogram to represent this data.

Distance travelled (km)	0–<2	2–<4	4–<6	6–<8	8–<10
Frequency	18	26	14	8	2

Questions 8 and 9 refer to the following.

Each student in a class has been assigned a newly planted tree to look after, and must provide a weekly report on its growth and condition. From the latest reports, the teacher recorded the height of each tree (in mm) and entered these in the stem-and-leaf plot shown below.

Key: 12|1 = 1210 mm
     12*|5 = 1250 mm

Stem	Leaf
12	1 2 4
12*	5 7 7 9 9
13	0 1 1 2 3 4 4
13*	5 6 6 7 9 9
14	0 2 3 4
14*	6 7

8. **MC** The class interval used in the stem-and-leaf plot is:
   A. 1   **B.** 10   **C.** 33   **D.** 50

9. **MC** The number of scores that have been recorded is:
   A. 21   **B.** 27   **C.** 33   **D.** 1210

Complex familiar

10. Consider the set of data in the stem plot shown.

Key: 0|1 = 1

Stem	Leaf
0	1
1	1 1 1 4 4 6 6 7 8
2	3 3 4 4 7 7 9

   a. Instead of grouping the data in 10s, the stems could be split in half to use groups of 5.
      Use the data from the original stem plot to complete the split stem plot.
   b. Comment on the effect of splitting the stem for the data in this question.

11. Organise the following data set into a frequency table justifying your choice of interval. Then draw a histogram to represent the data.

   5, 7, 14, 17, 13, 24, 22, 15, 12, 26, 17, 15, 14, 13, 15, 7, 8, 13, 17, 24,
   22, 7, 13, 20, 12, 15, 23, 20, 17, 15, 17, 16, 20, 23, 15, 16, 8, 17, 14, 15

12. Twenty transistors are tested by applying increasing voltage until they are destroyed. The maximum voltage that each could withstand is recorded below.

14.8 15.2 13.8 14.0 14.8 15.7 15.5 15.6 14.7 14.3
14.6 15.2 15.9 15.1 14.3 14.6 13.9 14.7 14.5 14.2

Prepare a stem-and-leaf plot of the data justifying your choice of interval and determine the percentage of transistors tested that could not withstand a voltage above 15 volts.

Fully worked solutions for this chapter are available online.

LESSON
11.4 Measures of centre

SYLLABUS LINKS

- Describe a graphical display in terms of the number of modes, shape (symmetric versus positively or negatively skewed), measures of centre and spread, and outliers, and interpret this information in the context of the data.
- Understand and calculate the mean, median, mode, range and interquartile range (IQR) of a dataset, with and without technology.
  - Mean: $\bar{x} = \dfrac{\sum x}{n}$
  - Median: $\left(\dfrac{n+1}{2}\right)^{\text{th}}$ data value

Source: General Mathematics Senior Syllabus 2024 © State of Queensland (QCAA) 2024; licensed under CC BY 4.0.

11.4.1 Describing distributions

The distribution of a set of data can be described in terms of a number of key features, including shape, modality, spread and outliers.

Shape

The shape of a numerical distribution is an important indicator of some of the key measures for further analysis and is one of the most important reasons for displaying the data in a graphical form. Shape will generally be described in terms of symmetry or skew.

Symmetrical data distributions have higher frequencies around their centres with a relatively evenly balanced spread to either side, whereas skewed distributions have the majority of their values towards one end.

Distributions with higher frequencies on the left side of the graph are positively skewed, whereas those with higher frequencies on the right side are negatively skewed.

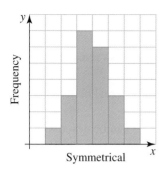

Symmetrical

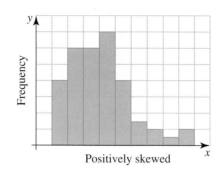

Positively skewed

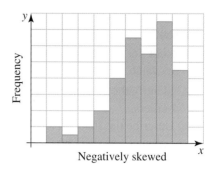

Negatively skewed

Modality

The mode of a distribution is the data value or class interval that has the highest frequency. This will be the column or row on the display that is the longest. When there is more than one mode, the data distribution is multimodal. This can indicate that there may be subgroups within the distribution that may require further investigation.

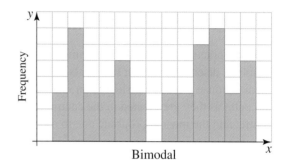

Bimodal

Bimodal distributions can occur when there are two distinct groups present, such as in data values that typically have clear differences between male and female measurements.

Spread

An awareness of how widely spread the data is can be an important consideration when conducting any further analysis. Common indicators of spread include the measures of **range** and the **standard deviation**. These measures of spread will be considered in more detail in the next lesson.

Outliers

An **outlier** is a data value that is an anomaly when compared to the majority of the sample. Sometimes outliers are just unusual readings or measurements, but they can also be the result of errors when recording the data.

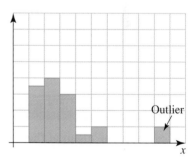

Outliers can have a significant effect on some of the measures that are used for further data analysis, and they are sometimes removed from the sample for those calculations. The graphical display of the data can alert us to the presence of potential outliers.

WORKED EXAMPLE 11 Describing a distribution

Describe the distribution of the data shown in the following histogram in terms of shape, modality and outliers.

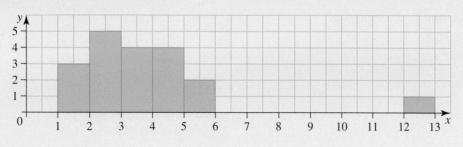

THINK	WRITE
1. Look for the mode and comment on its value.	The distribution has one mode with data values most frequently in the $2 \leq x < 3$ interval.
2. Identify the presence of any potential outliers.	There is one potential outlier in the interval between 12 and 13.
3. Describe the shape in terms of symmetry or skewness.	If we include the outlier, the data set can be described as positively skewed as it is clustered to the left. If we don't include the outlier, the distribution can be considered to be approximately symmetrical.

11.4.2 Measures of centre

In many practical settings, it is common to use a single measurement to represent an entire set of data. For example, discussions about fuel costs will often focus on the average price of petrol, while in real estate the median house price is considered an important measurement. These representative values are known as measures of centre as they are located in the central region of the data. The **mean**, **median** and **mode** are all measures of centre, and the most appropriate one to use depends on various characteristics of the data set.

The mean

The mean of a data set is what we commonly refer to as the average. It is calculated by dividing the sum of the data values by the number of data values. The symbol used for the mean is $\bar{x}$ (pronounced '*x*-bar').

Formula for mean

$$\bar{x} = \frac{\sum x_i}{n}$$

where $\sum$ (sigma) means 'the sum of', x_i are the data values and n is the number of data values.

---

eles-3082

WORKED EXAMPLE 12 Calculating the mean

Calculate the mean of the following data set, correct to 2 decimal places.

$$6, 3, 4, 5, 7, 7, 4, 8, 5, 10, 6, 10, 9, 8, 3, 6, 5, 4$$

THINK	WRITE
1. Calculate the sum of the data values.	$6 + 3 + 4 + 5 + 7 + 7 + 4 + 8 + 5 + 10 + 6 + 10 + 9 + 8 + 3 + 6 + 5 + 4 = 110$
2. Divide the sum by the number of data values.	$\bar{x} = \dfrac{\sum x_i}{n}$ $= \dfrac{110}{18}$ $= 6.111\ldots$
3. Write the answer.	The mean of the data set is 6.11.

Calculating the mean from a frequency table

To calculate the mean from a frequency table, add a new column and multiply each data value by its frequency. Sum the total value of this column and divide by the total frequency.

> **Formula for calculating the mean from a frequency table**
>
> $$\bar{x} = \frac{\sum x_i f}{\sum f}$$
>
> where $\sum$ means 'the sum of', f represents the frequencies and x_i represents the data values.

WORKED EXAMPLE 13 Calculating the mean from a frequency table

Calculate the mean of the data set displayed in the following frequency table. Give your answer correct to 2 decimal places.

Data value	Frequency
2	6
3	3
4	5
5	8
6	7
7	4

THINK

1. Add a column to the table and enter the product of the data values and the frequencies.

2. Calculate the totals of the frequency column and the xf column.

WRITE

Data value (x)	Frequency (f)	xf
2	6	12
3	3	9
4	5	20
5	8	40
6	7	42
7	4	28

Data value (x)	Frequency (f)	xf
2	6	12
3	3	9
4	5	20
5	8	40
6	7	42
7	4	28
	$\sum f = 33$	$\sum xf = 151$

3. Calculate the mean using the formula
$\bar{x} = \dfrac{\sum x_i f}{\sum f}$.

$$\bar{x} = \dfrac{\sum x_i f}{\sum f}$$

$$= \dfrac{151}{33}$$

$$= 4.55$$

4. Write the answer.

The mean of the distribution is 4.55.

Calculating the mean for grouped data

To calculate the mean from a table of data that has been organised into groups, we first need to calculate the midpoints of the intervals. To determine the midpoint, calculate the middle point of the corresponding interval. We then multiply the values of the midpoints by the corresponding frequencies and calculate the sum of these values. Finally, we divide this sum by the total of the frequencies.

Formula for calculating the mean of grouped data

$$\bar{x} = \dfrac{\sum x_m f}{\sum f}$$

where $\sum$ means the sum of, x_m represents the values of the midpoints and f represents the values of the frequencies.

eles-6401

WORKED EXAMPLE 14 Calculating the mean of a set of grouped data

Calculate the mean of the data set displayed in the following frequency table.

Interval	Frequency
0–< 5	3
5–< 10	12
10–< 15	3
15–< 20	2

THINK

1. Add a column to the table and calculate the midpoints for the corresponding intervals by adding the lowest and highest value of the interval and dividing by 2. For example, the midpoint for interval $0-<5 = \dfrac{0+5}{2} = 2.5$.

WRITE

Interval	Frequency (f)	Midpoint (x_m)
0–< 5	3	2.5
5–< 10	12	7.5
10–< 15	3	12.5
15–< 20	2	17.5

2. Add a column to the table and enter the product of the frequencies and midpoints (x_mf) for the corresponding intervals.

Interval	Frequency (f)	Midpoint (x_m)	x_mf
0–< 5	3	2.5	7.5
5–< 10	12	7.5	90
10–< 15	3	12.5	37.5
15–< 20	2	17.5	35

3. Calculate the totals of the f and x_mf columns.

Interval	Frequency	Midpoint (x_m)	x_mf
0–< 5	3	2.5	7.5
5–< 10	12	7.5	90
10–< 15	3	12.5	37.5
15–< 20	2	17.5	35
	$\sum f = 20$		$\sum fx = 170$

4. Calculate the mean using the formula $\bar{x} = \dfrac{\sum x_mf}{\sum f}$.

$$\bar{x} = \frac{\sum x_mf}{\sum f}$$
$$= \frac{170}{20}$$
$$= 8.5$$

5. Write the answer.

The mean of the distribution is 8.5.

11.4.3 The median

When considering a value that truly indicates the centre of a distribution, it would make sense to look at the number that is actually in the middle of the data set. The median of a distribution is the middle value of the ordered data set if there are an odd number of values when the data values are in ascending order.

If there are an even number of values, the median is halfway between the two middle values. It can be found using the following rule.

Calculating the median

When calculating the median, use the following steps:

1. **Arrange the data values in order (usually in ascending order).**

2. **The *position* of the median is the $\left(\dfrac{n+1}{2}\right)$th data value, where n is the total number of data values.**

Note: **If there is an even number of data values, then there will be two middle values. In this case, the median is the average of those data values.**

When there is an odd number of data values, the median is the middle value.

$$1 \qquad 1 \qquad 3 \qquad (4) \qquad 6 \qquad 7 \qquad 8$$

Median = 4

When there is an even number of data values, the median is the average of the two middle values.

$$2 \qquad 3 \qquad 3 \qquad (5) \qquad (6) \qquad 6 \qquad 7 \qquad 9$$

Median = $\dfrac{5+6}{2}$ = 5.5

eles-6402

WORKED EXAMPLE 15 Calculating the median

Calculate the medians of the following data sets.

a. $5, 3, 4, 5, 7, 7, 4, 8, 5, 10, 6, 10, 9, 8, 3, 6, 5, 4$

b. $16, 3, 4, 5, 17, 27, 14, 18, 15, 10, 6, 10, 9, 8, 23, 26, 35$

THINK	WRITE
a. 1. Put the data set in order from lowest to highest.	a. 3, 3, 4, 4, 4, 5, 5, 5, 5, 6, 6, 7, 7, 8, 8, 9, 10, 10
2. Identify the data value in the $\left(\dfrac{n+1}{2}\right)$ th position.	There are 18 data values, so the median will be in position $\left(\dfrac{18+1}{2}\right) = 9.5$, or halfway between the 9th and 10th data values.

3, 3, 4, 4, 4, 5, 5, 5, 5, 6, 6, 7, 7, 8, 8, 9, 10, 10

median = 5.5

3. Write the answer.	The median of the data set is 5.5.
b. 1. Put the data set in order from lowest to highest.	b. 3, 4, 5, 6, 8, 9, 10, 10, 14, 15, 16, 17, 18, 23, 26, 27, 35
2. Identify the data value in the $\left(\dfrac{n+1}{2}\right)$ th position.	There are 17 data values, so the median will be in position $\left(\dfrac{17+1}{2}\right) = 9$.

3, 4, 5, 6, 8, 9, 10, 10, 14, 15, 16, 17, 18, 23, 26, 27, 35

median = 14

3. Write the answer.	The median of the data set is 14.

11.4.4 Mode

The **mode** is the score that occurs most often. The data set can have no modes, one mode, two modes (bimodal) or multiple modes (multimodal).

466 Jacaranda Maths Quest 11 General Mathematics Units 1 & 2 for Queensland Second Edition

> **Calculating the mode**
>
> When determining the mode, use the following steps:
> 1. **Arrange the data values in ascending order (smallest to largest). This step is optional but helpful.**
> 2. **Look for the number that occurs most often (has the highest frequency).**

If no value in a data set appears more than once, then there is no mode. If a data set has multiple values that appear the most, then it has multiple modes. All values that appear the most are modes.

For example, the set 1, 2, 2, 4, 5, 5, 7 has two modes, 2 and 5.

WORKED EXAMPLE 16 Calculating the mode of a data set

Michelle is a real estate agent and has listed ten properties for sale in the last fortnight. She has listed the following numbers of bedrooms per house.

3, 2, 4, 5, 2, 2, 1, 2, 3, 2

Determine the mode for this data.

THINK	WRITE
1. Count the number of each score.	Score 1 (one bedroom) has a frequency of one. Score 2 (two bedrooms) has a frequency of five. Score 3 (three bedrooms) has a frequency of two. Score 4 (four bedrooms) has a frequency of one. Score 5 (five bedrooms) has a frequency of one.
2. Determine the highest frequency and the mode.	The highest frequency is 5, belonging to score 2 (two bedrooms).
3. Write the answer.	Therefore, the mode is 2.

11.4.5 Limitations of measures of centre

The mean

An important property of the mean is that it includes all the data in its calculation. As such, it has genuine credibility as a representative value for the distribution. On the other hand, this property also makes it susceptible to being adversely affected by the presence of extreme values when compared to the majority of the distribution.

Consider the data set: 3, 4, 5, 6, 7, 8, 9, which would have a mean of

$$\frac{3+4+5+6+7+8+9}{7} = 6.$$

Compare this to a data set with the same values with the exception of the largest one: 3, 4, 5, 6, 7, 8, 90, which has a mean of

$$\frac{3+4+5+6+7+8+90}{7} = 17.6.$$

As we can see, the mean has been significantly influenced by the one extreme value.

When the data is skewed or contains extreme values, the mean becomes less reliable as a measure of centre.

The median

In the previous example we saw that the mean is affected by an extreme value in the data set; however, the same cannot be said for the median. In both data sets the median will be the value in the fourth position.

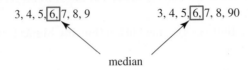

3, 4, 5, $\boxed{6,}$ 7, 8, 9   3, 4, 5, $\boxed{6,}$ 7, 8, 90

median

The median is therefore more reliable than the mean when the data is skewed or contains extreme values. Another potential advantage of median is that it is often one of the data values, whereas the mean often isn't. However, the median can be considered unrepresentative as it is not calculated by taking into account the actual values in the data set.

In most situations it is preferable to give both the median and the mean as a measure of centre, as between them they portray a more accurate picture of the data set. However, sometimes it is only possible to give one of these values to represent our data set.

Choosing between measures of centre

When choosing which measure of centre to use to represent a data set, take into account the distribution of the data. If the data has no outliers and is approximately symmetrical, then the mean is probably the better measure of centre to represent the data. If there are outliers, the median will be significantly less affected by these and would be a better choice to represent the data. The median is also a good choice to represent skewed data.

Also consider what each measure of centre tells you about the data. The values of the mean and the median can vary significantly, so choosing which one to represent the data set can be important, and you will need to justify your choice.

WORKED EXAMPLE 17 Determining the best measure of centre

The following histogram represents the IQ test results for a group of people.

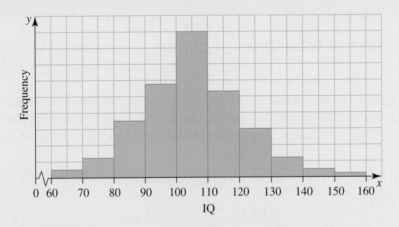

Determine which measure of centre is best to represent the data set.

THINK	WRITE
1. Look at the distribution of the data set.	The data set is approximately symmetrical and has no outliers.
2. If the data set is approximately symmetrical with no outliers, the mean is probably the better measure of centre to represent the data set. If there are outliers or the data is skewed, the median is probably the best measure of centre to use. Write the answer.	The mean is the better measure of centre to represent this data set.

 Resources

 Interactivity Mean, median, mode and quartiles (int-6496)

Exercise 11.4 Measures of centre

learn on

11.4 Exercise	**11.4 Exam questions** on

Simple familiar	Complex familiar	Complex unfamiliar
1, 2, 3, 4, 5, 6, 7, 8, 9, 10, 11, 12, 13, 14, 15	16, 17	18, 19, 20

These questions are even better in jacPLUS!
- Receive immediate feedback
- Access sample responses
- Track results and progress

Find all this and MORE in jacPLUS ▶

Simple familiar

1. **WE11** Describe the distribution of the data shown in the following histograms in terms of shape, modality and outliers.

a.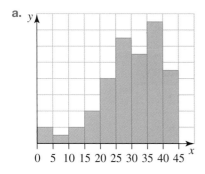

b.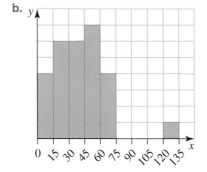

2. A group of 26 students received the following marks on a test:

6, 4, 3, 8, 6, 9, 5, 6, 9, 7, 7, 8, 5, 7, 4, 3, 8, 6, 5, 7, 9, 5, 6, 6, 7, 8

a. Construct a dot plot to display the data.
b. Describe the distribution in terms of shape, modality and outliers.

3. **WE12** Calculate the mean of the following data set.

108, 135, 120, 132, 113, 138, 125, 138, 107, 131,
113, 136, 119, 152, 134, 158, 136, 132, 113, 128

4. Calculate the means of the following data sets correct to 2 decimal places.

a. Key: $1|2 = 12$

Stem	Leaf
0	1 1 5 7
1	2 6
2	3 4 4 5
3	1 3
4	0 0 3
5	5
6	5

b.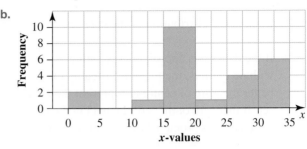

5. **WE13** Calculate the mean of the data set displayed in the following frequency table. Give your answer correct to 2 decimal places.

Data value	Frequency
6	4
7	1
9	3
11	6
12	2
13	4

6. Consider the following data set.

$$21, 22, 28, 21, 23, 24, 25, 21, 23, 24, 25, 28, 28, 21, 24, 22, 27, 21, 22, 24$$

a. Construct a frequency table for this data.
b. Calculate the mean of the data from the frequency table.

7. **WE14** Calculate the means of the data sets displayed in the following tables, giving your answer correct to 2 decimal places.

a.
Interval	Frequency
0–<5	12
5–<10	10
10–<15	1
15–<20	4

b.
Interval	Frequency
20–<35	12
35–<50	6
50–<65	13
65–<80	4

8. For each of the following sets of data, estimate the mean by creating a table using intervals that commence with the lowest data value and increase by an amount that is equal to the difference between the highest and lowest data value divided by 5. Give your answers correct to 2 decimal places.

a. 205, 203, 204, 205, 207, 216, 213, 218, 214, 220, 225, 220, 229, 228, 233, 238, 234

b. 5, 13, 24, 5, 27, 16, 13, 18, 24, 10, 5, 20, 30, 18, 13, 7, 14

9. **WE15** Calculate the medians of the following data sets.

   a. 15, 3, 54, 53, 27, 72, 41, 85, 15, 11, 62, 16, 49, 81, 53, 56, 75, 42

   b. 126, 301, 422, 567, 179, 267, 149, 198, 165, 170, 602, 180, 109, 85, 223, 206, 335

10. a. Calculate the median of the following data set.

    21, 22, 23, 24, 27, 26, 22, 27, 23, 21, 24, 20, 31, 25, 24, 28, 23

    b. Replace the highest value in the data set from part **a** with the number 96 and then calculate the median again.
    c. Describe how changing the highest value in the data set affects the median.

11. Calculate the medians of the following data sets.

    a.

    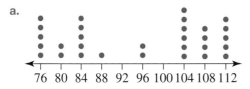

    b. 1.02, 2.01, 3.21, 4.63, 1.49, 3.45, 1.17, 1.38, 1.47, 1.70, 5.02, 1.38, 1.91, 8.54

12. **WE16** Determine the mode for the frequency distribution shown below.

    14, 14, 15, 13, 17, 13, 13, 14, 14, 14,
    16, 16, 17, 17, 16, 16, 16, 13, 14, 14,
    16, 17, 17, 14, 14, 15, 16, 16, 16, 16

13. Determine the modes of the following sets of data.

    a. 18, 21, 16, 19, 22, 21, 23, 21, 18, 18, 21, 19

    b. 144.1, 144.8, 144.5, 144.8, 144.9, 144.2, 144.9, 144.8, 144.7, 144.6, 144.3

14. **WE17** The following stem plot represents the lifespan of different animals at an animal sanctuary. Determine which measure of centre is better to represent the data set.

    Key: 1|2 = 12

Stem	Leaf
0	3 5 9
1	2 4 6 8
2	0 1 4 5 5 7 9
3	0 2 6
4	
5	
6	0 3

15. **a.** Calculate the mean (correct to 2 decimal places) and median for the following data set.

Average annual rainfall in selected Australian cities	
City	Rainfall (mm)
Sydney	1276
Melbourne	654
Brisbane	1194
Adelaide	563
Perth	745
Hobart	576
Darwin	1847
Canberra	630
Alice Springs	326

b. Determine which measure of centre would be the most appropriate to represent this data.

Complex familiar

16. The winning margins in the NRL over a particular period of time were as follows.

Winning margin	Frequency
2	4
4	12
6	8
8	5
10	4
12	4
16	1
20	1
34	1

a. Calculate the mean and the median.
b. Determine which measure of centre is the more appropriate measure for this data set and explain why.

17. Describe the distribution of the following data set in terms of shape, modality and outliers after drawing histograms with intervals of 10 commencing with the smallest values.

105, 70, 140, 127, 132, 124, 122, 125, 123, 126, 107, 105, 104, 113, 125, 70, 88, 103, 107,
124, 122, 76, 103, 120, 112, 115, 123, 120, 117, 115, 107, 106, 120, 123, 115, 74, 128, 119

18. The total number of games played by the players from two basketball squads is shown in the following stem plots.

Key: $0 \mid 1 = 1$ game played

Stem	Leaf
0	1
1	4 7
2	4 4 8
3	3 3 5 6
4	1 2 3
5	1 1
6	5
7	
8	
9	1

Key: $2 \mid 4 = 24$ games played

Stem	Leaf
2	4
3	1 2 6
4	3 4 5
5	2
6	
7	
8	2 5 7
9	3

Construct a back-to-back stem-and-leaf plot and compare the distributions. Use mathematical reasoning to justify your response.

19. Consider the following data set.

$$25, 23, 24, 25, 27, 26, 23, 28, 24, 20, 25, 20, 29, 28, 23, 27, 24$$

If the highest value in this data set was replaced with the number 79, describe how this would affect the mean. Use mathematical reasoning to justify your answer.

20. The following data set represents the salaries (in $000s) of workers at a small business.

$$45, 50, 55, 55, 55, 60, 65, 65, 70, 70, 75, 80, 220$$

When it comes to negotiating salaries, the workers want to use the mean to represent the data and the management want to use the median. Explain why this might be the case.

Fully worked solutions for this chapter are available online.

LESSON
11.5 Measures of spread

11.5.1 Measures of centre and measures of spread

Although measures of centre such as the mean or median give valuable information about a set of data, taken in isolation they can be quite misleading. Take for example the data sets {36, 43, 44, 59, 68} and {1, 2, 44, 80, 123}. Both groups have a mean of 50 and a median of 44, but the values in the second set are much further apart from each other.

Measures of centre tell us nothing about how variable the data values in a set might be; for this we need to consider the measures of spread of the data.

Range

In simplest terms, the spread of a data set can be determined by looking at the difference between the smallest and largest values. This is called the **range** of the distribution. Although the range is a useful calculation, it can also be limited. Any extreme values (outliers) will result in the range giving a false indication of the spread of the data.

> **Formula for range**
>
> **Range = largest value − smallest value**

Quartiles

A clearer picture of the spread of data can be obtained by looking at smaller sections. A common way to do this is to divide the data into quarters, known as **quartiles**. The word 'quartile' comes from the word 'quarter'.

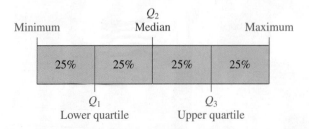

The **lower quartile** (Q_1) is the value that indicates the median of the lower half of the data.

The second quartile (Q_2) is the median of the distribution of data.

The **upper quartile** (Q_3) is the value that indicates the median of the upper half of the data.

When calculating the values of the lower and upper quartiles, the median should not be included. If the median is between values, then these values should be considered in your calculations.

The interquartile range

The **interquartile range** is found by calculating the difference between the third and first quartiles ($Q_3 - Q_1$), which gives an indication of the spread of the middle 50% of the data.

Formula for calculating the interquartile range (IQR)

Interquartile range (IQR) = upper quartile − lower quartile

$$= Q_{upper} - Q_{lower}$$
$$= Q_3 - Q_1$$

Note: The IQR is not affected by extremely large or extremely small data values (outliers), so in some circumstances the IQR is a better indicator of the spread of data than the range.

WORKED EXAMPLE 18 Calculating the interquartile range

Calculate the interquartile range of the following set of data.

23, 34, 67, 17, 34, 56, 19, 22, 24, 56, 56, 34, 23, 78, 22, 16, 15, 35, 45

THINK	WRITE
1. Put the data in order.	15, 16, 17, 19, 22, 22, 23, 23, 24, 34, 34, 34, 35, 45, 56, 56, 56, 67, 78
2. Identify the median.	There are 19 data values, so the median will be in position $\left(\dfrac{19+1}{2}\right) = 10$; that is, the median will be the 10th data value.

median
↓
15, 16, 17, 19, 22, 22, 23, 23, 24, (34)
The median is 34. |
| 3. Identify Q_1 by calculating the median of the lower half of the data. | There are 9 values in the lower half of the data, so Q_1 will be the 5th of these values.

Q_1
↓
15, 16, 17, 19, (22), 22, 23, 23, 24
$Q_1 = 22$ |
| 4. Identify Q_3 by finding the median of the upper half of the data. | There are 9 values in the upper half of the data, so Q_3 will be the 5th of these values.

Q_3
↓
34, 34, 35, 45, (56), 56, 56, 67, 78
$Q_3 = 56$ |

CHAPTER 11 Univariate data analysis **475**

5. Calculate the interquartile range using $IQR = Q_3 - Q_1$.

$$IQR = Q_3 - Q_1$$
$$= 56 - 22$$
$$= 34$$

6. Write the answer.

The interquartile range is 34.

11.5.2 Using technology

The measures of centre and spread that have been studied in this chapter can be determined with the use of technology. This can be particularly helpful where the data sets are larger and the individual data values are larger than or smaller than single digits.

WORKED EXAMPLE 19 Calculating the measures of centre, the range and the IQR using technology

Using technology, determine the mean, median, mode, range and IQR for the following data set.

$$1.83, 1.94, 1.98, 1.91, 1.88, 1.76, 2.12, 2.05, 2.11, 2.01,$$
$$2.04, 2.08, 2.07, 2.06, 2.05, 2.12, 1.94, 1.96, 2.12, 2.14$$

Give your answers correct to 2 decimal places.

THINK

1. First arrange the data in ascending order. Then use a scientific calculator to calculate the mean, median, mode, range and IQR.

WRITE

$1.76, 1.83, 1.88, 1.91, 1.94, 1.94, 1.96, 1.98, 2.01, 2.04,$
$2.05, 2.05, 2.06, 2.07, 2.08, 2.11, 2.12, 2.12, 2.12, 2.14$

```
(1.76+1.83+1.88▶
                  2.0085
```

```
(2.04+2.05)÷2
                  2.045
```

```
(1.94+1.94)÷2
                  1.94
```

```
(2.08+2.11)÷2
                  2.095
```

```
2.14-1.76
                  0.38
```

```
2.095-1.94
                  0.155
```

2. Write the answer.

Mean: 2.01
Median: 2.05
Mode: 2.12
Range: maximum value − minimum value
$= 2.14 - 1.76 = 0.38$
IQR: $Q_3 - Q_1 = 2.095 - 1.94 = 0.155$

11.5.3 Spread around the mean — the standard deviation

When the mean is used as a representative value for data, it makes sense to take note of how much the data varies in comparison to the mean. An indicator of the spread of data around the mean is the **standard deviation**. This measure applies to continuous numerical data. The larger the standard deviation, the more spread out the data is away from the mean.

Standard deviation

The standard deviation involves a high level of calculation, particularly with large groups of data. It can be easily computed using technology.

WORKED EXAMPLE 20 Calculating the standard deviation using technology

Using a scientific calculator, calculate the standard deviation for the following data set.
$$12, 23, 24, 16, 28, 21, 10, 18, 22, 29$$
Give your answers correct to 2 decimal places.

THINK	WRITE
1. Use a scientific calculator. Step 1: Use Mode 2 STAT to enter the values. Step 2: Enter all the values. Press touch AC, then press Shift STAT (above the 1) and select 4 Var and then 3 for the standard deviation.	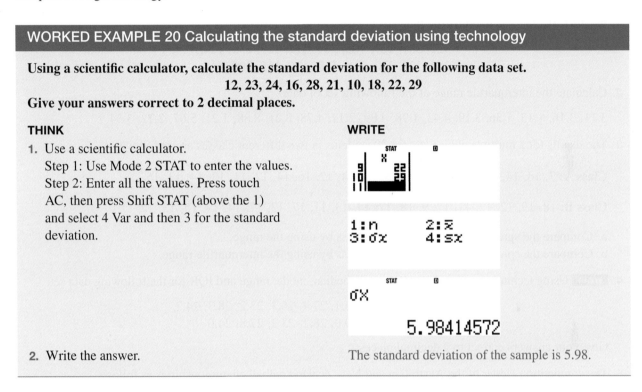
2. Write the answer.	The standard deviation of the sample is 5.98.

Preferred measures of spread

The standard deviation is generally considered the preferred measure of the spread of a distribution when there are no outliers and no skew, as all of the data contributes to its calculation. When there are outliers or the data is skewed, the interquartile range is a better option as it is not adversely influenced by extreme values.

As the interquartile range is calculated on the basis of just two numbers that may or may not be actual values from the data set, it could be considered to be unrepresentative of the data set.

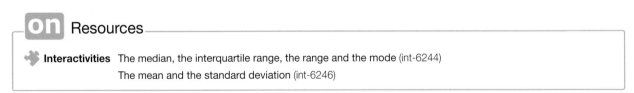

on Resources

Interactivities The median, the interquartile range, the range and the mode (int-6244)
The mean and the standard deviation (int-6246)

Exercise 11.5 Measures of spread

11.5 Exercise	11.5 Exam questions on

Simple familiar	Complex familiar	Complex unfamiliar
1, 2, 3, 4, 5, 6, 7, 8, 9	10, 11, 12, 13, 14	15

Simple familiar

1. **WE18** Calculate the interquartile range of the following set of data.

   421, 331, 127, 105, 309, 512, 129, 232, 124, 154,
   246, 124, 313, 218, 112, 136, 155, 305, 415

2. Calculate the interquartile range of the following set of data.

   3.11, 3.16, 1.13, 1.56, 3.19, 4.43, 1.98, 4.89, 2.12, 4.78, 3.21, 8.88, 1.21, 5.67, 2.22, 3.34

3. The results for a multiple choice test for 20 students in two different classes are as follows.

   Class A: 7, 13, 14, 13, 14, 14, 12, 8, 18, 13, 14, 12, 16, 14, 12, 11, 13, 14, 13, 15

   Class B: 18, 19, 12, 12, 11, 17, 9, 18, 17, 14, 13, 11, 17, 13, 17, 14, 14, 15, 13, 12

   a. Compare the spread of the marks for each class by using the range.
   b. Compare the spread of the marks for each class by using the interquartile range.

4. **WE19** Using technology, determine the mean, median, mode, range and IQR for the following data set.

   25.1, 23.2, 24.6, 25.1, 27.4, 26.3, 23.2, 28.3, 24.2,
   20.0, 25.6, 20.2, 29.0, 28.2, 23.2, 27.8, 24.0

   Give your answers correct to 2 decimal places.

5. The competition ladder of the Australian and New Zealand netball championship is as follows.

Position	Team	Win	Loss	Goals for	Goals against
1	Adelaide Thunderbirds	12	1	688	620
2	Melbourne Vixens	9	4	692	589
3	Waikato BOP Magic	9	4	749	650
4	Queensland Firebirds	9	4	793	691
5	Central Pulse	8	5	736	706
6	Southern Steel	6	7	812	790
7	West Coast Fever	5	8	715	757
8	NSW Swifts	4	9	652	672
9	Canterbury Tactix	2	11	700	882
10	Northern Mystics	1	12	699	879

   a. Using technology, calculate the spread for the 'Goals for' column by using the range.
   b. Using technology, calculate the spread for the 'Goals for' column by using the interquartile range.
   c. Compare the spread of the 'Goals for' column with the spread of the 'Goals against' column.

6. **WE20** Using a scientific calculator, calculate the standard deviation for the following data set.

$$30, 33, 45, 35, 26, 31, 47, 38, 39, 42, 30, 46$$

Give your answers correct to 2 decimal places.

7. Calculate the standard deviation for the following, correct to 3 decimal places.

$$1.32, 0.445, 3.102, 1.233, 0.344, 1.122, 0.998, 1.564, 0.213, 1.009$$

8. A survey of a large sample of people from particular areas of employment found the following average Australian salary ranges.

Employment area	Average minimum	Average maximum
Mining	$65 795	$262 733
Management	$66 701	$240 000
Engineering	$56 572	$233 451
Legal	$53 794	$193 235
Building and construction	$46 795	$186 412
Telecommunications	$47 354	$193 735
Science	$47 978	$211 823
Medical	$42 868	$228 806
Sales	$42 917	$176 783

a. Calculate the interquartile range for the average minimum salaries.
b. Calculate the interquartile range for the average maximum salaries.
c. Comment on the two interquartile ranges.

9. Answer the questions on the data in the following table. Where appropriate, give your answers correct to 2 decimal places.

Alcohol consumption per adult (litres)	
Country	Consumption per adult (litres)
Australia	10.21
Canada	10.01
France	12.48
Germany	12.14
Greece	11.01
Indonesia	0.56
Ireland	14.92
New Zealand	9.99
Russia	16.23
South Africa	10.16
Spain	11.83
Sri Lanka	0.81
United Kingdom	13.24
United States	9.7
Yemen	0.2

a. Calculate the interquartile range and standard deviation for the data set correct to 2 decimal places.
b. Calculate the interquartile range and standard deviation after removing the three lowest values, correct to 2 decimal places.
c. Compare the results from parts a and b.

Complex familiar

10. A survey of the number of motor vehicles that pass a school between 8:30 am and 9:30 am on 10 days during a term are as follows.

$$72, 89, 94, 78, 83, 84, 88, 97, 82, 88$$

If the lowest number is reduced by 10 and the highest value is increased by 10, describe how the standard deviation and the interquartile range will be affected by the change to the values. Use mathematical reasoning to justify your response.

11. Consider the following data.

Carbon dioxide emissions (million metric tons of carbon dioxide)						
Country	2001	2002	2003	2004	2005	2006
Australia	374.05	382.65	380.68	391.03	416.89	417.06
Canada	553.55	573.25	602.46	614.69	632.01	614.33
China	3107.99	3440.60	4061.64	4847.33	5429.30	6017.69
Germany	877.71	857.35	874.04	871.88	852.57	857.60
India	1035.42	1033.52	1048.11	1151.33	1194.01	1293.17
Indonesia	300.18	314.88	305.44	323.29	323.51	280.36
Japan	1197.15	1203.33	1253.29	1257.89	1249.62	1246.76
Russia	1571.14	1571.77	1626.86	1663.44	1698.56	1704.36
United Kingdom	575.19	563.89	575.17	582.29	584.65	585.71
United States	5762.33	5823.80	5877.73	5969.28	5994.29	5902.75

When comparing the Australian data with the data for India, China, the United Kingdom and the United States, identify which measure of spread is more appropriate — interquartile range or standard deviation.

12. A sample of crime statistics over a two-year period are shown in the following table.

Crime	Year 1	Year 2
Theft from motor vehicle	46 700	42 900
Theft (steal from shop)	19 800	20 600
Theft of motor vehicle	15 650	14 670
Theft of bicycle	4 200	4 660
Theft (other)	50 965	50 650

Apply your understanding of the interquartile range and standard deviation to explain the effect that removing the smallest category of crime for each year will have on the interquartile range and the standard deviation. Use mathematics to demonstrate by example.

13. Data collected on the number of daylight hours in Alice Springs is as shown.

10.3, 9.8, 9.6, 9.5, 8.5, 8.4, 9.1, 9.8, 10.0, 10.0, 10.1, 10.0, 10.1, 10.1, 10.6, 8.7, 8.8, 9.0, 8.0, 8.5, 10.6, 10.8, 10.5, 10.9, 8.5, 9.5, 9.3, 9.0, 9.4, 10.6, 8.3, 9.3, 9.0, 10.3, 8.4, 8.9

Calculate the range and interquartile range of the data, identify the difference between them, and explain what this indicates about the data.

14. The table shows the number of registered passenger vehicles in two particular years for the states and territories of Australia.

Number of passenger vehicles		
	Year 1	Year 2
New South Wales	3 395 905	3 877 515
Victoria	2 997 856	3 446 548
Queensland	2 138 364	2 556 581
South Australia	915 059	1 016 590
Western Australia	1 205 266	1 476 743
Tasmania	271 365	305 913
Northern Territory	73 302	91 071
Australian Capital Territory	191 763	229 060

Describe and contrast the effect of the removal of the three smallest values on the interquartile ranges and standard deviations.

Complex unfamiliar

15. The volume of wine ('000 litres) available for consumption in Australia for a random selection of months over a 10-year time period is shown in the following table.

38 595	41 301	44 212	39 362	38 914	38 273	39 456	38 823
41 123	42 981	44 567	41 675	41 365	42 845	43 987	41 583
39 347	42 673	44 835	39 773	38 586	38 833	39 756	39 095
42 946	46 382	44 892	41 038	41 402	42 587	43 689	41 209

Determine what percentage, correct to 2 decimal places, of the actual data values from the sample are within 1 standard deviation of the mean (i.e. between the number obtained by subtracting the standard deviation from the mean and the number obtained by adding the standard deviation to the mean).

Fully worked solutions for this chapter are available online.

LESSON
11.6 Review

11.6.1 Summary

doc-
41498

 Hey students! Now that it's time to revise this chapter, go online to:

Access the
chapter summary

 Review your
results

 Practise exam
questions

Find all this and MORE in jacPLUS

11.6 Exercise

learnon

| 11.6 Exercise | 11.6 Exam questions on |

These questions are
even better in jacPLUS!
- Receive immediate feedback
- Access sample responses
- Track results and progress

Find all this and MORE in jacPLUS

Simple familiar	Complex familiar	Complex unfamiliar
1, 2, 3, 4, 5, 6, 7, 8, 9, 10, 11, 12	13, 14, 15, 16	17, 18, 19, 20

Simple familiar

1. **MC** The interquartile range of the data distribution shown in the stem plot is:

Key: 2 | 6 = 26

Stem	Leaf
0	2
1	1 5
2	6 6 7 8
3	8 8 9
4	3 4
5	2

A. 41 **B.** 50 **C.** 28 **D.** 20.5

2. **MC** The mean of the data distribution shown in the table is:

Interval	Frequency (f)
0–<15	5
15–<30	7
30–<45	6
45–<60	2

A. 22.4 **B.** 26.25 **C.** 24.35 **D.** 25.65

3. **MC** Data gathered on the number of home runs in a baseball season would be classified as:

A. discrete
B. nominal
C. continuous
D. ordinal

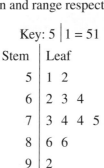

4. **MC** For the sample data set 2, 3, 5, 2, 3, 6, 3, 8, 9, 2, 8, 9, 2, 6, 7, the mean and standard deviation respectively would be closest to:

A. 5 and 6
B. 5 and 2.6
C. 2.6 and 5
D. 5 and 2.7

5. **MC** For the following stem plot, the median and range respectively are:

Key: $5 \mid 1 = 51$

Stem	Leaf
5	1 2
6	2 3 4
7	3 4 4 5
8	6 6
9	2

A. 73 and 41
B. 73.5 and 41
C. 71 and 39
D. 71 and 41

6. State whether each of the following data types is categorical or numerical.

a. The television program that people watch at 7:00 pm
b. The number of pets in each household
c. The amount of water consumed by athletes in a marathon run
d. The average distance that students live from school

7. For each of the numerical data types below, determine if the data are discrete or continuous.

a. The dress sizes of Year 11 girls
b. The volume of backyard swimming pools
c. The amount of water used in households
d. The number of viewers of a particular television program

8. A group of Year 11 students was asked to state the number of movies that they had downloaded in the last year. The results are shown below.

$$12, 1, 13, 20, 5, 22, 35, 12, 17, 20,$$

$$9, 5, 11, 0, 14, 25, 3, 8, 10, 9,$$

$$12, 6, 18, 7, 10, 9, 6, 23, 14, 19$$

a. Put the results into a table using the categories 0–4, 5–9, 10–14 etc.
b. Draw a column graph to represent the results.

9. The data below gives the number of errors made each week by 20 machine operators. Prepare a stem-and-leaf plot of the data.

$$6, 15, 20, 25, 28, 18, 32, 43, 52, 27,$$

$$17, 26, 38, 31, 26, 29, 32, 46, 13, 20$$

10. The time taken (in seconds) for a test vehicle to accelerate from 0 to 100 km/h is recorded during a test of 24 trials. The results are represented by the stem-and-leaf plot shown.
Using technology, determine the median of the data.

Key: $7 | 2 = 7.2$ s
$7* | 6 = 7.6$ s

Stem	Leaf
7	2 4 4
7*	5 5 7 9
8	0 0 1 2 4 4 4
8*	5 5 6 8 9
9	2 2 3
9*	5 7

11. The stem-and-leaf plot shown gives the exact mass of 24 packets of biscuits. Using technology, determine the mean and range of the data.

Key: $248 | 4 = 248.4$ g

Stem	Leaf
248	4 7 8
249	2 3 6 6
250	0 0 1 1 6 9 9
251	1 5 5 5 6 7
252	1 5 8
253	0

12. The frequency table below shows the crowds at football matches for a team over a season.

Interval	Midpoint	Frequency
5000–9999		1
10 000–14 999		5
15 000–19 999		9
20 000–24 999		3
25 000–29 999		2
30 000–34 999		2

a. Copy the frequency table and complete the midpoint column.
b. Show the information in a frequency histogram.

Complex familiar

13. Leesa ran a mean time of 10.5 seconds in her previous four trials for the one-hundred-metre sprint. If the times for three of the trials were 9.8 seconds, 11.1 seconds and 8.8 seconds, determine Leesa's time for the fourth trial.

14. The table below shows the number of sales made each day over a month in a car yard.

Number of sales	Frequency
0	2
1	5
2	12
3	6
4	2
5	0
6	1

Choose an appropriate graphical display with justification for this data and hence describe the distribution.

15. A teacher decided to display a set of students' scores out of a possible 50 marks on a stem-and-leaf graph and obtained the following graph.

Stem | Leaf
2 | 1 1 3 4 8 8 8 9 9
3 | 0 3 4 5 5 5 6 8 8 8 8 9
4 | 0 0 0 1 1 2 2 4 5 5 5 6 8

Describe the distribution in the context of the situation.
Determine a way of displaying the data that would give the teacher a better understanding of the students' results. Describe this new graph.

16. If the median of a set of data is the 35th data value, determine how many data values are there in the set.

Complex unfamiliar

17. The histogram shown displays the annual earnings of a group of professional tennis players.

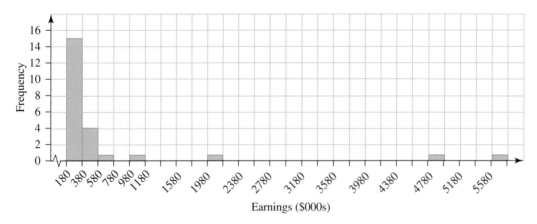

Determine the approximate positions of the mean and median on this histogram, using mathematical reasoning to justify your answer.

18. The histogram shown displays the body mass index (BMI) for a group of people.

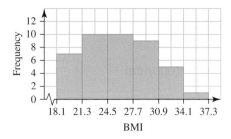

Determine the mean BMI for this group and describe the shape of the histogram.

19. The formula for the standard deviation of a set of data is

$$s = \sqrt{\frac{\sum (x_i - \overline{x})^2}{n - 1}}$$

For a particular set of data, $s = 5.2$ and $\sum (x - \overline{x})^2 = 1352$. Determine the number of data values in the set.

20. Consider the following data set.

$$4, 18, 35, 26, 12, 25, 21, 34, 43, 37,$$

$$6, 25, 25, 23, 34, 38, 37, 22, 36, 31$$

Determine what percentage of the data set lies within 2 standard deviations of the mean.

Fully worked solutions for this chapter are available online.

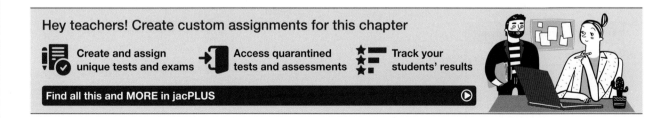

Answers

Chapter 11 Univariate data analysis

11.2 Classifying data and displaying categorical data

11.2 Exercise

1. a. Numerical
   b. Categorical
   c. Categorical
   d. Numerical
   e. Numerical

2. Nominal

3. Categorical, ordinal

4. a. Numerical and continuous
   b. Numerical and continuous
   c. Numerical and discrete
   d. Categorical

5. a. Continuous   b. Discrete
   c. Ordinal   d. Nominal

6.

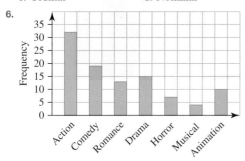

Favourite movie genre

7.

Favourite movie genre

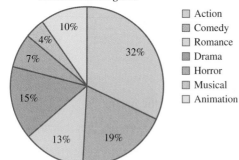

- Action
- Comedy
- Romance
- Drama
- Horror
- Musical
- Animation

8.

Favourite pizza	Frequency
Margherita	7
Pepperoni	11
Supreme	9
Meat feast	14
Vegetarian	6
Other	13

9. a. Nominal categorical

   b.

Transport method	Frequency
Bus	9
Walk	3
Train	4
Car	7
Bicycle	3

c.

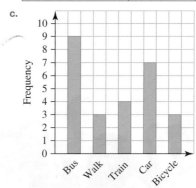

Transport method

10.

	Data	Type	
a	Wines rated as high, medium or low quality	Categorical	Ordinal
b	The number of downloads from a website	Numerical	Discrete
c	The electricity usage over a three-month period	Numerical	Continuous
d	A volume of petrol sold by a petrol station per day	Numerical	Continuous

11. a.

Favourite animal	Frequency
Dog	7
Cat	8
Rabbit	4
Guinea pig	3
Rat	2
Ferret	1

b.

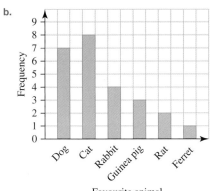

Favourite animal

c. Cat

12. a. Flat white   **b.** 70

13. a.

Result	Frequency
A	3
B	2
C	8
D	4
E	3

b.

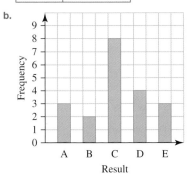

Result

c. Ordinal categorical

14.

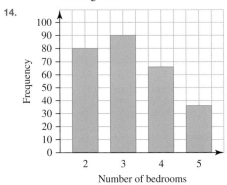

Number of bedrooms

15. a. This is categorical data, so a pie graph is a suitable diplay.

Type of music

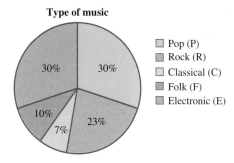

- ☐ Pop (P)
- ▨ Rock (R)
- ☐ Classical (C)
- ▨ Folk (F)
- ▨ Electronic (E)

b. The most popular music is pop. 30%

16. a. Ordinal categorical

b. The data should be in order from 'Strongly agree' through to 'Strongly disagree'.

c.

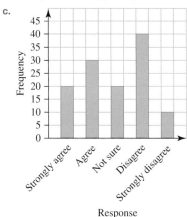

Response

17. 15%

18. 64%

19. a.

Temperature	Frequency
Above average	3
Average	9
Below average	3

b.

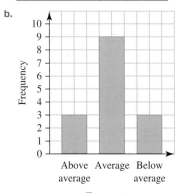

Temperature

Ordinal categorical

20. a.

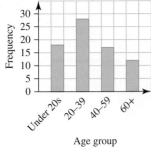

Age group

b. Yes, the modal category is now 20–39.

11.3 Dot plots, stem-and-leaf plots, column charts and histograms

11.3 Exercise

1. a.

Number of wickets

b.

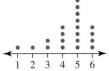

Hours checking email

2. a.

Round score

b.

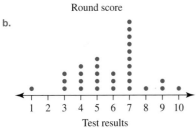

Test results

3. a. Key: 1*|9 = $19

```
Stem | Leaf
1*   | 9
2    | 1 1 2 2 2 2 2 2 3 3 4 4
2*   | 5 6
```

b. Key: 0*|6 = 6 hours

```
Stem | Leaf
0*   | 6 7 9
1    | 3 4 4
1*   | 7 7
2    | 0 0 1 1 1 3 3 4 4
2*   | 5 5 6
```

4.

Mark on quick quiz

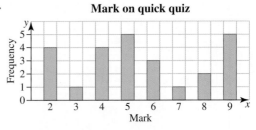

Mark

5. a.

Mark	Tally	Frequency
40–49	|	1
50–59	||	2
60–69	|||| ||||	9
70–79	|||| |||	8
80–89	|| |	3
90–99	||	2

b.

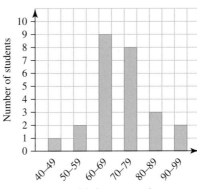

Maths exam mark

6.

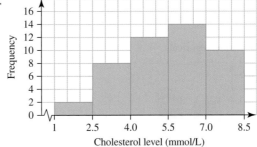

Cholesterol level (mmol/L)

7.

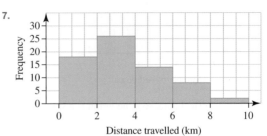

Distance travelled (km)

8. D

9. D

10. a. Key: 0|1 = 1

Stem	Leaf
0	1
0*	
1	1 1 1 4 4
1*	6 6 7 8
2	3 3 4 4
2*	7 7 9

b. Splitting the stem for this data gives a clearer picture of the spread and shape of the distribution of the data set.

11.

Class interval	Frequency
5– < 10	6
10– < 15	9
15– < 20	15
20– < 25	9
25– < 30	1

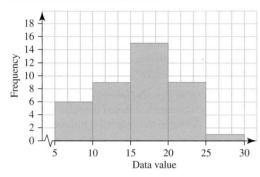

12. Key: 14|3 = 14.3 V   14*|8 = 14.8 V

Stem	Leaf
13*	8 9
14	0 2 3 3
14*	5 6 6 7 7 8 8
15	1 2 2
15*	5 6 7 9

The percentage of transistors tested that could not withstand a voltage above 15 volts is 65%.
Intervals of 0.5 were chosen as intervals of 1.0 would not have shown as clear a picture of the spread of data.

11.4 Measures of centre

11.4 Exercise

1. a. The distribution has one mode with data values that are most frequent in the 35– < 40 interval. There are no obvious outliers, and there is a negative skew to the distribution.

b. The distribution has one mode with data values that are most frequent in the 45– < 60 interval. There are potential outliers in the 120– < 135 interval, and the distribution is either symmetrical (excluding the outliers) or has a slight positive skew (including the outliers).

2. a.

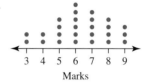

Marks

b. The distribution has one mode with a value of 6. There are no obvious outliers and there is a slight negative skew to the distribution.

3. 128.4

4. a. 26.18   **b.** 21.67

5. 10

6. a.

Data value	Frequency
21	5
22	3
23	2
24	4
25	2
27	1
28	3

b. 23.7

7. a. 6.94   **b.** 46.36

8. a. See the table at the bottom of the page.*

---

*8. a

Interval	Frequency (f)	Midpoint (x)	xf
203– < 210	5	206.5	1032.5
210– < 217	3	213.5	640.5
217– < 224	3	220.5	661.5
224– < 231	3	227.5	682.5
231– < 238	2	234.5	469
238– < 245	1	241.5	241.5
	$\sum f = 17$		$\sum fx = 3727.5$

$\bar{x} = 219.26$

b. See the table at the bottom of the page.*

9. a. 51   b. 198

10. a. 24

    b. 24

    c. The median is unchanged.

11. a. 100   b. 1.805

12. 16

13. a. 21   b. 144.8

14. The median, as the data set has two clear outliers

15. a. Mean $= 867.89$ mm, median $= 654$ mm

    b. The median, as it is not affected by the extreme values present in the data set.

16. a. Mean $= 7.55$, median $= 6$

    b. The median would be the preferred choice due to the extreme value of 34.

17.

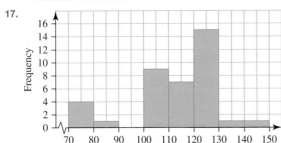

The distribution has one mode with data values that are most frequent in the 120–< 130 interval. There are potential outliers in the 70–< 80 interval, and there is a negative skew to the distribution.

18. Please see the worked solutions available online.

19. As the highest value increased, the mean increased significantly.

20. It would be in the workers' interest to use a higher figure when negotiating salaries, whereas it would be in the management's interest to use a lower figure.

11.5 Measures of spread

11.5 Exercise

1. 186

2. 2.555

3. a. Class A $= 11$, Class B $= 10$

    b. Class A $= 2$, Class B $= 5$

4. Mean $= 25.02$
   Median $= 25.10$
   Mode $= 23.2$
   Range $= 9.0$
   IQR $= 4.4$

5. a. 160

    b. 57

    c. Goals against: range $= 293$, interquartile range $= 140$
    The 'Goals against' column is significantly more spread out than the 'Goals for' column.

6. Standard deviation $= 7.07$

7. 0.822

8. a. \$16 327.50

    b. \$46 902

    c. There is a much larger spread in the maximum salaries than the minimum salaries.

9. a. Interquartile range $= 2.78$, standard deviation $= 5.04$

    b. Interquartile range $= 2.78$, standard deviation $= 2.10$

    c. The interquartile range has stayed the same value, while the standard deviation has reduced significantly.

10. The standard deviation increased by 4.12, while the interquartile range was unchanged.

11. Please see the worked solutions available online.

12. Please see the worked solutions available online.

13. The range is slightly more than double the value of the interquartile range $(2.9 > 1.25)$. This indicates that the data is bunched with no outliers.

14. Please see the worked solutions available online.

15. 59.38%

*8. b

Interval	Frequency (f)	Midpoint (x)	xf
5–< 10	4	7.5	30
10–< 15	5	12.5	62.5
15–< 20	3	17.5	52.5
20–< 25	3	22.5	67.5
25–< 30	1	27.5	27.5
30–< 35	1	32.5	32.5
	$\sum f = 17$		$\sum fx = 272.5$

$\bar{x} = 16.03$

11.6 Review

11.6 Exercise

1. D
2. B
3. A
4. D
5. B
6. a. Categorical   b. Numerical   c. Numerical
   d. Numerical
7. a. Discrete   b. Continuous   c. Continuous
   d. Discrete
8. a.

Number of DVDs	Tally	Number of students
0–4	||	3
5–9	ЖЖ ||||	9
10–14	ЖЖ ||||	9
15–19	||	3
20–24	||||	4
25–29	|	1
30–34		0
35–39	|	1

b.

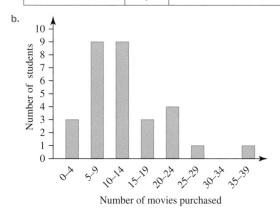

9. Key: 0 | 6 = 6 errors

Stem	Leaf
0	6
1	3 5 7 8
2	0 0 5 6 6 7 8 9
3	1 2 2 8
4	3 6
5	2

10. 8.4 s
11. Mean = 250.65 g, range = 4.6 g

12. a.

Class	Class centre	Frequency
5000–9999	7 500	1
10 000–14 999	12 500	5
15 000–19 999	17 500	9
20 000–24 999	22 500	3
25 000–29 999	27 500	2
30 000–34 999	32 500	2

b.

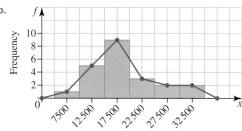

No. of people at a football match

13. 12.3 s
14. Please see the worked solutions for sample responses.
15. The distribution has a negative skew, with most of the scores closer to 50.

Stem	Leaf
2	1 1 3 4
2*	8 8 8 9 9
3	0 3 4
3*	5 5 5 6 8 8 8 8 9
4	0 0 0 1 1 2 2 4
4*	5 5 5 6 8

This stem-and-leaf plot is slightly negatively skewed but shows that most of the scores lie between 35 and 44 marks.

16. 69
17. Please see the worked solutions for sample responses.
18. The histogram has two modes and is near symmetrical, with a slight positive skew.
   The mean BMI for this group is 25.95.
19. 51
20. 95%

12 Univariate data comparisons

LESSON SEQUENCE

Fully worked solutions for this chapter are available online.

EXAM PREPARATION

Access exam-style questions in every lesson, available online.

 **Resources**

Solutions	Solutions — Chapter 12 (sol-1253)
Exam questions	Exam question booklet — Chapter 12 (eqb-0269)
Digital documents	Learning matrix — Chapter 12 (doc-41500)
	Chapter summary — Chapter 12 (doc-41501)

LESSON
12.1 Overview

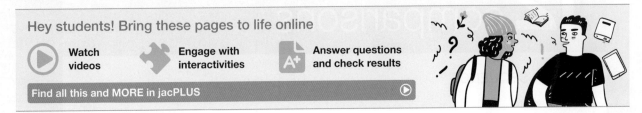
12.1.1 Introduction

Imagine that a new drug for the relief of cold symptoms has been developed. To test the drug, 40 people were exposed to a cold virus. Twenty patients were then given a dose of the drug, while the other 20 patients were given a placebo. (In medical tests a control group is often given a placebo drug. The subjects in this group believe that they have been given the real drug, but their dose contains no drug at all.) All participants were then asked to indicate the time when they first felt relief of symptoms. The number of hours from the time the dose was administered to the time when the patients first felt relief of symptoms was recorded. How would these two sets of data be compared?

In this chapter we will look at one method of data comparison called parallel boxplots, which are based on the work completed in Chapter 11.

12.1.2 Syllabus links

Lesson	Lesson title	Syllabus links
12.2	**Constructing boxplots**	○ Construct boxplots (Prerequisite skill).
12.3	**Outliers and fences**	○ Construct and use parallel boxplots, including identifying possible outliers, to compare datasets in terms of median, spread (range and IQR) and outliers to interpret and communicate the differences observed in the context of the data. • outliers (identifying): $Q_1 - 1.5 \times \text{IQR} \le x \le Q_3 + 1.5 \times \text{IQR}$ where Q_1 is lower quartile and Q_3 is upper quartile
12.4	**Parallel boxplots**	○ Construct and use parallel boxplots, including identifying possible outliers, to compare datasets in terms of median, spread (range and IQR) and outliers to interpret and communicate the differences observed in the context of the data. • outliers (identifying): $Q_1 - 1.5 \times \text{IQR} \le x \le Q_3 + 1.5 \times \text{IQR}$ where Q_1 is lower quartile and Q_3 is upper quartile
		○ Compare datasets in terms of mean, median, range, IQR and standard deviation, interpret the differences observed in the context of the data, and report the findings in a systematic and concise manner.

Source: General Mathematics Senior Syllabus 2024 © State of Queensland (QCAA) 2024; licensed under CC BY 4.0.

LESSON
12.2 Constructing boxplots

SYLLABUS LINKS

- Construct boxplots (Prerequisite skill).

Source: General Mathematics Senior Syllabus 2024 © State of Queensland (QCAA) 2024; licensed under CC BY 4.0.

12.2.1 Boxplots

We use a **box-and-whisker plot**, commonly referred to as a **boxplot**, to represent the five-number summary in a graphical form.

> **Boxplot: five-number summary**
>
> **The five-number summary gives five key values that provide information about the spread of a data set. These values are:**
> 1. **the lowest score (X_{min})**
> 2. **the lower quartile (Q_1)**
> 3. **the median (Q_2)**
> 4. **the upper quartile (Q_3)**
> 5. **the highest score (X_{max}).**

The boxplot is often displayed either above or below a number line, which allows easy identification of the key values.

Boxplots usually consist of both a central box and 'whiskers' on either side of the box. The box represents the interquartile range (IQR) of the data set, with the distance between the start of the first whisker and the end of the second whisker representing the range of the data. If either the lowest score is equal to the lower quartile or the highest score is equal to the upper quartile, then there will be no whisker on that side of the boxplot.

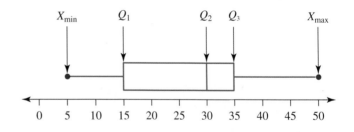

The shape of boxplots

The shape of a boxplot will mirror the distribution of the data set.

For example, a boxplot with a small central box and large whiskers indicates that the majority of the data is clustered around the median, whereas a boxplot with a large central box and small whiskers indicates that the data is spread more evenly across the range.

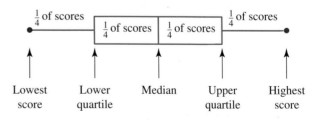

Positively skewed data will have the central box on the left-hand side of the boxplot with a large whisker to the right, whereas negatively skewed data will have the central box on the right-hand side of the boxplot with a large whisker to the left.

Learning to interpret the shape of boxplots will help you to better understand the data that the boxplot represents.

Some of the shapes of boxplots are shown here.

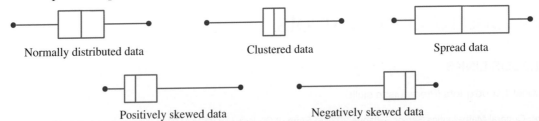

Normally distributed data   Clustered data   Spread data

Positively skewed data   Negatively skewed data

WORKED EXAMPLE 1 Developing a five-number summary

For the following set of scores, develop a five-number summary.
12 15 46 9 36 85 73 29 64 50

THINK	WRITE
1. Re-write the list in ascending order.	Med 9 12 [15] 29 36 | 46 50 [64] 73 85
2. Write the lowest score.	Lowest score $= 9$
3. Calculate the lower quartile. (There are 10 data values, the median falls between the 5th and 6th values, and Q_1 is the middle value of the lowest 5 values.)	Lower quartile $= 15$
4. Calculate the median.	Median $= \dfrac{36 + 46}{2}$ $= 41$
5. Calculate the upper quartile. (There are 10 data values, the median falls between the 5th and 6th value and Q_3 is the middle value of the upper 5 values.)	Upper quartile $= 64$
6. Write the highest score.	Highest score $= 85$
7. Combine the data into a five-number summary and write the answer.	Five-number summary $= 9, 15, 41,$ $64, 85$

WORKED EXAMPLE 2 Determining the IQR from a boxplot

The boxplot shows the marks achieved by students on their end of year exam.
a. State the median.
b. Obtain the interquartile range.
c. Determine the highest mark in the class.

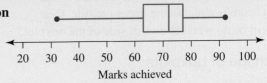

20 30 40 50 60 70 80 90 100
Marks achieved

THINK	WRITE
a. The mark in the box shows the median (72).	a. Median $= 72$ marks
b. 1. The lower end of the box shows the lower quartile (63).	b. Lower quartile $= 63$ marks
2. The upper end of the box shows the upper quartile (77)	Upper quartile $= 77$ marks
3. Subtract the lower quartile from the upper quartile.	Interquartile range $= Q_3 - Q_1$ $= 77 - 63$ $= 14$
4. Write the answer.	Interquartile range $= 14$ marks
c. 1. The top end of the whisker gives the highest mark (92).	c. Top mark $= 92$ marks
2. Write the answer.	The highest mark in the class is 92.

eles-6403

WORKED EXAMPLE 3 Constructing a boxplot from a five-number summary

After analysing the speed (in km/h) of motorists through a particular intersection, the following five-number summary was developed.

The lowest score is 82 km/h.
The lower quartile is 84 km/h.
The median is 89 km/h.
The upper quartile is 95 km/h.
The highest score is 114 km/h.
Show this information in a boxplot.

THINK

1. Draw a number line from 70 to 120 using
   1 cm = 10 km/h and label the axis.

2. Draw the box from 84 to 95, representing Q_1 to Q_3.

3. Mark the median at 89 (Q_2).

4. Draw the whiskers to 82 (X_{min}) and 114 (X_{max}).

WRITE

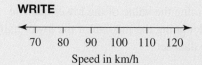

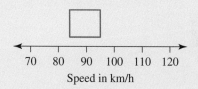

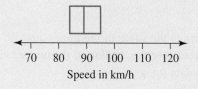

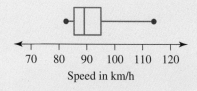

WORKED EXAMPLE 4 Constructing a boxplot from a stem-and-leaf plot

Construct a boxplot for the data contained in the following stem-and-leaf plot, which shows the number of coffees sold by a café each day over a 21-day period.

Key: 6|3 = 63 coffees

Stem	Leaf
6	3 5 8
7	0 2 4 5 7 9
8	1 1 3 6 8
9	0 1 5 6 7
10	1 4

THINK	WRITE
1. Determine the median of the data, recalling the median formula $\left(\dfrac{n+1}{2}\right)$ th data value.	There are 21 values, so the median is the $\left(\dfrac{21+1}{2}\right) = 11$th data value.

$$\text{median}$$
$$63, 65, 68, 70, 72, 74, 75, 77, 79, 81, \boxed{81}$$
$$Q_2 = 81$$

| 2. Determine the value of the lower quartile by calculating the median of the lower half of the data set. | $\left(\dfrac{10+1}{2}\right) = 5.5$ |

There are 10 values in the lower half of the data, so Q_1 will be between the 5th and 6th values.

$$Q_1$$
$$63, 65, 68, 70, 72, 74, 75, 77, 79, 81$$
$$Q_1 = \dfrac{72+74}{2}$$
$$= 73$$

| 3. Determine the value of the upper quartile by calculating the median of the upper half of the data set. | $\left(\dfrac{10+1}{2}\right) = 5.5$ |

There are 10 values in the upper half of the data, so Q_3 will between the 5th and 6th values.

$$Q_3$$
$$83, 86, 88, 90, 91, 95, 96, 97, 101, 104$$
$$Q_3 = \dfrac{91+95}{2}$$
$$= 93$$

4. Write the five-number summary.	$X_{\min} = 63$
	$Q_1 = 73$
	$Q_2 = 81$
	$Q_3 = 93$
	$X_{\max} = 104$

5. Rule a suitable scale for your boxplot that covers the full range of values. Draw the central box first (from Q_1 to Q_3, with a line at Q_2) and then draw in the whiskers from the edge of the box to the minimum and maximum values.
Remember to label your axis.

Coffees sold

Exercise 12.2 Constructing boxplots

learn on

12.2 Exercise	**12.2 Exam questions** on

Simple familiar	**Complex familiar**	**Complex unfamiliar**
1, 2, 3, 4, 5, 6, 7, 8, 9, 10, 11, 12, 13, 14, 15	N/A	N/A

Simple familiar

1. Copy and complete the following sentences.

   a. When you want to calculate the _____ of a data set, the first thing you must do is put the data in order from _____ to _____.

   b. When the median falls between two values, you need to calculate the _____ of those two values.

   c. _____, the _____ and _____ divides the data into _____.

2. Put these words in order from the one with the smallest value to the one with the largest value.
   Upper quartile; Minimum; Median; Maximum; Lower quartile.

A calculator can be used for many of the following questions.

3. **WE1** For the following data set, develop a five-number summary.

   $$15, \ 17, \ 16, \ 8, \ 25, \ 18, \ 20, \ 15, \ 17, \ 14$$

4. For each of the following data sets, develop a five-number summary.

   a. 23, 45, 92, 80, 84, 83, 43, 83
   b. 2, 6, 4, 2, 5, 7, 1
   c. 60, 75, 29, 38, 69, 63, 45, 20, 29, 93, 8, 29, 93

5. **WE2** From the five-number summary 6, 11, 13, 16, 32, identify:

   a. the median
   b. the interquartile range
   c. the range.

6. From the five-number summary 101, 119, 122, 125, 128, identify:

   a. the median
   b. the interquartile range
   c. the range.

7. **WE3** A five-number summary is given below.
   Lowest score = 39.2   Upper quartile = 52.3
   Lower quartile = 46.5   Highest score = 57.8
   Median = 49.0
   Construct a boxplot of the data.

Questions 8, 9 and 10 refer to the boxplot shown.

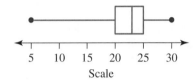

8. **MC** The median of the data is:

    A. 5 **B.** 20 **C.** 23 **D.** 25

9. **MC** The interquartile range of the data is:

    A. 5 **B.** 20 **C.** 25 **D.** 20 to 25

10. **MC** Identify the statement that is *not* true of the data represented by the box-and-whisker plot.

    A. One-quarter of the scores are between 5 and 20.
    B. One-half of the scores are between 20 and 25.
    C. The lowest quarter of the data is spread over a wide range.
    D. Most of the data are contained between the scores of 5 and 20.

11. **WE4** The following stem plot shows the ages of 25 people when they had their first child.

    Key: $1^*|7 = 17$ years old

Stem	Leaf
1*	7 8 8
2	0 2 3 3 4
2*	5 6 6 7 8 9
3	0 0 1 2 2 4
3*	6 8 9
4	1 3

    Construct a boxplot of the data.

12. The number of sales made each day by a salesperson is recorded over a fortnight:

    $$25, 31, 28, 43, 37, 43, 22, 45, 48, 33$$

    a. Write a five-number summary of the data.
    b. Construct a box-and-whisker plot of the data.

13. The data below show monthly rainfall in millimetres.

Jan.	Feb.	Mar.	Apr.	May	June	July	Aug.	Sept.	Oct.	Nov.	Dec.
10	12	21	23	39	22	15	11	22	37	45	30

    a. Provide a five-number summary of the data.
    b. Construct a box-and-whisker plot of the data.

14. The boxplot shows the distribution of final points scored by a football team over a season's roster.

    a. Identify the team's greatest points score.
    b. Identify the team's smallest points score.
    c. Identify the team's median points score.
    d. Calculate the range of points scored.
    e. Calculate the interquartile range of points scored.

    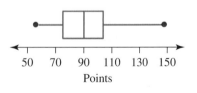

15. The boxplot shows the distribution of data formed by counting the number of honey bears in each of a large sample of packs.

For any pack, determine:

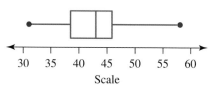

a. the largest number of honey bears
b. the smallest number of honey bears
c. the median number of honey bears
d. the range of numbers of honey bears
e. the interquartile range of honey bears.

Fully worked solutions for this chapter are available online.

LESSON
12.3 Outliers and fences

SYLLABUS LINKS

- Construct and use parallel boxplots, including identifying possible outliers, to compare datasets in terms of median, spread (range and IQR) and outliers to interpret and communicate the differences observed in the context of the data.
  - outliers (identifying): $Q_1 - 1.5 \times IQR \leq x \leq Q_3 + 1.5 \times IQR$ where Q_1 is lower quartile and Q_3 is upper quartile

Source: General Mathematics Senior Syllabus 2024 © State of Queensland (QCAA) 2024; licensed under CC BY 4.0.

12.3.1 Identifying possible outliers

If there is an outlier (an extreme value) in the data set, rather than extending the whiskers to reach this value, we extend the whiskers to the next smallest or largest value and indicate the outlier value with an 'x', as shown in the diagram.

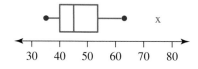

We can identify whether a value in our data set is an outlier or not by calculating the lower and upper fences of our data set.

Lower and upper fences

To calculate the **lower fence** and **upper fence** of the data set, we first need to calculate the interquartile range (IQR). Once this has been calculated, the lower and upper fences are given by the following rules.

> **Identifying outliers**
>
> $Q_1 - 1.5 \times IQR \leq x \leq Q_3 + 1.5 \times IQR$
>
> **Lower fence** $= Q_1 - 1.5 \times IQR$
>
> **Upper fence** $= Q_3 + 1.5 \times IQR$

If a data value lies outside the lower or upper fence, then it can be considered an outlier.

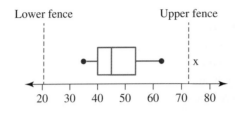

Lower fence   Upper fence

20 30 40 50 60 70 80

WORKED EXAMPLE 5 Constructing a boxplot with outliers

eles-3084

The following stem plot represents the time taken (in minutes) for 25 students to finish a maths puzzle.

Key: $1|5 = 1.5$ minutes

Stem	Leaf
1	5
2	8
3	4 6 6 9
4	0 0 2 3 5 5 7 8 8
5	0 2 4 4 5 8 9
6	0 4
7	
8	5

a. Calculate the values of the lower and upper fences.
b. Identify any outliers in the data set.
c. Construct a boxplot to represent the data.

THINK

a. 1. To calculate the upper and lower fences, we must first determine the median (Q_2) by recalling the formula $\left(\dfrac{n+1}{2}\right)$ th data value.

2. The lower quartile (Q_1) can then be calculated.

WRITE

a. There are 25 values, so the median is the $\left(\dfrac{25+1}{2}\right) = $ 13th data value.

median

1.5, 2.8, 3.4, 3.6, 3.6, 3.9, 4.0, 4.0, 4.2, 4.3, 4.5, 4.5, (4.7)

$Q_2 = 4.7$

$\left(\dfrac{12+1}{2}\right) = 6.5$

There are 12 values in the lower half of the data, so Q_1 will be between the 6th and 7th values.

Q_1

1.5, 2.8, 3.4, 3.6, 3.6, 3.9, 4.0, 4.0, 4.2, 4.3, 4.5, 4.5

$Q_1 = \dfrac{3.9 + 4.0}{2}$

$= 3.95$

3. The upper quartile (Q_3) can then be calculated.

$$\left(\frac{12+1}{2}\right)=6.5$$

There are 12 values in the upper half of the data, so Q_3 will be between the 6th and 7th values.

Q_3

4.8, 4.8, 5.0, 5.2, 5.4, 5.4, 5.5, 5.8, 5.9, 6.0, 6.4, 8.5

$$Q_3 = \frac{5.4+5.5}{2}$$
$$= 5.45$$

4. Calculate the IQR.

$$\begin{aligned} IQR &= Q_3 - Q_1 \\ &= 5.45 - 3.95 \\ &= 1.5 \end{aligned}$$

5. Calculate the values of the lower and upper fences by recalling the formulas:
Lower fence $= Q_1 - 1.5 \times IQR$
Upper fence $= Q_3 + 1.5 \times IQR$

$$\begin{aligned} \text{Lower fence} &= Q_1 - 1.5 \times IQR \\ &= 3.95 - 1.5 \times 1.5 \\ &= 3.95 - 2.25 \\ &= 1.7 \end{aligned}$$

$$\begin{aligned} \text{Upper fence} &= Q_3 + 1.5 \times IQR \\ &= 5.45 + 1.5 \times 1.5 \\ &= 5.45 + 2.25 \\ &= 7.7 \end{aligned}$$

b. 1. Identify whether any values lie below the lower fence or above the upper fence.

b. Values below the lower fence (1.7): 1.5
Values above the upper fence (7.7): 8.5

2. Write the answer.

There are two outliers: 1.5 and 8.5.

c. 1. Write the five-number summary, giving the minimum and maximum values as those that lie within the lower and upper fences.

c. $X_{min} = 2.8$
$Q_1 = 3.95$
$Q_2 = 4.7$
$Q_3 = 5.45$
$X_{max} = 6.4$

2. Rule a suitable scale for your boxplot to cover the full range of values. Draw the central box first (from Q_1 to Q_3, with a line at Q_2) and then draw in the whiskers from the edge of the box to the minimum and maximum values. Mark the outliers with an 'x'. Remember to label your axis.

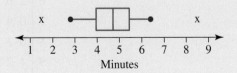

Exercise 12.3 Outliers and fences

12.3 Exercise	12.3 Exam questions on

Simple familiar	Complex familiar	Complex unfamiliar
1, 2, 3, 4, 5, 6, 7, 8, 9, 10, 11	12, 13, 14	15

Simple familiar

1. a. Calculate the upper and lower fences for the data contained in the following stem plot, which shows the number of sandwiches sold by a café per day over a 21-day period.

Key: $2|9 = 29$ sandwiches

Stem	Leaf
2	9
3	1 3 6 8 9
4	2 4 5 5 6 7 7 8
5	0 0 3 5 8
6	1 2

b. Construct a boxplot to represent the data.

2. The boxplot shows the temperatures in Brisbane over a 23-day period.
   a. Determine the median temperature.
   b. Determine the range of the temperatures.
   c. Determine the interquartile range of temperatures.
   d. Identify whether there are any outliers.

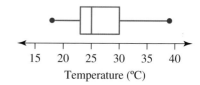

15 20 25 30 35 40
Temperature (°C)

3. **WE5** The following stem plot represents the time taken (in minutes) for 25 students to finish a logic problem.

Key: $4|4 = 4.4$ minutes

Stem	Leaf
4	4
5	
6	2 6 9
7	0 4 7 7 8
8	0 3 3 5 6 8 9
9	1 2 4 6 7
10	2 4 4
11	5
12	

a. Calculate the values of the lower and upper fences.
b. Identify any outliers in the data set.
c. Construct a boxplot to represent the data.

4. The boxplot represents the scores made by an Australian football team over a season.

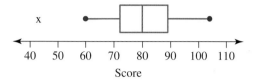

a. Determine the highest amount of points the team scored in the season.
b. Determine the lowest amount of points the team scored in the season.
c. Determine the range of points scored.
d. Determine the interquartile range of points scored.

5. **MC** The five-number summary for a data set is 45, 56, 70, 83, 92.
Identify which of the following statements is definitely *not* true.

A. There are no outliers in the data set.
B. Half of the scores are between 56 and 70.
C. The range is 47.
D. The value of the lower fence is 15.5.

6. **MC** The formula for the lower fence of a set of data is:

A. $Q_1 - 1.5 \times IQR$
B. $Q_1 + 1.5 \times IQR$
C. $Q_2 - 1.5 \times IQR$
D. $Q_1 - 2.5 \times IQR$

7. **MC** The formula for the upper fence of a set of data is:

A. $Q_1 - 1.5 \times IQR$
B. $Q_1 + 1.5 \times IQR$
C. $Q_3 - 1.5 \times IQR$
D. $Q_3 + 1.5 \times IQR$

8. **MC** For the five-number summary 15, 24, 33, 42, 51, identify which of the following is true.

A. The range is 35.
B. The value of the upper fence is 70.
C. The data is symmetrical.
D. The IQR is 22.

9. Determine whether the following statements are true or false.

a. You can always determine the median from a boxplot.
b. A stem plot contains every piece of data from a data set.
c. Boxplots show the complete distribution of scores within a data set.

10. Answer the following for the data set 12, 18, 21, 16, 9, 15, 21, 32, 15, 18, 27, 24, 19, 24, 30.

a. Calculate the five-number summary of the data.
b. Calculate the IQR.
c. Calculate the upper and lower fences.
d. Identify the outliers, if there are any.

11. **MC** The formula for the range of values that excludes outliers is:

A. $Q_3 - 1.5 \times IQR \le x \le Q_1 + 1.5 \times IQR$
B. $Q_1 - 1.5 \times IQR \le x \le Q_1 + 1.5 \times IQR$
C. $Q_3 - 1.5 \times IQR \le x \le Q_3 + 1.5 \times IQR$
D. $Q_1 - 1.5 \times IQR \le x \le Q_3 + 1.5 \times IQR$

Complex familiar

12. The five-number summary of a data set is 15, 29, 43, 57, 96.
Identify whether there is an outlier in the data set. (Outliers were not considered when calculating the five-number summary.)

13. Explain why outliers are considered an obstacle to making estimates of data and how this might be overcome.

14. If an outlier is added to the top range of a data set and included in the calculation of the five-number summary, describe the effect on each of the five numbers.

15. The lower and upper fences for a set of data are 12 and 72 respectively. If the IQR for the set of data is 10, determine Q_1 and Q_3.

Fully worked solutions for this chapter are available online.

LESSON
12.4 Parallel boxplots

SYLLABUS LINKS

• Construct and use parallel boxplots, including identifying possible outliers, to compare datasets in terms of median, spread (range and IQR) and outliers to interpret and communicate the differences observed in the context of the data.
  • outliers (identifying): $Q_1 - 1.5 \times \text{IQR} \leq x \leq Q_3 + 1.5 \times \text{IQR}$ where Q_1 is lower quartile and Q_3 is upper quartile

• Compare datasets in terms of mean, median, range, IQR and standard deviation, interpret the differences observed in the context of the data, and report the findings in a systematic and concise manner.

Source: General Mathematics Senior Syllabus 2024 © State of Queensland (QCAA) 2024; licensed under CC BY 4.0.

12.4.1 Comparing data sets

In some situations you will be required to compare two or more data sets. We can use two different graphical representations to easily compare and contrast data sets.

Back-to-back stem plots

As mentioned in Chapter 11, back-to-back stem plots are plotted with the same stem, with one of the plots displayed to the left of the stem and the other plot to the right.

```
                Key: 7|5 = 7.5
        Leaf | Stem | Leaf
          7 3 |  6  |
        9 6 2 |  7  | 5 7
        5 4 2 |  8  | 0 1 4 8
    8 6 6 0 0 |  9  | 2 6 6 9 9
        9 5 3 | 10  | 3 5 7
              | 11  | 1 4
              | 12  | 2
```

Remember to start the numbering on both sides with the smallest values closest to the stem and increasing in value as you move away from the stem. Also include a key with your back-to-back stem plot.

After drawing back-to-back stem plots, you can easily identify key points such as:
• which data set has the lowest and/or highest values
• which data set has the largest range
• the spread of both data sets.

Parallel boxplots

Parallel boxplots are plotted with one of the boxplots above the other. Both boxplots share the same scale.

Parallel boxplots allow us to easily make comparisons between data sets, as we can see the key features of the boxplots in the same picture. The position and size of the interquartile ranges of the data sets can be seen, as well as their range.

However, while parallel boxplots do display information about the general distribution of the data sets they cover, they lack the detail about this distribution that a histogram or stem plot provides.

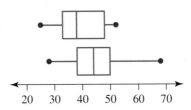

eles-3171

WORKED EXAMPLE 6 Constructing and comparing parallel boxplots

The following back-to-back stem plot shows the size (in kg) of two different breeds of dog.
a. **Construct parallel boxplots of the two sets of data.**
b. **Use the boxplots, including the medians, ranges, interquartile ranges and outliers, to compare and contrast the two sets of data.**

Key: $2|6 = 26$ kg

Breed X			Breed Y
Leaf	Stem	Leaf
9 8 7 7 6 4 4	1	
8 7 5 4 3 3 1 0	2	6 9
	3	3 5 5 7 8
	4	0 2 4 5 6 9
	5	1 3

THINK

a. 1. Calculate the five-number summary for the first data set (Breed X).

WRITE

a. $X_{min} = 14$, $X_{max} = 28$
There are 15 pieces of data, so the median is
the $\left(\dfrac{15+1}{2}\right) = $ 8th data value.
$Q_2 = 20$
There are 7 pieces of data in the lower half, so Q_1 is the 4th data value.
$Q_1 = 17$
There are 7 pieces of data in the upper half, so Q_3 is the 4th data value.
$Q_3 = 24$
Five-number summary: 14, 17, 20, 24, 28

2. Calculate the five-number summary for the second data set (Breed Y).

$X_{\min} = 26$, $X_{\max} = 53$

There are 15 pieces of data, so the median is the $\left(\dfrac{15+1}{2}\right) = $ 8th data value.

$Q_2 = 40$

There are 7 pieces of data in the lower half, so Q_1 is the 4th data value.

$Q_1 = 35$

There are 7 pieces of data in the upper half, so Q_3 is the 4th value.

$Q_3 = 46$

Five-number summary: 26, 35, 40, 46, 53

3. Use the five-number summaries to plot the parallel boxplots. Use a suitable scale that will cover the full range of values for both data sets.

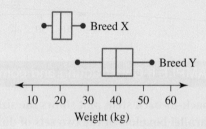

b. Compare and contrast the data sets, looking at where the key points of each data set lie. Comment on any noticeable differences in the centre and spread of the scores, as well as the shape of the distributions.

b. From a comparison of the boxplots it can be seen that Breed X has a much smaller median weight of 20 kg compared to 40 kg for Y. Almost every one of the Breed Y dogs are larger than Breed X, as the largest data value for Breed X is 28 and the smallest of Breed Y is 26.

The IQR for Breed X is 7 and the range is 14, whereas the IQR for Breed Y is 11 and the range is 27, so Breed Y has a much greater spread of sizes within the breed than Breed X, but overall Breed Y is a heavier breed of dog than Breed X.

12.4 Exercise	12.4 Exam questions on

Simple familiar	Complex familiar	Complex unfamiliar
1, 2, 3, 4, 5, 6	7, 8, 9, 10, 11, 12	13, 14

Simple familiar

1. **WE6** The following back-to-back stem plot shows the amount of sales (in $000s) for two different high street stores.

Key: 3|4 = $3400

Store 1 Leaf	Stem	Store 2 Leaf
	3	4 7 9
7 4	4	2 4 6 8
6 2 1	5	1 2 5 5 9
8 8 5 5	6	3 5 7
5 3 2	7	
6 1	8	
0	9	

   a. Construct a parallel boxplot of the two sets of data.
   b. Compare and contrast the two sets of data.

2. The prices of main meals at two restaurants that appear in the *Good Food Guide* are shown in the following back-to-back stem plot.

Key: 1|8 = $18

Restaurant A Leaf	Stem	Restaurant B Leaf
	1	8 9
9 9 8 5 5 4	2	2 5 5 7
8 6 5 5 2	3	0 0 2 5 5 8
2 0 0	4	0 3 6
	5	
9	6	

   a. Identify any outliers in either set of data.
   b. Prepare five-number summaries for the price of the meals at each restaurant.
   c. Construct a parallel boxplot to compare the two data sets.
   d. Compare and contrast the cost of the main meals at each restaurant.

3. The following parallel boxplot shows the weekly sales figures of three different mobile phones across a period of six months.

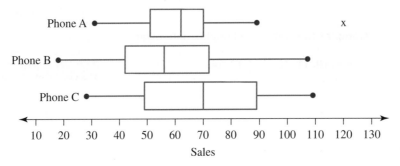

a. Determine which phone had the highest weekly sales overall.
b. Determine which phone had the most consistent sales.
c. Determine which phone had the largest range in sales.
d. Determine which phone had the largest interquartile range in sales.
e. Determine which phone had the highest median sales figure.

4. The following parallel boxplot details the amount of strawberries harvested in kg at two different farms for the month of March over a 15-year period.

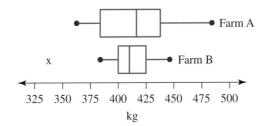

Decide whether the following statements are true or false.

a. Farm A produced a larger harvest of strawberries in March than Farm B more often than not.
b. The strawberry harvest at Farm B in March is much more reliable than the strawberry harvest at farm A.
c. Farm A had the highest producing month for strawberries on record.
d. Farm A had the lowest producing month for strawberries on record.

5. **MC** The lower quartile of Group B is:

Key: 12|2 = 122

Group B Leaf	Stem	Group A Leaf
	12	2
	13	3 8
6	14	0 4 4 6
8 5 4 2 2	15	2 3 5 7 8
8 5 5 3 0 0	16	2 4 4 5
7 4 4 1 0	17	2 6
1 1	18	1

A. 156.5.
B. 144.
C. 155.
D. 152.

6. **MC** For the stem plot shown in question **5**, identify which of the following statements is false.

   A. The data from Group A shows less consistency than the data from Group B.

   B. The data from Group B has a lower interquartile range.

   C. Group B has a greater median.

   D. None of the above. (All of the statements are true.)

Complex familiar

7. The following table displays the number of votes that two political parties received in 15 different constituencies in the local elections.

Party A	425	630	813	370	515	662	838	769
Party B	632	924	514	335	748	290	801	956

Party A	541	484	745	833	497	746	651
Party B	677	255	430	789	545	971	318

   Compare the performance of the two political parties. Ensure you comment on the distribution of each data set.

8. The following back-to-back stem plot shows the share prices (in $) of two companies from 18 random months out of a 10-year period.

Key: 1*|7 = $17

Company A Leaf	Stem	Company B Leaf
4 2	1	
9 7 5	1*	7 9
4 4 1	2	0 3 4
9 8 6 6	2*	7 8 8
3 3 2 0	3	1 2 4
8 6	3*	6 9
	4	0 1 2
	4*	5 6

   Comment on the distributions of both data sets.

9. The parallel boxplot shows the difference in grades (out of 100) between students at two schools. Determine which school has achieved better results. Use statistics to justify your answer.

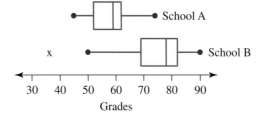

10. The parallel boxplot shows the performance of two leading brands of battery in a test of longevity.

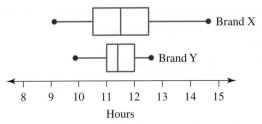

Identify the better brand. Justify your decision and refer to statistics in your response.

This data set is for questions **11** and **12**. The back-to-back stem plot displays the rental price (in $) of one-bedroom apartments in two different suburbs.

Key: $25|0 = \$250$

Suburb A Leaf	Stem	Suburb B Leaf
	25	0
	26	5 9
5 5	27	0 0 5
9 9 5 0	28	5 9 9
5 5 0	29	0
5 5 0 0 0 0	30	0 0 0
5 5 0 0	31	0 5 5
	32	9 9
	33	
	34	0 0
0	35	

11. Compare and contrast the rental price in the two suburbs.

12. The rental prices in a third suburb, Suburb C, were also analysed, with the data having a five-number summary of 280, 310, 325, 340, 375.
    Compare the rent in the third suburb with the rent in the other two suburbs.

Complex unfamiliar

13. The five-number summaries for the amount collected by three different charities in collection tins over a series of weeks are as follows.

    Charity 1: 225, 310, 394, 465, 580
    Charity 2: 168, 259, 420, 493, 667
    Charity 3: 262, 312, 349, 388, 445

    Compare and contrast the amount collected by the three charities, referring to the boxplots to justify your reasoning.

14. The following data sets show the daily sales figures for three new drinks across a 21-day period.

    Drink 1: 35, 51, 47, 56, 53, 64, 44, 39, 50, 47, 62, 66, 58, 41, 39, 55, 52, 59, 47, 42, 60
    Drink 2: 48, 53, 66, 51, 37, 44, 70, 59, 41, 68, 73, 62, 56, 40, 65, 77, 74, 63, 54, 49, 61
    Drink 3: 57, 49, 51, 49, 52, 60, 46, 48, 53, 56, 52, 49, 47, 54, 61, 50, 33, 48, 54, 57, 50

    Compare and contrast the sales of the three drinks by plotting a parallel boxplot. Use mathematical reasoning to justify your answer.

Fully worked solutions for this chapter are available online.

LESSON
12.5 Review

12.5.1 Summary

doc-41501

Hey students! Now that it's time to revise this chapter, go online to:

Access the chapter summary

Review your results

Practise exam questions

Find all this and MORE in jacPLUS

12.5 Exercise

learn on

12.5 Exercise	12.5 Exam questions on

Simple familiar	Complex familiar	Complex unfamiliar
1, 2, 3, 4, 5, 6, 7, 8, 9, 10, 11, 12	13, 14, 15, 16	17, 18, 19, 20

These questions are even better in jacPLUS!
- Receive immediate feedback
- Access sample responses
- Track results and progress

Find all this and MORE in jacPLUS

Simple familiar

1. **MC** For the parallel boxplots shown, identify the correct statement.

Group A

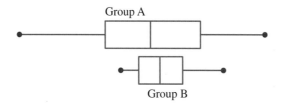

Group B

A. Group A has a smaller IQR than Group B.
B. Group B has a greater range than Group A.
C. Group A has a higher median than Group B.
D. 25% of Group B is greater than Group A's Q_1 and less than its median.

2. **MC** For the data set 789, 211, 167, 321, 432, 222, 234, 456, 456, 234, the five-number summary is:

A. 167, 222, 321, 456, 789.
B. 167, 222, 277.5, 456, 789.
C. 167, 234, 432, 456, 789.
D. 167, 234, 432, 456, 789.

3. **MC** The boxplot shown would best be described as:

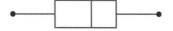   x

A. positively skewed.
B. symmetrical with an outlier.
C. positively skewed with an outlier.
D. negatively skewed with an outlier.

4. **MC** For the parallel boxplots shown, identify the statement that is **not** correct.

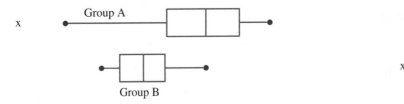

A. 75% of Group A is larger than 75% of Group B.
B. 25% of Group A has a larger spread than 75% of Group B.
C. The median of Group A is equal to the highest value of Group B.
D. Group A is negatively skewed and Group B is positively skewed.

5. **MC** For the data set 21, 56, 110, 15, 111, 45, 250, 124, 78, 24, the number of outliers and the value of $1.5 \times$ IQR will respectively be:

A. 0 and 87.　　　　　**B.** 1 and 87.　　　　　**C.** 1 and 111.　　　　　**D.** 1 and 130.5.

6. For the following data set, give a five-number summary.
$$24, 53, 91, 57, 29, 69, 29, 15, 84, 6$$

7. For the box-and-whisker plot shown:

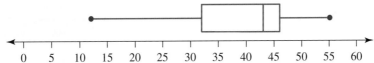

a. state the median
b. calculate the range
c. calculate the interquartile range.

8. The number of babies born each day at a hospital over a year was tabulated, and the five-number summary is given below.
Lowest score = 1, Lower quartile = 8, Median = 14, Upper quartile = 16, Highest score = 18
Show this information in a box-and-whisker plot.

9. Construct parallel boxplots for the following pair of five-number summaries.
Group X: 14, 18.5, 21.5, 27.5, 33
Group Y: 11, 17.5, 21, 26.5, 35

10. The following back-to-back stem-and-leaf plot gives the age at marriage of a group of 10 women and a group of 10 men.

Key: $1|8 = 18$ years old

Men Leaf	Stem	Women Leaf
8 7	1	8 8
9 8 7 5 1	2	0 2 3 4 4 5
6 3	3	0 1
0	4	

a. Construct side-by-side boxplots of the data.
b. Make comparisons about the distribution of the sets of data.

Questions 11 and 12 refer to the following boxplots.

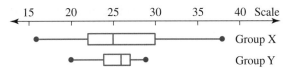

11. **MC** Identify the statement that is a correct comparison of the data.

    A. Group X has a higher median and shows more variability than Group Y.

    B. Group X has a lower median and shows more variability than Group Y.

    C. Group X has a higher median and shows less variability than Group Y.

    D. It is impossible to make comparisons like this without seeing the data displayed on a stem-and-leaf plot.

12. **MC** Identify the statement that is not true in regards to the boxplots.

    A. One-quarter of all Group X data is greater than any of Group Y data.

    B. The median of Group X is 25.

    C. The interquartile range of Group X is 25.

    D. The range of Group Y is 9.

Complex familiar

13. The stem-and-leaf plot shown represents the number of typing errors recorded by a class of students in 1 page of typing.

Key: $1 \mid 2 = 12$ $1* \mid 5 = 15$

Stem	Leaf
0*	0 1 4
0*	6 7 8 9
1*	0 0 1 1 2 3 3 4
1*	5 6 8 9
2*	3

Construct a boxplot of the data.

14. The box-and-whisker plots show the sales of two different brands of washing powder at a supermarket each day.

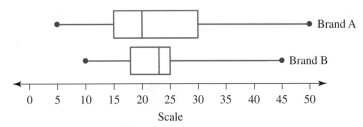

Describe the spread of the sales for each brand of washing powder. Use statistics to justify your answer.

15. The number of words in each of the first 12 sentences is counted in each of 3 different types of book: a children's book, a Year 12 Geography text, and a major daily newspaper. The results are shown in the table.

Children's book	6	8	12	15	6	8	10	8	5	11	10	8
Geography text	16	18	25	13	10	25	29	18	7	22	28	22
Newspaper	12	6	8	14	18	7	12	10	21	17	16	8

Make comparisons about the sentence length of each type of publication. Use statistics in your answer.

16. The following table shows the top 10 tourist destinations as collected by a tourism officer.

Country	Nights in country (× 1000)
Spain	—
Italy	158 527
France	98 700
United Kingdom	80 454
Austria	72 225
Germany	54 097
Greece	46 677
Portugal	25 025
Netherlands	25 014
Czech Republic	17 747

The tourism officer forgot to include the nights in Spain. The officer knows that Spain is both the most popular tourist destination and is an outlier in this set of data. Determine the minimum number of nights spent in Spain.

Complex unfamiliar

17. The stem-and-leaf plot shown gives the batting scores of two cricket players — Smith and Jones — who share the responsibility of 'opening the batting' for their side.

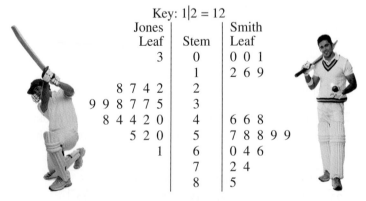

Key: 1|2 = 12

Jones Leaf	Stem	Smith Leaf
3	0	0 0 1
	1	2 6 9
8 7 4 2	2	
9 9 8 7 7 5	3	
8 4 4 2 0	4	6 6 8
5 2 0	5	7 8 8 9 9
1	6	0 4 6
	7	2 4
	8	5

Determine which player is the best batsman. Use graphs and statistics to justify your response.

18. The biggest winning margins in AFL Grand Finals up to the year 2013 are shown in the following table.

Winning margin	Year	Winning team	Winning score	Losing team	Losing score
119	2007	Geelong	163	Port Adelaide	44
96	1988	Hawthorn	152	Melbourne	56
83	1983	Hawthorn	140	Essendon	57
81	1980	Richmond	159	Collingwood	78
80	1994	West Coast	143	Geelong	63
78	1985	Essendon	170	Hawthorn	92
73	1949	Essendon	125	Carlton	52
73	1956	Melbourne	121	Collingwood	48
63	1946	Essendon	150	Melbourne	87
61	1995	Carlton	141	Geelong	80
61	1957	Melbourne	116	Essendon	55
60	2000	Essendon	135	Melbourne	75

Complete the sentences below.

a. 50% of the losing scores were less than __% of the winning margins.

b. 25% of the losing scores were greater than ___% of the winning scores.

19. The following table shows the AFL Grand Final statistics for a sample of players who have kicked a total of 5 or more goals from the clubs Carlton and Collingwood.

Player	Team	Kicks	Marks	Handballs	Disposals	Goals	Behinds
Alex Jesaulenko	Carlton	23	11	9	32	11	0
John Nicholls	Carlton	29	3	1	30	13	1
Wayne Johnston	Carlton	78	19	17	95	5	7
Robert Walls	Carlton	19	9	5	24	11	1
Craig Bradley	Carlton	61	11	37	98	6	2
Mark MacLure	Carlton	34	16	14	48	5	4
Stephen Kernahan	Carlton	44	26	8	52	17	5
Ken Sheldon	Carlton	36	5	12	48	5	2
Syd Jackson	Carlton	13	3	1	14	5	1
Rodney Ashman	Carlton	25	4	10	35	5	2
Greg Williams	Carlton	30	6	29	59	6	4
Alan Didak	Collingwood	46	17	24	70	6	2
Peter Moore	Collingwood	42	22	13	55	11	7
Ricky Barham	Collingwood	42	15	16	58	5	5
Travis Cloke	Collingwood	26	16	9	35	5	4
Ross Dunne	Collingwood	17	6	6	23	5	2
Craig Davis	Collingwood	27	8	8	35	6	3

Complete the sentences below.

a. ___% of Carlton players kicked more goals than 100% of Collingwood players if the outlier for Collingwood is not considered.

b. There were 50% more kicks per player for both clubs than ____% handballs per player for both clubs.

20. The following table shows some key nutritional information about a sample of fruits and vegetables.

Food	Calcium (mg)	Serve weight (g)	Water (%)	Energy (kcal)	Protein (g)	Carbohydrate (g)
Avocado	19	173	73	305	4.0	12.0
Blackberries	46	144	86	74	1.0	18.4
Broccoli	205	180	90	53	5.3	10
Cantaloupe	29	267	90	94	2.4	22.3
Carrots	19	72	88	31	0.7	7.3
Cauliflower	17	62	92	15	1.2	2.9
Celery	14	40	95	6	0.3	1.4
Corn	2	77	70	83	2.6	19.4
Cucumber	4	28	96	4	0.2	0.8
Eggplant	10	160	92	45	1.3	10.6
Lettuce	52	163	96	21	2.1	3.8
Mango	21	207	82	135	1.1	35.2
Mushrooms	2	35	92	9	0.7	1.6
Nectarines	6	136	86	67	1.3	16.0
Peaches	4	87	88	37	0.6	9.6
Pears	19	166	84	98	0.7	25.1
Pineapple	11	155	86	76	0.6	19.2
Plums	10	95	84	55	0.5	14.4
Spinach	55	56	92	12	1.6	2.0
Strawberries	28	255	73	245	1.4	66.1

a. Using technology or otherwise, compare and comment on the data for serve weight with the weight of water content for the samples given.

b. Using technology or otherwise, compare and comment on the data for protein and carbohydrate content as percentages of the total serve weight for the samples given.

Fully worked solutions for this chapter are available online.

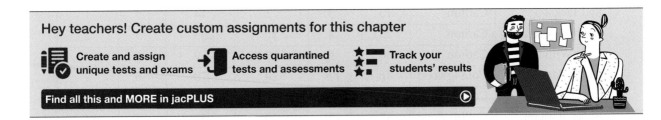

Hey teachers! Create custom assignments for this chapter

Create and assign unique tests and exams

Access quarantined tests and assessments

Track your students' results

Find all this and MORE in jacPLUS

Answers

Chapter 12 Univariate data comparisons

12.2 Constructing boxplots

12.2 Exercise

1. a. median, smallest, largest
   b. mean
   c. Q_1, median, Q_3, four

2. Minimum, Lower quartile, Median, Upper quartile, Maximum

3. 8, 15, 16.5, 18, 25

4. a. 23, 44, 81.5, 83.5, 92
   b. 1, 2, 4, 6, 7
   c. 8, 29, 45, 72, 93

5. a. 13   b. 5   c. 26

6. a. 122   b. 6   c. 27

7.
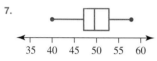

8. C

9. A

10. D

11.
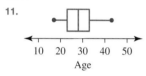

12. a. 22, 28, 35, 43, 48
    b.

13. a. 10 mm, 13.5 mm, 22 mm, 33.5 mm, 45 mm.
    b.
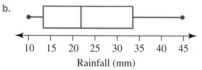

14. a. 147   b. 56   c. 90   d. 91   e. 28

15. a. 58   b. 31   c. 43   d. 27   e. 8

12.3 Outliers and fences

12.3 Exercise

1. a. Lower fence = 19, upper fence = 71
   b.
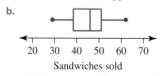

2. a. 25°C   b. 21°C   c. 7°C   d. No

3. a. Lower fence = 4.625, upper fence = 12.425

b. 4.4 is an outlier.

c.
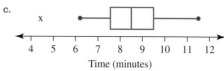

4. a. 104   b. 43   c. 61   d. 17

5. B

6. A

7. D

8. C

9. a. True   b. True   c. False

10. a. Five-number summary: 9, 15, 19, 24, 32
    b. IQR = 9
    c. Lower fence = 1.5
       Upper fence = 37.5
    d. There are no outliers since the 9 and 32 lie within the lower and upper fences.

11. D

12. Lower fence = −13
    Upper fence = 99
    Since the lowest and highest values of 15 and 96 fit within the upper and lower fences of −13 and 99, there are not any outliers in the data.

13. Outliers can unfairly skew data, so they can dramatically alter the five-number summary. Identify and remove any outliers from the data before determining the five-number summary.

14. The minimum number will not be increased as it will remain as the minimum.

15. 27, 57

12.4 Parallel boxplots

12.4 Exercise

1. a.

b. On the whole, store 2 has fewer sales than store 1; however, the sales of store 2 are much more consistent than store 1's sales.
   The sales of store 1 have a negative skew, whereas the sales of store 2 are symmetrical. There are no obvious outliers in either data set.

2. a. $69 in Restaurant A is an outlier
   b. Restaurant A: 24, 28, 35, 40, 42
      Restaurant B: 18, 25, 30, 38, 46
   c.
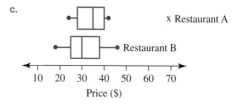

d. The meals in Restaurant A are more consistently priced but are also in general higher priced. The distribution of prices at Restaurant A has a positive skew, whereas the distribution of prices at Restaurant B is nearly symmetrical.

3. a. Phone A   b. Phone A
   c. Phone B   d. Phone C
   e. Phone C

4. a. True   b. True   c. True   d. False

5. C

6. D

7. The spread of votes for Party B is far larger than it is for Party A. Party A polled more consistently and had a higher median number of votes. Party A had a nearly symmetrical distribution of votes, whereas Party B's votes had a slight negative skew.

8. On the whole, the share price of Company B is greater than the share price of Company A. However, the share price of Company A is more consistent than the share price of Company B. The share price of Company A has a negative skew, whereas the share price of Company B has a nearly symmetrical distribution.

9. School B has a larger range than school A and a larger interquartile range, which means the scores for school B are a little more spread. However, 75% of the results for school B are 70 or above, whereas 75% of the results for school A are below 60. Therefore, school B has achieved better results.

10. The medians for both brands are similar. Brand Y does not have the best or worst battery, but it has the most consistent battery life, so Brand Y is the better performing brand.

11. The rental prices in Suburb A are far more consistent than the rental prices in Suburb B. There is one outlier in the data sets ($350 in Suburb A). Although Suburb A has a higher median rental price, you could not say that it was definitely more expensive than Suburb B.

12. Suburb C has a higher average rental price than either Suburb A or B. The spread of the prices in Suburb C is more similar to those in Suburb B than Suburb A.

13.

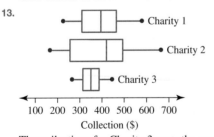

Collection ($)

The collections for Charity 3 were the most consistent of the three charities. Charity 2 collected more money on average than the other charities, but also had the poorest performing week in total. There are no outliers in any of the data sets.

14. The sales of Drink 3 are by far the most consistent, although overall Drink 2 has the highest sales. Drink 2's sales are also the most inconsistent of all the drinks. There is one outlier in the data sets (33 in Drink 3).

12.5 Review

12.5 Exercise

1. D

2. B

3. B

4. C

5. D

6. 6, 24, 41, 69, 91

7. a. 43   b. 43   c. 14

8.

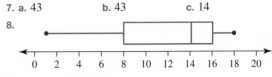

9.

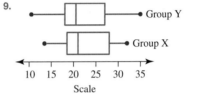

10. a.

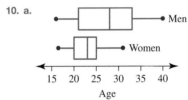

b. Women had a smaller range of 13 compared to men with the range of 23. The interquartile range for women was 5 compared to 12 for men. The median age was smaller for women (23.5) compared to men (27.5). Hence, there is less variability in the age at marriage for women.

11. B

12. C

13.

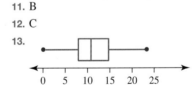

14. The number of daily sales for Brand A varies more than those for Brand B. The lowest score for Brand A is lower than that for Brand B, and the highest score for Brand A is higher than that for Brand B.

15. The longest sentence was in the Geography text book with 29 words. The newspaper's longest sentence was 21 words, while the children's book's longest sentence was 15 words. The shortest sentence went in the order: children (5), newspaper (6) and the Geography text (7). The variability was greatest in the Geography text (range = 22, IQR = 10.5). Overall the Geography text also had a larger central value (median of 20) than the others (children (8) and newspaper (12)).

16. 210 000

17. Jones's results are more clustered than Smith's results, indicating less variability in scores obtained. Both players have obtained large ranges overall. Jones's consistency may indicate that he is better as an opening batsman, whereas a big hitter like Smith might be better at the end.

18. a. 100%   **b.** 0%

19. a. 50%   **b.** 100%

20. a. The boxplots appear to indicate that there are only slight differences between the serve weights and water weights of the samples. The distributions are very similar in shape, with the water weights being slightly less overall.

b. Carbohydrate for this sample of foods is much greater but more variable than protein, as indicated by the larger range and IQR. The protein amounts are all less than the Q_1 for the carbohydrate amounts, with the exception of two upper outliers for protein that lie between the Q_1 and median for carbohydrate. Carbohydrate is positively skewed with one upper outlier.

GLOSSARY

adjacency matrix a matrix that represents the number of connections between objects in a network or the number of edges that connect the vertices of a graph

adjacent the side next to the angle used for reference in a right-angled triangle

allowance an extra payment made to a worker for working in unfavourable conditions

angle of depression the angle measured down from the horizontal line (through the observation point) to the line of vision

angle of elevation the angle measured up from the horizontal line (through the observation point) to the line of vision

annulus the area between two circles with the same centre. The formula is $A = \pi(R^2 - r^2)$, where R is the radius of the outer circle and r is the radius of the inner circle.

area the two-dimensional space taken up by an object

area scale factor the ratio of the corresponding areas of similar objects. It is equal to the linear scale factor raised to the power of two.

bar (column) charts displays with the categories of data on one axis (usually the horizontal axis) and the frequency of the data on the other (usually the vertical axis)

BIDMAS the order of operations to evaluate an expression: brackets, indices, division, multiplication, addition and subtraction

box-and-whisker plot or **boxplot** a graphical representation of the five-number summary

break-even point the point at which revenue begins to exceed the cost of production

budget a list of all planned income and costs

capacity the amount of fluid an object can contain

categorical data data that can be organised into groups or categories and is often an 'object', 'thing' or 'idea'. Examples include brand names, colours, general sizes and opinions.

circumference the perimeter of a curved figure, the circumference of a circle is $C = 2\pi r = \pi D$

coincident two lines that lie one on top of the other

column matrix a matrix that has only one column

commission payment made to a salesperson. A commission is usually paid as a percentage of sales.

cone a solid object in which one end is circular and the other end is a single vertex. Its cross-section is a series of circles that gradually get smaller as they approach the vertex.

congruent objects that are the same size and shape

cosine the ratio of the adjacent side and hypotenuse in a right-angled triangle

cylinder a solid object with ends that are identical circles and a cross-section that is the same along its length

discount the amount of money which the price of an item is reduced by

dividend yield (percentage dividend) the dividend paid per share as a percentage of the share price,
$$\text{dividend yield} = \frac{\text{dividend}}{\text{shareprice}} \times 100$$

DMS degrees, minutes, seconds

dot plot a plot in which every data value is represented by a dot, used to identify the most common values

double time a penalty rate that pays the employee twice the normal hourly rate

elements entries in a matrix

elimination the process of simplifying a mathematical expression by removing a variable; common when solving simultaneous equations

evaluate find a numerical answer for an equation

floor plan a plan showing the floor dimensions of a structure and detailed dimensions of features such as doors, windows, wall thicknesses and stairs

formula or **rule** an equation showing the relationship between two or more quantities

frequency tables displays that tabulate data according to the frequencies of predetermined groupings

gradient also known as the slope; determines the change in the *y*-value for each change in *x*-value. This measures the steepness of a line as the ratio $m = \dfrac{\text{rise}}{\text{run}}$. If (x_1, y_1) and (x_2, y_2) are two points on the line, $m = \dfrac{(y_2 - y_1)}{(x_2 - x_1)}$.

goods and services tax (GST) a consumption tax charged on most goods and services

Heron's formula a way of calculating the area of the triangle if you are given all three side lengths, $A = \sqrt{s(s-a)(s-b)(s-c)}$, where $s = \dfrac{a+b+c}{2}$

histogram a display of continuous numerical data similar to a bar chart, in which the width of each column represents a range of data values and the height of the column represents that range's frequency.

hypotenuse the longest side of a right-angled triangle. The hypotenuse is opposite the right angle.

identity matrix a square matrix in which all the elements on the diagonal line from the top left to bottom right are 1s and all of the other elements are 0s

inflation the rate at which the price of goods and services increase

interquartile range the difference between the upper and lower quartiles of a data set

linear equation an equation in which the highest power of any variable is one

linear relationship a relation between up to two variables of degree 1 that produces a straight line

linear scale factor the ratio of the corresponding side lengths of similar objects

lower fence the lower boundary beyond which a data value is considered to be an outlier: $Q_1 - 1.5 \times \text{IQR}$

lower quartile (Q_1) the median of the lower half of an ordered data set

matrix a rectangular array of rows and columns that is used to store and display information

mark-up the amount or percentage increase added to the costs of goods or services

mean commonly referred to as the average; a measure of the centre of a set of data. The mean is calculated by dividing the sum of the data values by the number of data values.

median the middle value of the ordered data set if there are an odd number of values, or halfway between the two middle values if there are an even number of values, $\left(\dfrac{n+1}{2}\right)^{\text{th}}$ data value

mode the category or data value(s) with the highest frequency. It is the most frequently occurring value in a data set.

net a 2-dimensional plan of the surfaces that make up a 3-dimensional object

networks an arrangement of interconnecting lines which shows the pathways between points

nominal data categorical data that has no natural order or ranking

non-linear relationship a relation between variables that forms a shape or curve rather than a straight line

numerical data data that can be counted or measured

opposite the side opposite to the angle used for reference in a right-angled triangle

order the indice or power of a number expressed as a base number and an indice

ordinal data categorical data that can be placed into a natural order or ranking

outlier an extreme value or unusual reading in the data set, generally considered to be any value beyond the lower or upper fences

overtime when a person earns more than the regular hours each week

parallel boxplots displays in which two or more boxplots share the same scale to enable comparisons between data sets

penalty rate a higher rate of pay made to a person who is working overtime

per cent the amount out of 100, or per hundred; for example, 50 per cent (or 50%) means 50 out of 100 or $\dfrac{50}{100}$

percentage discount the discount of the price of an item expressed as a percentage of the original cost

perimeter the distance around an object

pie charts graphs that use sectors of circles to represent categories of data

piecewise graphs continuous graphs formed by two or more linear graphs that are joined at points of intersection

piecework payment for the amount of work completed

polygons 2-dimensional shapes consisting of at least three straight sides

price-to-earnings ratio (or **P/E ratio**) the market price of a company's share price divided by its earnings or profit per share. It gives an indication of how much shares cost per dollar of profit earned.

prism a solid object that has identical ends that are joined by flat surfaces, and a cross-section that is the same along its length

profit the positive difference between what a product is sold for less than what it cost

proportional when two quantities have the same ratio; therefore, they always have the same size in relation to each other

pyramid a solid object whose base is a polygon and whose sides are triangles that meet at a single point

Pythagoras' theorem in a right-angled triangle, the square on the hypotenuse is equal to the sum of the squares on the other two sides, $c^2 = a^2 + b^2$

Pythagorean triads (or **Pythagorean triples**) sets of three numbers which satisfy Pythagoras' theorem

quartiles these divide a set of data into quarters. The lower quartile (Q_1) is the median of the lower half of an ordered data set. The upper quartile (Q_3) is the median of the upper half of an ordered data set. The middle quartile (Q_2) is the median of the whole data set.

range a measure of the spread of a numerical data set determined by calculating the difference between the smallest and largest values

rate a measure of how one quantity is changing compared to another

retainer a fixed payment usually paid to someone receiving commission. They receive the retainer regardless of the number of sales made.

row matrix a matrix that has only one row

salary a form of payment where a person is paid a fixed amount to do their job. A salary is usually based on an annual amount divided into weekly or fortnightly instalments.

scalar multiplication each element of the matrix is multiplied by the same number, called a 'scalar'. A scalar quantity can be any real number, such as negative or positive numbers, fractions or decimal numbers.

scale the ratio of the length on a drawing to the actual length

scale factor a number by which the side lengths on the first of two similar figures is multiplied by to obtain the measurements on the second of the figures

sectors fractions of a circle. The area of a sector can be calculated using $A = \dfrac{\theta}{360}\pi r^2$.

similar objects that are the same shape but have different sizes

simple interest interest calculated as a percentage of the amount of money invested or borrowed. It remains constant for the term of the investment or loan.

simultaneous equations equations belonging to a system of equations in which the solutions for the values of the unknowns must satisfy each equation

sine the ratio of the opposite side and the hypotenuse in a right-angled triangle

sine rule the ratios of a side length with the sine of the angle opposite it are equal throughout a triangle: $\dfrac{a}{\sin(A)} = \dfrac{b}{\sin(B)} = \dfrac{c}{\sin(C)}$. It's used to find the unknown side length or angle in non-right-angled triangles.

site plan a plan showing the boundaries of a block of land and the position of the structure on the lot

slope–intercept form the form of a linear equation expressed as $y = mx + c$, where m is the slope and c is the y-intercept

sphere a solid object that has a curved surface such that every point on the surface is the same distance (the radius of the sphere) from a central point

square matrix a matrix that has the same number of rows and columns

standard deviation the most common measure of the spread of data around the mean; found by taking the square root of the variance, $s = \sqrt{\dfrac{\sum (x_i - \bar{x})^2}{n-1}}$

stem-and-leaf plots or **stem plots** arrangements used for numerical data in which data points are grouped according to their numerical place values (the 'stem') and then displayed horizontally as single digits (the 'leaf')

step graphs discontinuous graphs formed by two or more linear graphs that have zero gradients

substituted when a variable in a formula is replaced with an equivalent value or expression

substitution replacement of the pronumeral with its corresponding value in order to work out the value of an algebraic expression

surface area the combined total of the areas of each individual surface that forms a solid object

survey plan a plan showing all boundaries of blocks of land and the position of roadways

tangent the ratio of the opposite side and the adjacent side in a right-angled triangle

time and a half a penalty rate where the employee is paid 1.5 times the normal hourly rate

transpose to rearrange an expression or formula

trigonometry a branch of mathematics in which sides and angles of triangles are calculated

trigonometric ratio a ratio of the lengths of two sides of a right-angled triangle. The three most common trigonometric ratios are sine, cosine and tangent.

unit cost for cost of a single unit of an item, this may be a standard weight or volume, or individual item

upper fence the upper boundary beyond which a data value is considered to be an outlier: $Q_3 + 1.5 \times \text{IQR}$

upper quartile (Q_3) the median of the upper half of an ordered data set

variables a quantity that can take on a range of values depending on its relationship to other values; typically represented by pronumerals

volume the amount of space that is taken up by any solid or 3-dimensional object

volume scale factor the ratio of the corresponding volumes of similar objects. This is equal to the linear scale factor raised to the power of 3.

wage a form of payment that is based on an hourly rate

x-intercept the point where the graph of an equation crosses the x-axis. This occurs when $y = 0$.

y-intercept the point where the graph of an equation crosses the y-axis. This occurs when $x = 0$.

zero matrix a square matrix that consists entirely of '0' elements